Village Information System

Vol. One
Innovation of Natural Infrastructure

VILLAGE INFORMATION SYSTEM

VOLUME ONE: Innovation of Natural Infrastructure
VOLUME TWO: Innovation of Human Infrastructure
VOLUME THREE: Development Planning

About the Author

Hridai R. Yadav obtained degrees of M.Phil., Ph.D. from Jawaharlal Nehru University, New Delhi, worked with CSIR, Government of India as Pool Scientist for eight years, having authored 10 Books and 73 Research Papers. He has been the Member of the National Wastelands Development Boards, and Nominated as U.G.C. Observer for the selection of Professor in the Universities and U.G.C. Nominee in the Board of Management of Autonomous Institute of Engineering and Technology (GBTU), Lucknow. H.R. Yadav has been the members of various Government and Autonomous Institutions/Universities besides being Founder Member-Secretary-Director of the Institute of Wastelands Reclamation and Rural Development (1990) and Institute of Science and Technology for Rural Development (1994) and U.G.C. accredited RIWARD Post Graduate College, Hanumanganj, Sultanpur. Presently he is engaged in the implementation of Agro-Afforestation on the Patta Banjar Lands, Women Empowerment and Rural Health Mission Programmes in Amethi.

Village Information System

Vol. One
Innovation of Natural Infrastructure

[With 86 Coloured Maps]

Prof. (Dr.) Hridai R. Yadav

CONCEPT PUBLISHING COMPANY PVT. LTD.,
NEW DELHI-110059

This Study has been supported by NRDMS, Department of Science and Technology, Government of India, New Delhi

ISBN-13: 978-81-8069-940-5 (Set)
ISBN-13: 978-81-8069-941-2 (Vol. I)

First Published 2013

Published and Printed by

Concept
Since 1974

Concept Publishing Company Pvt. Ltd.
A/15-16, Commercial Block, Mohan Garden
New Delhi-110059 (India)
T: 25351460, 25351794, *F*: 091-11-25357109
E: publishing@conceptpub.com, W: www.conceptpub.com
Editorial Office: H-13, Bali Nagar, New Delhi-110 015, India.

Cataloging in Publication Data--Courtesy: D.K. Agencies (P) Ltd. <docinfo@dkagencies.com>

Yadav, Hridai Ram, 1957-
Village information system / Hridai R. Yadav.
v. 1 cm.
Includes bibliographical references (p.) and index.
Contents: v. 1. Innovation of natural infrastructure.
ISBN 9788180699405 (Set)
ISBN 9788180699412 (Vol. 1)

1. Rural development projects—Geographic information systems—India—Amethi. 2. Infrastructure (Economics) Geographic information systems—India—Amethi. 3. Natural resources—Geographic information systems—India—Amethi. I. Title.

DDC 307.141209542 23

Preface

The soul of India lives in the villages, where more than 76 per cent population survives on the natural land, water, infrastructure in the remote rural areas. Development programmes for the remote rural areas has been pursued with genuine concern for human values. The development should ensure the betterment and welfare of all the deprived resourceless multitudes living in the remote rural areas.

The existing process of development has neither been true, nor good, nor beautiful, but it does not mean that everything is wrong. Really the development lies in the techno-scientific application for quantity and quality production for realizing the better quality of life by the common people.

The process of development should be meant equally, effectively, efficiently with the people's participation, by the people and for the people for whom such development programmes have been formulated.

The formulation of the plan based on the natural resources planning has been adopted but at the grassroots level or village level or field level the formulated and implemented plans did not yield the desired results due to inequities amongst the people and disparities between rich and poor. Tremendous pressure on land, water and forest, natural infrastructure and its overexploitation has led to deforestation, land degradation and environmental crisis and socio-economic crises.

The nature has given in bounty to us, land, water, forest resources, agro-climatic conditions and eco-system of the remote villages must be utilized optimally but not exploited in an unplanned system. The development planning process for the appropriate management and optimum utilization of available water, land, forest, natural resources has always been challenges to the policy makers and planners.

Thus, there is, an urgent need to evolve a planning strategy through evolving village information system and innovation of Natural-Social-Economic Institutional Resources and the assessment of the ongoing development programmes and need based people participatory village panchayat planning.

In view of the above Natural Resources Data Management System (NRDMS), Department of Science and Technology, Government of India sanctioned a project on "Developing Village Information System in Amethi" which has been successfully implemented at village panchayat level in the Amethi Block.

The outcome of the implemented project has been properly documented and an effort has been made to publish the achievement of this study in the Book Form in three Volumes:

Vol. I: Village Information System (Innovation of Natural Infrastructure)
Vol. II: Village Information System (Innovation of Human Infrastructure)
Vol. III: Village Information System (Development Planning)

The 1st Volume of this study of the Village Information System is an innovation of the natural infrastructure, viz. land, water, forest and agro-climatic situations of village panchayat level covering 82 revenue villages and 377 purwas/tolas/hamlet villages based on the revenue records, i.e. Village Khasra Maps, Khasara Register and direct field level survey work has been carried out. In this volume an attempt has been made to study soil, climate, water resources, land use, wastelands, land capability and agro-afforestation etc. at village panchayat level in the Amethi Block.

I hope this study will serve as useful guide for policy planners while implementing Village Development Programmes at grassroots level. This study will also be useful for academicians, research scholars and grassroots level programme implementing institution and social workers, equally to understand field level or village level through people participatory development planning process and also the implementation of various integrated rural development programmes.

Prof. (Dr.) Hridai R. Yadav

Acknowledgements

Natural Resources Data Management System (NRDMS), Department of Science and Technology, Government of India, sanctioned a project entitled "Developing Village Information System in Amethi" to the Institute of Wastelands Reclamation and Rural Development (RIWARD), Payagipur, Sultanpur, U.P. An attempt has been made to document the outcome of the sanctioned project in the form of Book "Village Information System (Innovation of Natural Infrastructure) Volume-I for which Techno-Scientific assistance of Dr. Bhoop Singh, Director, NRDMS, Department of Science and Technology, Government of India has been highly appreciable because without their support this study could not have been successfully implemented.

Sincere gratitude are recorded to the revenue authorities of the Amethi Tehsil especially all the village Lekhpals, Kanoongos, Registrar Kanoongo, Tehsildar and S.D.M. Amethi and the District Magistrate, Amethi.

Neeraj Kumar and Dinesh Kumar deserve heartiest thanks for supporting the collection and collation of the information at village panchayat level in Amethi.

The tedious typing of the data and manuscript and finalization of the Maps etc. has been very carefully managed by Sri Kamlesh Chandra Verma, who deserves thanks.

I thank to all those, who directly or indirectly supported in the completion of this study. I acknowledge my sincere gratitude to Sri Ashok Kumar Mittal, CMD, Concept Publishing Co. Pvt. Ltd., New Delhi for the publication of this study.

Last, but not the least, I am grateful and indebted to Baba Ji and my wife Reeta and lovely son Sahit Hridai, without whose support writing and completion of this study would not have been possible.

Prof. (Dr.) Hridai R. Yadav

Contents

List of Maps

List of Tables and Figures

TABLES

FIGURES

1

Introduction

1. INTRODUCTION

Integrated economic development has been realized and commendable advances have been made in the diverse field of rural development programmes, would in reality be incomplete, if such rural development programmes fail to translate into better and dignified life for the majority of people who live in the remote rural areas because the benefit of development must reach to the down-trodden, deprived rural resourceless poor multitudes of the society **equally, effectively** and **efficiently**.

The planning is an organized process by which the people achieve their developmental goals or the planning is a means to achieve the development, but when the objectives and intentions of development are distorted then the planning process also becomes a distorted exercise. Planning is a systematic procedure which objects at the implementation of several specific aims and objectives—

- To specify potential variants of reaching development targets, i.e. variants of optimum utilization of various resources;
- To find out resources available, reaching such targets;
- Optimum utilization of available resources at grassroots level;
- To find out, which variants is optional (in the sense of allowing the best set of targets to be reached within the available resources);
- To innovate the available resources/infrastructure available at grassroots level which will ensure operational implementation of variants;
- To develop village information system especially for the Natural-Socio-Economic and institutional infrastructure at grassroots level before initiation of need based village development planning process; and
- To supervise the planning process for the implementation of developmental programmes in order to find out eventual discrepancies and to set in motion the adaptive achievements.

Presently in our country the planning at Macro-Meso-Micro-level is done without assessing the available resources of infrastructure at various levels and ignoring the participatory willing consent of the local people. In some of the cases the policy planners interpret people's need and aspirations for developing a plan for the people, by the people and with the people in which the planners can provide back up support to achieve their planning goals.

Thus, the policy planners will have to provide integrated development planning techno-scientific backup support at the grassroots level with the involvement of people's representatives, villagers and officials of the implementating institutions involved in the village development.

The planners should not only be a technocrat to provide projections but they should study and innovate the available resources/infrastructure at village level before formulation of the village development planning. The planners should assist villagers as a social scientist who can easily understand the basic grassroots level need of the people, so that according to the requirements and available village infrastructure village planning is formulated and village development programmes are being implemented with the peoples participation and to achieve the planned target of village development.

Development is perused with genuine concern for human values. Development means evolution of the people to acquire better perception and best control over its natural and human infrastructure at National-State-District-Block-Village level perspectives.

The present development process is neither true, nor good, nor beautiful but it does not mean that everything is wrong. Really the development lies in the development of Science and Technology and its application by the common people. The progress of the people in the Techno-Socio-Economic front has been made more destructive than constructive approaches of the development for each and every section of the society.

There is a quest for evolving a model of development planning. The approach for macro-level planning in our country did not yield the desired results due to inequities amongst the people and disparities at regional level. The environmental conditions continued to degrade leading to severe stress on forest cover, land and water resources. The sustainable and an integrated development of the area has been ensured during late 1970s. The conceptual changes in the planning practices had been adopted in macro-meso-micro level process of planning. The main objective of the planning was **to increase the land productivity** and **to increase the employment generation with the focus on creation of assets. The poverty alleviation and provision of common minimum amenities and infrastructural facilities** had been the main objectives of the then planning process.

73rd and 74th Constitutional Amendment Acts, (1992 and 1993), the State Governments were empowered from the institution of local self-governance right from the **Village level to Blcok level** and up to the **District level Panchayat** and within the municipal jurisdiction, i.e. **Zila Panchayat-Block Panchayat-Village Panchayat** and **Nagar Palikas** in the context of rural and urban development respectively.

The eleventh and twelfth schedules of the Constitution, while specifying the sectors for the Village-Block-District Panchayat and Nagar Palikas have also defined the role for the multi level institution of governance in the planning and implementation of the integrated development programmes.

Accordingly, the state government have constituted such local level bodies with definite mandate, to collect revenues, draw up and to implement integrated development programmes at various levels. It needs capacity building at the Panchayat Raj Institutions of Local Self-Governance to prepare and implement local area specific rural development and integrated rural development schemes through assessing and utilizing the available natural and human resources at grassroots level, i.e. village level.

The formulation of planning process and the implementation of integrated rural development programmes are an information intensive task. The available natural resources and human infrastructure resources information system at grassroots level, i.e. village level-block level-district level are not properly arranged while such available infrastructural information needs to be integrated for the developmental planning process.

The conventional methods of data collection, collation, storage of the information are not amendable for quick updation, retrieval and holistic analysis for the formulation of planning process. The several problems which are existing in the conventional system of planning process are as follows:

- In compatible and non-standard data sets instituting integration of sectoral indicators of the data sets.
- In efficient inter-sectoral data flow at various levels.
- Lack of information of quality, completeness and linkage of the data.

Integrated approach to planning process requires a detailed knowledge of the inter-relations and inter-dependencies between various sectoral data to resolve often conflicting demands. It leads to a requirement for appropriate data management system, analyzing tools, technologies and large matrix of sectoral indicators data, in a digital format on natural resources, demography, socio-economic, infrastructural information system and invites for specific scientific and technological interactions.

- Creation of need base data and information system.
- Development and upgradation in the data handling and analysis tools.
- Efficient distribution of data base and information system to the user institutions and developmental planning and programmes implementing agencies/institutions.

Thus the innovation of available natural, social, economic and institutional infrastructure at village level and developing Village Information System at grassroots level has become a challenging task for the policy makers and planners to develop the need based village development planning and also to assess the progress of on-going implementation of various rural development programmes being implemented at village level. The people's participatory planning process and implementation of rural development programmes through the people's involvement can be an asset for successful implementation of the programmes for realizing the aspirations of the people for better quality of life.

2. REVIEW OF THE LITERATURE

There are insignificant studies on innovation of infrastructure and developing Village Information System at grassroots level. Development is afterall a multifaceted phenomenon, which can neither be understood, nor promoted unless it is seen from interdisciplinary and holistic perspectives. Moreover development is basically a human enterprise and therefore, it requires a combined efforts of all systems of knowledge, be they physical, biological, social, human to comprehend and articulate it. It is more, so with respect to planned development. Planning must draw upon all relevant disciplines including physical on the extreme and linguistics on the other has been explained by R.P. Mishra in the changing desceptions of development problems.

Oufer has analyzed the planning in Intervening Opportunities, a theory relating to mobility and distance. The policy makers and planners should not only be a technocrat to provide projections but they should assist the villagers as a social-scientist who can easily understand the basic grassroots level need of the village people.

R.P. Mishra has analyzed the regional planning and tried to explain the various aspects of regional level planning process. H.R. Yadav has explained the Village Development Planning at village level in Mangraura Block of Pratapgarh District. In this volume various Natural, Social, Economic and Institutional resources information has been analyzed, various rural development programmes being implemented at village level has been explained and village level planning of Mangraura Block of the district Pratapgarh has been analyzed for the integrated village development planning.

In view of the increasing demands of the policy, planners the development in the areas of Information and Communication Technologies (ICT) and Geo-spatial Data Management and Analysis Techniques like—Remote Sensing and Geographical Information System (GIS)—certain initiatives have been taken at the national level to create awareness on the usefulness of these technologies among policy makers, planners and decision makers.

Natural Resources Data Management System (NRDMS) Programme of Department of Science and Technology has been one such pioneering effort, while the others are being:

- National (Natural) Resources Information System (NRIS) of the Department of Space, Government of India.
- Integrated Mission for Sustainable Development (IMSD), Department of Space, Government of India.
- Geographic Information System (GIS/NIC) and District Information System (DIS/NIC) of the National Informatics Centre, Government of India.
- National Spatial Data Infrastructure (NSDI), Department of Science and Technology, Government of India.

The Department of Science and Technology, Government of India, started, during 1982, programme on Natural Resources Data Management System (NRDMS) which is multi-disciplinary and multi-institutional research and development programme with a vision to provide appropriate techno-scientific inputs for operationalizing the concept of decentralized development planning at various levels.

This is longest known tradition of systematically collecting spatial data through various organizations at National level in our country. Different line departments are engaged for collection of non-spatial data in strong traditional system. Hanumantha Rao Committee (1984) has emphasized a management of the spatial data for efficient local level planning.

An effort has been made to develop scientific data based approach to support the local level planning. At the end of the first decade of the programme, 73rd and 74th Constitutional Amendments came into being and with the entry of computers at District level and Geographic Information System (GIS) in India during late 1980s the goals of NRDMS programmes were evolved to develop spatial data management technologies for integrated rural development planning and its demonstration in the local level decision making. In the context of Natural Resources which is important for local area development is relevant for the study. Natural Resources Data Management System (NRDMS) has been supporting several academic and R&D Institutions through Grants-in-Aid programmes.

Natural Resources Data Management System has been focusing on technological and institutional capacity building in a multi-level framework so that the people are able to utilize the local level resources. In this regard NRDMS develops and provides tools, and technologies for assessing local level natural resources endowment, their utilization of potential and generating alternative scenarios for the development.

NRDMS operates through District Data Base Centre as its hubs. There are 45 pilot District Centres located in Karnataka, Andhara Pradesh, Orissa, West Bengal, Bihar, Uttar Pradesh, Uttarakhand, Haryana, Gujarat and North-Eastern Region.

In spite of the national efforts, a large segment of end-areas yet to explore the full potential of the system and use in most of their planning activities due to lack of technical, organizational and institutional capacity. To reduce the impediments in the widespread use of the technologies and make it an effective tool in the process of local level integrated village development planning. NRDMS is focusing its research on simplifying the methods and techniques of data collection, organization and analysis, devise suitable strategies to transfer the techniques for its adoption by the end of the use in their planning process.

Recent trends in the development of spatial data technologies, the emerging changes in the system governance of policy planning and institutionalization of National Spatial Data Infrastructure (NSDI) tried to need the aspirations of the users and in support of the overall transformation in the concept of information sharing with the other user institutions. The need felt for focusing the Research and Development thrusts of the programes are as follows:

- Spatial Data Management and Analysis Technologies;
- Advanced Technology Application in large scale spatial data gathering;
- Land and Water Management; and
- Technology Transfer and Capacity Building.

The basic main vision of the programme has been to enable people, institutions and communities for self-governance with spatial data technologies for sustained participation of the local people in the decision making process for the integrated development of the local areas.

The important goals of the programmes have been to strengthen the Science and Technology institutions for integrated resources management through development of appropriate tools and techniques in the data collection, organization and analysis. Technical capacity building at various levels of administration for enabling the adoption of the developed technologies for an integrated development planning.

The major objectives of the Natural Resources Data Management System has been to promote and demonstrate the use of Spatial Data Technologies for micro level planning under diverse terrain conditions.

To provide software support for the data management modelling and operation research.

To promote Research and Development in spatial data technology.

Technology transfer and capacity building of potential users.

Forage linkages with users at different levels.

To provide Science and Technology inputs for framing policies related to spatial data technologies.

To develop and demonstrate pilot scale infrastructure and provide research support to National Spatial Data Infrastructure and documentation and dissemination.

In order to implement at least some of the activities to realize the vision of Natural Resource Data Management System support from a large segment of Scientists, entrepreneurs are required. Natural Resources Data Management System is expected to stimulate the imagination of the research community to relate specific problems in their areas of interest and evolve suitable research proposal to pressure line of investigation to arrive at meaningful solutions to the problems as well as in advancing the existing knowledge in certain sphere of activity thereby aiding the overall development of the area.

3. BACKGROUND OF THE STUDY

It has been argued that an integrated and faster development is required to bridge the gap between **urban** to **rural** and **rurban** areas. An appropriate faster development of rural areas are required to fulfil the increasing demand of even growing population, so that the disparities between rural to urban can be bridged to ensure integrated development.

It has been realized that the development should be participatory and it should be planned according to the local needs of the deprived rural multitudes. Development should be planned through tapping the village infrastructure, viz. natural and human resources to ensure successful implementation of the development programmes self-sustainable.

Several rural development programmes are being implemented aiming at the sustainable integrated development of the rural area with the focus on enhancing the livelihood of weaker and vulnerable sections of the society. In conformity with the commitment to accord foremost priority to sustainable village development a number of steps have been initiated to make the implementation of ongoing rural development programmes more effective and accelerated.

The formulation and implementation of rural development programme at village level in an intensive, innovative, informative and infrastructure (Natural and Human) oriented activity. The natural and human resources/infrastructure, information system at village level are not properly arranged but such available information needs to be integrated and innovated for integrated village development planning. The conventional methods of information collection, storage, and collation of the collected data are not appropriate for quick updation, retrieval and holistic analysis and several problems are existing in the conventional system.

Participatory planning and development should be initiated and sustained through the participation of rural people at the grassroots level, i.e. village level.

It has also been mentioned in the Constitution of our country that government should evolve methods to unite the Village Panchayat and provide such powers and rights which can be helpful to work as an unit of Self-Government.

The father of the Nation Mahatma Gandhi stated that, "*In the Panchayati Raj System, each village should be self-sustained and independent, the implementation of the Welfare programmes of rural masses can be implemented in a better manner in the villages of vast rural India.*"

In view of the above, as per policy of the Government of India, the State Governments established Self-Government to strengthen the Panchayati Raj Institutions in the rural areas. Historical and revolutionary decisions have been taken to allot the work and to allocate the financial resources to implement the planned programmes and also allocated financial, administrative rights to provide the Self-Governance of Village Panchayats-Block Panchayat-District Panchayat.

The real meaning of democracy is important to the people and Self-Government is the main sole of rural development, management and financial and administrative control of village panchayat and continued participation, cooperation of the local people to establish village panchayat for Self-Governance—**For the people, by the people and with the people** to ensure participatory planning and successful implementation of rural development programmes for self-reliance.

The optimum utilization of available Natural-Social-Economic-Institutional infrastructural resources to meet the increasing demand of growing cattle and human population for Fuel Wood, Fodder, Fibre, Fruits, Fisheries, Food grains (F^6) and vegetables should be ensured through self-government at grassroots level, i.e. Village Panchayat level to ensure the **Self-Government-Panchayati Raj-Lok Raj-Gram Raj-Ram Raj** in which scientific technological infrastructure facilities are procured in an innovative manner to generate employment and increase the income of deprived rural multitudes surviving in the remote rural areas, to improve their socio-economic conditions for an integrated sustainable development.

In view of the above, an attempt has been made to point out certain innovative, infrastructural resources, i.e. natural and human resources (socio-economic-institutional) and the implementation of various rural development programmes being implemented under the various schemes of the Government/State Government/Non-Government organizations. It has also been attempted to explain the need based village development planning through peoples participation for an integrated village development.

4. OBJECTIVES OF THE STUDY

The main objectives of the study are as follows:

(1) To innovate the available natural resources/infrastructure at village level in Amethi Block.

(2) To study the Physiographic, social, climate, land use, wastelands, land capability and agro-afforestation and natural vegetation at village level for all 44 village panchayats of the Amethi Block.

(3) To develop the village information system for the available natural infrastructure at village level for all 44 village panchayats of the Amethi.

(4) To develop Geographic Information System of the natural infrastructure based on the each village khasra map of all 44 village panchayats of Amethi.

(5) To analyze and develop geo-referencing of the available natural infrastructure Village Information System based on the village khasra map of each 44 village panchayats and analyze the natural resources/infrastructure of each village separately at grassroots level in Amethi.

The innovation of available village infrastructure and developing Information System has been analysed in three volumes:

Volume-I: Village Information System (Innovation of Natural Infrastructure).
Volume-II: Village Information System (Innovation of Socio-Economic Institutional Infrastructure).
Volume-III: Village Information System (Development Planning).

5. STUDY AREA

District Amethi of Eastern Uttar Pradesh has been recently created, taking Tiloi and Salon Tehsils of the Rae Bareli district and Amethi, Gauriganj, Musafirkhana Tehsils of the Sultanpur District of Uttar Pradesh.

The Rae Bareli district was created by the British rulers in 1858, it was named after Rae Bareli town as its headquarters, which is a part of the Ganga plain. The Sultanpur

District has been named after its headquarters town Sultanpur. The Rae Bareli District was also called Kosala-Vishaya, Kosala-Mahainapada or the Kosala country. The river Sai formed the southern boundary of Kosala. Rae Bareli district has civilized and settled life since the medieval period of history. It is believed that Rae is corrupt form of Rahia nearby village and Barelli that of Bharaula. In the ancient times the Sultanpur district Headquarters town was known as Kusapura or Kush Bhawanpur after the name of its founder Kusha son of Rama. The town was held by Bhars till the end of the 13th Century, when two horse dealers Saiyid Muhammad and Saiyid-Ala-ud-din preferred some horses for the sale of the Chief of Bhars. He seized the horses and killed both of them. After about one year unsuccessful seize, the old town of Kush Bhawanpur was reduced to ashes and new town called Sultanpur came up in the existence which was named from the rank of victor.

The newly created district Amethi, geographically lies between 81°02′E to 81°52′E longitudes and 25°41′ North to 26°40′North latitude. The district Amethi is bounded by Barabanki and Faizabad districts in north, Pratapgarh and Rae Bareli districts in the south and west, Sultanpur district in the east and north (**Map 1.1**).

The district Amethi is predominantly rural base in nature, agriculture is the main occupation of the people, over 85 per cent of the population are engaged in the agricultural sector of economy. The industrial and tertiary sectoral economy engaged about 15 per cent of the population of the district. Thus the district Amethi can be easily called an agrarian area.

The district Amethi is divided into five tehsils and the headquarters of the tehsils are Amethi, Gauriganj, Musafirkhana, Salon and Tiloi (**Map 1.2**). There are 16 blocks in the district.

Fig. 1.1: District Amethi

Sl. No.	*Tehsil*	*Blocks*
1.	Salon	1. Salon, 2. Dih, 3. Chhatoh
2.	Tiloi	4. Tiloi, 5 Singhpur, 6. Bahadurpur
3.	Musafirkhana	7. Shukul Bazar, 8. Jagdishpur, 9. Musafirkhana
4.	Gauriganj	10. Gauriganj, 11. Jamo, 12. Shahgarh
5.	Amethi	13. Amethi, 14. Sangrampur, 15. Bhnetua, 16 Bhadar.

There are 44 Village Panchayats (**Map 1.4**) and 82 revenue villages (**Map 1.3**) and 372 hamlet villages/purwa/tola in the Amethi block of the District Amethi of the Uttar Pradesh. The detailed physiographic location and topographical variations are depicted in **Map 1.5**, which has been derived from the Survey of India Toposheets. The District Amethi is in the south of the erstwhile province of Oudh and is the southern-east portion of Lucknow. The list of the Village Panchayats, revenue villages and hamlet villages/purwa/tola are mentioned below:

Table 1.1: District Amethi—Village Panchayat/Revenue/Hamlet Villages

Sl. No.	*Village Panchayat*	*Revenue Villages*	*Hamlet Villages/Purwa/Tola*
1.	**Katara Maharani**	**1. Katara Maharani**	1. Katara Maharani, 2. Munshiganj, 3. Chauraha, 4. Usarapur.
		2. Laukipur	1. Laukipur
		3. Matiyar Baghaura	1. Mariyar Baghaura

(Contd.)

Table 1.1 (*Contd.*)

Sl. No.	*Village Panchayat*	*Revenue Villages*	*Hamlet Villages/Purwa/Tola*
2.	**Sarai Khema**	**1. Sarai Khema**	1. Sarai Khema Khas, 2. Munshiganj, 3. Basuapur, 4. Halapur, 5. Beni Madhavpur, 6. Kutia, 7. Ahmad Nagar
		2. Dandupur	1. Dandupur Khas, 2. Pure Kutur, 3. Jagapur
		3. Ram Nagar Mafi	1. Pure Pandit, 2. Mainapur
3.	**Trilokpur**	**1. Trilokpur**	1. Pure Subedar, 2. Pure Bhagwandin, 3. Trilokpur, 4. Pure Darshan, 5. Dubepur.
		2. Teri	1. Teri, 2. Kutur Ka Purwa.
4.	**Ramdaipur**	**1. Ramdaipur**	1. Ramdaipur, 2. Chhawani, 3. Punnapur.
		2. Bhupatipur	1. Bhupatipur.
		3. Naugnawa	1. Naugnawa.
		4. Bahapur	1. Bahapur, 2. Sitarampur, 3. Rajan Ka Pura, 4. Gaudawa.
5.	**Jangal Ram Nagar**	**1. Jangal Ramnagar**	1. Chhawani, 2. Pure Mannuram, 3. Pure Durga Pasi, 4. Ram Nagar, 5. Ram Nagar Pachhimi, 6. Katara, 7. Shiv Darshan Nagar, 8. Hanumat Nagar, 9. Amriti, 10. Uttari, 11. Van Saţahi, 12. Pure Swayambar, 13. Pure Pathak, 14. Sitarampur Usari, 15. Pure Jhilcai, 16. Pure Pandey, 17. Dhamrawa, 18. Pure Bodhan, 19. Derkha Chhatolia, 20. Pure Gayasi, 21. Baramati, 22. Pure Parsada, 23. Pure Bhwal, 24. Pure Digarihan, 25. Pure Sheetal, 26. Pure Hira Simar, 27. Pure Pandit, 28. Pure Belaura, 29. Semsahawa, 30. Pure Bhagan Ahir, 31. Bhusupar, 32. Pure Ram Prasad Kurmi, 33. Ranbir Nagar Uttari Bhag, 34. Ranbir Nagar Dakshini Bhag, 35. Gahari, 36. Gondwa, 37. Pure Murai, 38. Pure Mistri, 39. Karbala, 40. Devi Patam, 41. Kakarahi, 42. Amriti. 43. Mahua Tali. 44. Chhiwala Tali, 45. Walada. 46. Pure Dutta, 47. Usarahia, 48. Pure Banaha.
6.	**Benipur**	**1. Benipur**	1. Benipur, 2. Benipur New
7.	**Parsanwa**	**1. Parsanwa**	1. Pure Bhatani, 2. Parsanwa, 3. Pasanwa Khalawa, 4. Parsanwa Dihwa.
		2. Ram Nathpur Chhota	1. Ram Nathpur Chhota.
8.	**Mahmoodpur**	**1. Mahmoodpur**	1. Mahmoodpur, 3. Ramnathpur Bada, 3. Munshiganj Road, 4. Lakharia Khera Ke Pichhe, 5. Awas Vikas Gopalpur.
		2. Marafapur	1. Marafapur, 2. Mainapur.
		3. Valipur	1. Valipur Khas, 2. Valipur Aam.
9.	**Raipur Phulwari**	**1. Raipur Phulwari**	1. Pure Ram Prasad Pandit, 2. Baniyani, 3. Salamat Ali Ka Purwa, 4. Ojha Ka Purwa, 5. Datta Ka Purwa, 6. Kakawa Mod, 7. Raipur Phulwari.
		2. Katara Raja Himmat Singh	1. Katara Raja Himmat Singh
10.	**Umapur Ganapatti**	**1. Umapur Ganapatti**	1. Bhagwanpur, 2. Pure Guru Baksh, 3. Pure Prem, 4. Hati, 5. Pure Ichha, 6. Pure Nidhi, 7. Sri Rampur.

(*Contd.*)

Table 1.1 (*Contd.*)

Sl. No.	Village Panchayat	Revenue Villages	Hamlet Villages/Purwa/Tola
11.	**Bhaganpur**	**1. Bhaganpur**	1. Bhaganpur, 2. Durgapur, 3. Pure Lal, 4. Pure Nand Mishra, 5. Sringarpur, 6. Pure Bhojai.
		2. Jangal Tikari	1. Jangal Tikari, 2. Dhamranwa.
12.	**Tala**	**1. Tala**	1. Tala, 2. Lodhani Patti, 3. Bhaisapur, 4. Pure Dina Mankata, 5 Baripur, 6. Mahadewan.
13.	**Kushi Tali**	**1. Kushi Tali**	1. Patakhapur, 2. Pure Nardutt, 3. Kushi Tali, 4. Pure Ram Lal.
		2. Haripur	1. Haripur
14.	**Loharata**	**1. Loharata**	1. Loharata, 2. Mali Ka Purwa.
		2. Sompur Mankanth	1. Pure Shiv Gulam, 2. Lala Ka Purwa, 3. Kalu Tewari Ka Purwa, 4. Chak Belan.
		3. Ahirawal	1. Ahirawal, 2. Pure Mod.
15.	**Darakha**	**1. Darakha**	1. Pure Chandika, 2. Pure Shiv Lal, 3. Pure Bariyar, 4. Pure Dutta, 5. Darakha Diti, 6. Pure Jagannath, 7. Naya Purwa, 8. Pure Shital.
16.	**Mochawa**	**1. Mochawa**	1. Mochawa, 2. Maheshpur.
		2. Sarai Hirmati	1. Sarai Hirmati, 2. Barsati
		3. Jhaunei	1. Jhaunei, 2. Pure Daya Mishra
		4. Sultanpur	1. Sultanpur, 2. Pure Daljit, 3. Pure Dashratn, 4. Pure Galhin, 5. Pure Daya Mishra.
17.	**Nuwawan**	**1. Nuwanwa**	1. Nuwanwa Khas, 2. Dubepur, 3. Kodari, 4. Saidapur, 5. Teliyani, 6. Pure Gopal, 7. Pure Yamati.
18.	**Loniapur**	**1. Loniapur**	1. Mattha Shukul Ka Purwa, 2. Teliyani, 3. Badhai Ka Purwa, 4. Loniapur.
		2. Bharathipur	1. Bharathipur
19.	**Kherauna**	**1. Kherauna**	1. Kherauna, 2. Baramati, 3. Kewalapur, 4. Hardew Nagar.
		2. Sarai Raj Shah	1. Sarai Raj Shah
		3. Sarwanpur	1. Sarwanpur, 2. Keshawa Nagar, 3. Gayatri Nagar, 4. Gangaganj, 5. Sringarpur, 5. Pure Bhojai.
20.	**Katara Phulkunwar**	**1. Katara Phulkunwar**	1. Katara Phulkunwar, 2. Bahorapur, 3. Tewaripur, 4. Kewalapur.
		2. Jagipatti	1. Jagapatti
		3. Baghwaria	1. Baghwaria
		4. Parsauli	1. Parsauli
21.	**Hathkila**	**1. Hathkila**	1. Hathkila, 2. Dewaraka Tola, 3. Dewatar, 4. Baripur.
22.	**Rebha**	**1. Rebha**	1. Rebha, 2. Pure Pothi
		2. Mandharpatti	1. Mandharpatti
23.	**Raidaipur**	**1. Raidaipur**	1. Raidaipur
		2. Chachakapur	1. Chachakapur
		3. Birahimpur	1. Birahimpur Kundwa.
24.	**Korari Girdhar Shah**	**1. Korari Girdhar Shah**	1. Pure Matadin, 2. Ghatampur, 3. Pure Bhagwan, 4. Mahadewan, 5. Pure Roshani, 6. Baba Ka Purwa, 7. Pure Telin, 8. Pure Shoschiman.
		2. Mangalpur	1. Mangalpur

(*Contd.*)

Table 1.1 (*Contd.*)

Sl. No.	*Village Panchayat*	*Revenue Villages*	*Hamlet Villages/Purwa/Tola*
		3. Dariyapur	1. Tewari Ka Purwa, 2. Gohanwa, 3. Dariyapur, 4. Bikham Bharpur, 5. Manharia.
25.	**Dedhpasar**	**1. Dedhpasar**	1. Tulapur, 2. Parashuram, 3. Dedhpasar, 4. Sarai Kodha, 5. Pure Dakshin.
		2. Jamalpatti	1. Pure Uttar, 2. Jamalpatti.
26.	**Mahaso**	**1. Mahaso**	1. Mahaso, 2. Basupur Tola, 3. Dewatar, 4. Paripar.
		2. Gyan Chandpur	1. Pandit Ka pura, 2. Banktawa, 3. Pahia.
27.	**Gangauli**	**1. Gangauli**	1. Pandit Ka Purwa, 2. Subedar Ka Purwa, 3. Gangauli Matera, 4. Pure Diggaj Ka Purwa, 5. Bari Ka pura, 6. Tulsi Pandey Ka Purwa, 7. Agaha, 8. Pure Kurai, 9. Dina ka Purwa, 10, Soiya.
28.	**Dhandhudhar**	**1. Dhandhudhar**	1. Dhandhudhar, 2. Pure Upadhyaya
		2. Pithipur	1. Pure Hanuman Tewari, 2. Pure Ojhan, 3. Pure Mishra.
29.	**Gaderi**	**1. Gaderi**	1. Gaderi, 2. Sadwa, 3. Banpurwa, 4. Pure Hardatt, 5. Pure Duban, 6. Pure Kalika, 7. Pure Tularam, 8. Rampur, 9. Pure Gangu, 10. Pure Bal, 11. Pure Bidhasia, 12. Samari, 13. Pure Bhagwandin, 14. Pure Hawaldar, 15. Pure Chandra Mani, 16. Bangarwa.
30.	**Bhusahari**	**1. Bhusahari**	1. Bhusahari Khas, 2. Barain Tola, 3. Pure Gaderian, 4. Karnaipur, 5. Makhdumpur, 6. Patipur, 7. Jham Singh Ka Purwa.
31.	**Purabgaon**	**1. Purabgaon**	1. Purabgaon.
32.	**Chaturbhujpur**	**1. Chaturbhujpur**	1. Pure Khuklan, 2. Hanumanganj, 3. Jagat Gaon, 4. Chaturbhujpur, 5. Pure Harsahai, 5. Pilepur.
33.	**Mahrajpur**	**1. Maharajpur**	1. Maharajpur Khas, 2. Kurmiyan Ka Purwa, 3. Khusadi.
34.	**Agahar**	**1. Agahar**	1. Agahardih, 2. Pure Ram Sewak, 3. Pure Bhawani, 4. Pure Gosnai, 5. Pure Patoli.
35.	**Saidpur**	**1. Saidpur**	1. Pure Shivalal, 2. Pure Dala, 3. Pure Vir Singh, 4. Saidpur
		2. Dalshahpur	1. Dalshahpur, 2. Ram Nagar Pur, 3. Sadakpur.
36.	**Kakawa**	**1. Kakawa**	1. Kuwnarganj, 2. Kakawa, 3. Pure Ganesh Baksh Singh, 4. Pure Ram Sahai, 5. Pure Girdhari, 6. Shahadara, 7. Kakarahia, 8. Voda Nagar, 9. Pure Valdi Tewari.
		2. Vidhasiya	1. Pure Gadarian, 2. Vidhasiya, 3. Pure Dhandauia, 4. Pure Valdu, 5. Pure Tada Tewari.
37.	**Saraiya Duban**	**1. Saraiya Duban**	1. Saraiya Duban, 2. Jaggubari, 3. Pure Lal Ram Prasad, 4. Gudiyan Tola, 5. Pure Man Kunwar, 6. Pure Brahmachari, 7. Murarpur.
38.	**Kohara**	**1. Kohara**	1. Harakhpur, 2. Dodhai Ka Purwa, 2. Dihawan, 4. Vishyapur, 5. Devigarh, 6. Kohara Khas, 7. Ahiranki Saraiya, 8. Aditya Nagar (Chamrauti), 9. Pure Jogdhar, 10. Ulapur, 11. Pure Shivdas, 12. Pure

(*Contd.*)

Table 1.1 (*Contd.*)

Sl. No.	*Village Panchayat*	*Revenue Villages*	*Hamlet Villages/Purwa/Tola*
			Hanuman Mishra, 14. Bhawani Baksh (Chaubepur), 14. Pure Jagannath, 15. Baniyan Ka Purwa, 16. Ganga Mishra Ka Purwa, 17. Pure Shiv Bhikh, 18. Malin Ka Purwa, 19. Ojha Ka Purwa, 20. Murai Ka Purwa, 21. Pure Bhadosian, 22. Mukunpur, 23. Pure Ghisiya.
39.	**Mahmadpur**	**1. Mahmadpur**	1. Mahmadpur, 2. Shankarpur, 3. Dudhai Ka Purwa, 4. Kalu Ka Purwa, 5. Sitaram Nagar, 6. Chaubepur, 7. Kurmi Ka Purwa, 8. Pure Baksh Singh, 9. Hulasi Ka Purwa, 10. Bhawani Baksh Ka Purwa, 11. Pure Pandit Ka Purwa, 12. Pure Hari Baksh.
40.	**Naraini**	**1. Naraini**	1. Narainidih, 2. Pure Pachhim, 3. Pure Matlu, 4. Pure Wakta, 5. Dadi, 6. Pure Gangeran, 7. Pure Ayodhya, 8. Pure Khojhwa.
41.	**Nainaha**	**1. Nainaha**	1. Nainaha Khas, 2. Pure Tewari, 3. Pure Bishen, 4. Dhanapur, 5. Pure Satya Shukul.
		2. Bartali	1. Bartali Khas, 2. Dhaksahiya.
		3. Rajapur Kohara	1. Pure Jangali, 2. Damgarha, 3. Rajapur, 4. Ajyab Sinda, 5. Sadarpur.
42.	**Ramgarh**	**1. Ramgarh**	1. Badisoiya, 2. Chhoti Soiya, 3. Pure Bhoop, 4. Rampur, 5. Bahni, 6. Mahoka, 7. Saraiya Purwa, 8. Pure Kotwa, 9. Ramgarh Khas.
43.	**Himmatgarh**	**1. Himmatgarh**	1. Pure Bhoop, 2. Madaripur, 3. Pure Ajit, 4. Jhali, 5. Visheshganj, 6. Pure Upadhyaya, Himmatgarh.
44.	**Dehra**	**1. Dehra**	1. Dehra Khas, 2. Pure Darjiyan, 3. Pure Shivpal, 4. Pure Pandey, Shanichara, 6. Gulabgarh, 7. Shukla Ka Purwa.
		2. Kudwa Mohamadpur	1. Daya Garha, 2. Pure Jhingur, 3. Kudwa Mohamadpur.
Total	**44**	**82**	**377**

Villages Panchayats, Revenue Villages/Hamlet Villages and Village Panchayatwise location of all 44 village panchayats of Amethi block is depicted in **Map 1.4** of the District Amethi. The villages of the District Amethi resemble those of the rest of the southern Audh and are collection of mud built huts. In some places hamlet villages of Tola or purwas, large and small are scattered all over the face of the newly created District. Although elsewhere the number of inhibited sites are also existing which is one of the major characteristics of the District Amethi. Some of the hamlet villages or purwa or tolas are very small and contain less houses, their origin being the desire of a careful cultivators to live close to his fields, which has resulted appearing in high cultivation and flourishing crops and hence increasing agricultural productivity, while the tenants reside in the main settlement area (*Abadi*) due to which the outlying crops are usually found very poor in terms of agricultural productivity. The topographical and specific general description of all 44 village panchayats of Amethi Block as per topographical sheets of Survey of India has been depicted in **Map 1.5.**

6. TOPOGRAPHY

The topography of the district is irregular in shape but fairly compact tract of undulating land. The general slope is from north-west to south-east. The highest points of the crowns of the water sheds of the different drainage channels. The land is fairly flat and gently undulating, the highest points being the crown of the watersheds. The slope of the land from the north-west to south-east varies as per the direction in which the rivers drain. The elevation varies from about 11 metres above sea level in the north-east to about eight metres at the extreme south-east corner. The relief and slope gradient of the district is depicted in **Map 1.6**.

The land of the District Amethi is generally plain except near ponds and adjacent to lakes. The river Gomati enters in the district in the north-west corner of the Shukul Bazar Block and runs in the well defined course towards the south-east through Jagdishpur and Musafirkhana Block of the District Amethi. Finally, the Gomati River flows through Sultanpur District into Jaunpur District.

Towards the south of the river Gomati the land can be classified into different three belts, viz.:

(i) A riverain tract along the bank of the river;
(ii) A central belt of cultivated villages; and
(iii) Picturesque landscape and a belt of rice lands with large arid plains of usar land and swampy *jhils*.

Many parts of the District Amethi are ill-drained. There are numerous swamps and lakes in the district, while some of the swamps and lakes are seasonal and are apt to dry off in summer. Such lakes and swamps can be exploited to develop for fisheries particularly when the state government has been going ahead in a big way in establishing fisheries estates in every district utilizing waterlogged land especially in the swampy land, lakes and the land along the canal banks. Such fisheries estate has been established in Bhojpur village of the Amethi Tehsil and Block Bhadar of the District Amethi.

The central track of Sai upland on the northern clay tract has been divided due to the crown of the watersheds of the streams and lakes. There are large number of marshes and irregular inland water bodies while most of the water bodies do not serve any useful land scape and it can be developed for the fisheries culture.

7. GEOLOGY

It has been recorded that the highest points are the crowns of the watershed of various drainage systems. Geologically the District Amethi forms the part of the Gangetic plain which is of the recent origin according to the geological chronology and it reveals ordinary Gangetic alluvium containing the modules of Calcareousness or *Kankar*. It has been noted that there are deposit of the red and brick earth, on the basis of the geology and topography of the southern part of the district can be divided into the Sai catchment area which comprises Salon and Tiloi Tehsils of the District. The northern tract of the District is lying in the south of the river Gomati comprising Amethi, Gauriganj and Musafirkhana Tehsils, which can be further divided into the reverine tract of the gornatic and the lake tract. Geologically the District Amethi does not reveal anything striking except the ordinary Gangetic alluvium. The *kankar* is only the mineral found along the bed of the river Gomati. It lies at the depth upto about one metre from the surface. *Reh* and the *Multani Mitti* (Clay soil) is also found in some parts of the District Amethi. The *reh* is used for washing the clothes and *Multani Mitti* is used for the pottery purpose.

Older alluvial, saline, alkaline soils and ravines are found in the District Amethi. The sedimentary (unconsolidated) rocks and minerals are depicted in **Map 1.7**.

Thus, geologically there is no noteworthy feature in the physiognomy of the District Amethi and geologically it is not important. It is characterized by unrelieved monotony of alluvial deposits which is of several thousand metres. The solid geology of the area is totally obscured underneath the mantle. Rock and Minerals of the District Amethi is depicted in Map 1.7. The *Kankarili* land and ravine land is found in the northern part of the district. The soils are sandy loamy, clayey and subdivided into *Khadar* and newer alluvium. The land of the district is formed of rich alluvial soil which is of great agricultural importance but *kankar* or calcareous lime stones are common. Various geologists have developed different hypothesis to explain the geological evolution of the plain.

Geologically, there is no striking importance except the ordinary gangetic alluvium in the District Amethi. *Kankar pan*, *Reh* and *Multani Mitti* is found in the different parts of the district. Other alluvial soils and saline, alkaline and saline-alkaline soils are also found in the different parts of the District Amethi.

8. HISTORICAL BACKGROUND

The Gazetteer of the district published in 1903 A.D., throws some light on the historical background of the area. It has been observed that the chief land owning families of the past were the Rajputs of the different clans. These Rajputs were possessing 76.16 per cent land of the total area of the district. The Raj Kumars alone held over one-fourth of the land of the district, while their kinsmen, the Bachhgotis and Rajwaras owned 11.4 per cent and 3.4 per cent land respectively. The Raj Kumars of the Bachhgotis was the Raja of Kurwar while the Taluqdar of Samrathpur represented another branch of the family. The chief of the Rajwaras was the Taluqdar of Pratappur. The another member of the Rajwaras family was the Raja of Hasanpur. The families of the Maniarpur and Gangeo and between them they owned the large portion of the land area. Next to the Bachhgotis and their kinsman comes the Bandhalgotis, who owned almost the whole of the Amethi Pargana. Raja of Amethi was their head, while the Taluqdar of Shahgarh belonged to the same clan. The other Rajputs with large properties in the District Amethi where the Bhale Sultans owned 4.72 per cent, the Kanhpurias with 4.7 per cent and the Bais with 2.8 per cent. The half of the Bhale Sultan Rajputs were belonging to Hindus while half of them were Muslims and they were dwelling in the northern parts of the district in the Musafirkhana and Jagdishpur Blocks. The Kanhpuria Rajputs were chiefly confined to Pargana Gaura Jamo, almost all of which belong to them. The Bais Rajputs were scattered about the small groups.

During the beginning of nineteenth century Babu Madho Singh, eleventh in descent from Jioharain was the rural of the estate which consisted of 100 villages. Babu Madho Singh had been remembered as successful leader and managed his property well and died in 1823 and he had been succeeded by his widow, namely Thakurain Dario Kunwar, a most remarkable woman who through toils and turmoil, not only bravely held her own, but added to her estates than her husband had done in his life time. With the death of the Thakurain's husband, Babu Madho Singh the direct line of succession had ended. Babu Rustam Shah was the next male collateral whom Thakurain disliked. Babu Rustam Shah was in the service of Maharaja Man Singh the Nazim of the day and with his help succeeded in capturing Thakurain and made her to write a deed in his favour. A formidable woman, whose pride was hurt grieved for a few months and later on died. Babu Rustam Shah was given the possession of the property by the Nazim. Later on Rustam Shah came to know that the Nazim had ulterior motives in helping him. A Pandit advised Babu Rustam Shah that the time was not propitious to fight and kill the Nazim. Rustam Shah sought asylum across the British border and was made the Taluqdar of Dera which consisted of 336 villages. Babu Rustam Shah rendered excellent services during the Mutiny. Babu Rustam Shah died in 1877 and he was succeeded by his nephew Raja Rudra Pratap Singh. Bariyar Singh youngest brother of Rustam Shah received an estate of 20 villages and there Patti's in the Pargana of Baraunsa and Aldemau in return for services rendered during the Mutiny.

Amethi' has also strong historical background. Amethi was ruled by Manohar Singh, who had six sons, who was the only son of the first of the Bandhalgotis. The six sons of Manohar Singh divided the estate between themselves. The Raj Singh, the younger among six brothers succeeded in adding to his share those of his brother, Ram Singh and Kunwar Singh who died childlessly. The fourth descendant from Raj Singh, Sri Ram Deo had two brothers, viz. Shyamlal and Dharambir who owned Barna. Tikar Estate on the extreme west and Tikeri estate on the extreme east. Ram Sahai the grandson of Sri Ram Deo was given as his share of the estate Kasrawan on the northern portion of Amethi. The great grandson Sultan Saheb obtained Shahgarh on intermediate between Kasrawan and the older estates. Thus the hold of Banghalgotis seemed to have spread over the entire Amethi.

Gurdatt Singh was the next of the Bandhalgotis who defied in 1743 the local authorities and ultimately he had to flee to the neighbouring Jungles of Ram Nagar. His fort of Rajpur was destroyed in the seize of eighteen days and his estate was taken over. Digpal Singh, the son of Gurdatt Singh, recovered the estate and from his time dates the present Taluqa of Amethi. The property had been called as Udaiwan. Gurdatt Singh was sometimes self-styled Raja and some times Babu. The Raja title was hereditary but it is not known how long it had been adopted by the head of family. Gurdatt Singh has two sons, namely Harchand Singh who became a separate proprietor of Kannu Kasrawan. Harchand Singh obtained the bulk of his fathers possession and Jaichand Singh became a separate proprietor of Kannu Kasrawan. Excepting Raghipur, Harchand Singh owned the whole of Amethi Pargana. During 1810 Harchand Singh was defeated by Saadat Ali Khan. Raja Harchand Singh was left only 44 villages. Dalpat Singh son of Raja Harchand Singh abdictedly recovered during 1813 all his father's property and original estates. Raja Dalpat Shah died in 1815 and he was succeeded by his son Bisheshar Singh and died without child in 1842. Raja Dalpat Shah was succeeded by his nephew Madho Singh who tried to expand his estate and he has faced hostilities from the Nazim of Sultanpur. During 1845 Maharaja Man Singh was given on lease the whole pargana with the exception of a few villages due to in decisive, negotiations. Raja Madho Singh died in 1891 and just after his death his son succeeded by an adopted heir, Raja Bhagwan Baksh Singh, son of Babu Sheodarshan Singh, a relative of the late Raja. The estate of Amethi Pargana consisted of 314 villages and four pattis.

The Brother of Bikkam Shah of Amethi Sultan Shah founded the house of Shahgarh and he built a fort and derived its name and called after himself. The Shahgarh estate consisted of 121 villages. Shahgarh estate was with the rest of the pargana from 1803 to 1810 which has been leased to Harchand Singh and later on taken away and comprised 40 villages which has been increased up to 60 villages during 1846 when it was again given back to Amethi estate.

The Kanhpurias of the district are decedants of Rahas who was the second son of Kanha and who was the founder of the Clan. Prashad Singh who was the seventh descent from Rahas who had three sons, viz. Janga Singh, who obtained Tiloi and Man Singh received Atheha and Madan Singh achieved Semrauta estate. The Udaibhan of Tiloi and Gulal Shah of Shah Mau was the fourth descendant of Janga Singh. In the beginning of twentieth century Raja Tiloi was the descendant of Udaibhan and his property consisted of Suratgarh and Naudand pargana of Gaura Jamo.

The other descendant of Kanhpurias was Indrajit Singh, who was the great grandson of Janga Singh, Balbhadra Singh son of Indrajit Singh had four sons, viz. Pratap Bahadur Singh Raja of Katari who owned 13 villages in the Gaura Jamo pargana and he was the grandson of Balbhadra Singh. Raja Shah the second son of Balbhadra Singh founded the house of Jamo. Mahabir Baksh Singh of Jamo owned 17 villages of the descendant pargana of Raja Shah in the tenth generation. Raghuraj Singh another descendant of Baraulia owned 13 villages in Gaura Jamo. Babu Devi Baksh Singh the son of Tribhuwan Shah and the third son of Balbhadra Singh was the Taluqdar of Bhawan Shahpur owned 12 villages belonging to one patti in the north of Pargana Amethi. The old pargana of Jais was also included in the possession of Kanhpurias which was broken during 1775 into four parganas—Jais, Maharajganj, Semrauta and Gaura Jamo.

Thus there is no systematic archaeological excavation has been carried out in the district accidental finds of ruins of ancient monuments reveal their antiquity. There are remains of a mediaeval **Dargah** known as **Alha-Udar-Ki-Baithak** which had been well known to Banaphar heroes during 12th Century. Ikshavaku a member of the Solar race of the ancient Indian Kshatriyas was earlier known as the ruler of the area. During the 19th descendant in the Solar Dynasty Mandhata was the first emperor under whom the kingdom of Ayodhya became a imperial power. Ikshavaku successors were weak. The Sagara, 38th king in the line retrieved the lost prestige of his fore fathers and became the second emperor in the dynasty, which has been mentioned in the Mahabharata. Well known next king was Bhagiratha. After few generations Dilipa-II came in exhistance and the area was under him and his immediate successors that the areas came to acquire the name **Kosala**. Raghu a great conqueror was the next successor who gave the name of dynasty as Raghuvansh. The reign of Rama was characterized by plenty and prosperity. Rama was most widely known king of Kosala and his rule was known as **Ram Rajya** symbol of ideal rule. Rama divided his kingdom towards the end of his life among his sons and the area represented by Rae Bareli came to the dominion of Kusha, Rama's eldest son who succeeded to the throne and had his capital at Ayodhya and the later on the area was ruled by the Kosalan Kings. The son of Maha-Kosala, Prasenjit was the contemporary of Mahavira and Buddha, who was the last great monarch of Kosala. In the Buddhist and Jain literature the Kosala Kingdom figured as one of the sixteen Maharajan Padasie great kingdom.

The fifth descendant of Prasenjit, Mahapadma the Nand King of Magadha defeated Sumitra during the middle of the fourth century B.C. and annexed the entire territory of Kosala including the other area presently knwon as Rae Bareli. The Maurya emperor of Magadha succeeded Nandas. The Kanishka, the Kushana Emperor (78-102 A.D.) extended his dominion over the whole of northern India including Kosala. The Kushana dominion came into end towards the end of the second century A.D. Several dynasties had been established after the ruins of Kushana empire was that of Bharshivas, which was a branch of Nagas. It had been observed that a considerable number of Bharshivas has infiltrated into this region which had been the forebears of the later day Bhars of this area.

During the emperor Harsha's period the Chinese pilgrim Hiuen Tsang visited India in 636 A.D. Hiuen Tsang refers in his travelogue to A-Yu-te, which means 'yielded good crops and luxuriant in fruits, vegetables and flowers and has genial climate. The people at that time were found of good works, had agreeable ways and determined to the practical learning'. After the death of King Harsha, the Yashoverman of Kannauj ruled the area during the 8th century.

During the 9th century the area formed part of the dominion of the Gurjara Parthiharas. Rajput Raja Argal had relationship with the Gurjara. Parthiharas who were holding over the area but could not control the Chief of the Bhar who has consolidated strength of the Bhars.

The invasion by Mahmood Ghajani during the end of the 1018 A.D. marked the end of Gurjara Parthihara dynasty, which was followed by the establishment of the Gahadwala dynasty. Gahadawala dynasty came to an end due to defeat in 1193 A.D. of their king Jaichand by Shihab-ud-din Ghauri. Taking the advantage of local political conditions the Bhar Chief occupied the area. Various old mounds called **Dihs** or **Kherus** are supposed to contain ruins of strong Bhar castles, or fortresses, such as those at Inhauna, Jais, Parshadepur, Salon etc. During 1206 after the death of Shiab-ud-din Ghauri, Qutub-ud-din Aibak assumed supreme authority over the Gauri's Indian dominions and thus founded the Slave dynasty of the Sultans of Delhi. Kanhpuria Rajput clan during 13th century was the progenitor of the clan founded the village Kanhapur which had given its name to clan. During the end of the 14th century. **Malik Sarwar Khwaja-i Jahan** was appointed Subedar of the Awadh and established himself at Jaunpur as an independent ruler and Sharqi dynasty was founded by him. The Bhar chief became dominant taking advantage of the change but Sharqi kings resented the Bhars and Sharqi kings army routed the Bhars to the last man in the area. The Kanhpuria, Bais, and other Rajputs helped Ibrahim Shah Sharqi in his campaigns against the Bhars and as a result of the latters annihilation had gained considerable strength and possession in spite of that king's efforts to keep them under control, drove out a majority of the Muslim Zamindars of the area.

During 1457 Mahmud Shah was succeeded by his son who was killed by his brother after few months. Husain Shah Sharqi ascended the throne of Jaunpur in 1458 and restored the ousted Muslim to their power and possessions in the area and changed the name of town Rae Bareli to Husainabad which did not survive its originator's effort. Sharqi King was finally defeated by Buhlul Lodhi in 1486 and the local Rajputs strengthen their position. During 1556 the great Mughal emperor, Akbar ascended the throne of Agra and during the 1559 Khan Zaman an imperial officer was sent to oust the Afghans from Jaunpur and connected territories including this area.

Under the administrative arrangements of the Akbar's empire the area was divided between the Sirkars of Awadh and Lucknow of the Subah of Awadh. The management of Mahals or parganas then existing is not clear altogether but with the aid of **Ain-i-Akbari** and local tradition it had been more or less satisfactory. The Sirkar of Manikpur made 12 Mahals out of which 9 Mahals lay in this region. The northern most Mahal of Bhalol or Bhilwal of their Sirkar included the northern part of the pargana Kumhrawan. The large Mahal included the parganas of the same name contained a brick fort on the Sai and a Military contingent, the Chief land holders being the Daurias and Rajputs.

The Mahal of Salon was possessed a brick fort which was held by the Kanhpurias and Bisen Rajputs. To the north of this Mahal was the Mahal of Nasirabad which was included into the present pargana of Parshadepur and the old pargana of Rokha. To its north lay the Mahal of Jais which was included in the old pargana of Jais, and the pargana of Mohanganj and a position of pargana Simrauta was held by Kanhpurias and the area was dominated by them.

A number of villages of the Salon Tehsil was included in the two Mahals, Qaryat Guzarah and Quaryat Paegah.

The Mahal of Anhonah (Inhauna) of the Sirkar and Subah of Awadh corresponded to the present pargana of Inhauna of this District Amethi.

Ahmad Beg Khan nephew of Nur Jahan held the fief of Jais and several fine buildings, such as tombs and palaces were built. During the reign of Aurangzeb, 1974, Sheikh Abdul Karim built a mosque at Jais and during 1678, the local emperor ordered the repair and restoration of the Masjid.

The Kanhpuria Raja of Tiloi has been the most prominent Rajput Chief of the region and during the last quarter of the eighteenth century, the history of the area reflects the rise of local Bais and Kanhpuria chiefs. The end of the Akbar's reign had marked a season of great vitality among these Rajput families.

The Saadat Khan the Governor of the province of Agra was given the charge of the Subah of Awadh during 1722 and by 1724, he had firmly established himself, although nominally subedars of the imperial government were virtually independent rulers of the Kingdom of the Avadh. Saadat Khan made a tour of the area in order to secure the submission of the powerful local Rajput Chiefs who cherished their traditional freedom and sovereignty and were very reluctant to yield to the Nawab. Particularly Mohan Singh of Tiloi put up a gallant resistance to the Nawab's efforts.

During 1738, Saadat Khan was succeeded by his nephew and son-in-law Safdarjang (1739-54). Mohan Singh of Tiloi continued to give trouble to the Nawab as well and had to pay for it with this life. His son Prem Singh, succeeded to the estate had been a man of peaceful disposition but his brother Nawal Singh gave unremitting trouble and vexation to the Nawab. Prem Singh was succeeded in 1748 by his son. Balbhadra Singh another famous barrier of his family and his whole life was spend in fighting often beyond the borders of his area.

During 1754 Safdarjung died and he was succeeded by his son Shiraj-ud-daulah. Muhammad Quli Khan unsuccessfully tried to displace the Nawab from the Manikpur area along with Hindu Chiefs did not let him remain in peaceful possession of his responsibilities. Balbhadra Singh of Tiloi rose in rebellion and was suppressed with difficulty by Najaf Khan, the Nawab's Commandant at Pratapgarh. Muhammad Quli Khan did not help in punishing the local Taluqdars or in reducing their turbulence. During 1759, Shuja-ud-daulah siezed the territory of Mahammad Quli Khan and tried to restore the estate but he was prevented for doing so due to the incursions of the Marathas into Awadh region. During 1760 Shuja-ud-daulah confiscated the rent free tenures granted by the former rulers which resulted in the considerable distress to the people of the area.

The Suja-ud-daulah died on 26th January, 1775 and he was succeeded by his son Asaf-ud-daulah. During the year 1783 Asaf-ud-daulah assigned to his mother Bahu Begum, the pargana of Salon, Jais and Nasirabad in Jagir. Some of the villages of Salon Pargana belonged to the Kanhpurias who were displaced and their villages were given to Pir Zadas. The Salon was also made the headquarters of a **Chakala**. Finally, the the area was divided between the Chakalas and Salon of the Nizamat of Baiswara. After the reign of the Asaf-ud-daulah, more than half a century history of the area was simply a record of constant fighting between officials of the Nawab and Chief of the Local Rajputs.

The Kanhpuria Rajputs of Nain were in constant opposition to the famous Nazim Darshan Singh who held Salon during the 1827 to 1836. During 1837 Kundan Lal Pathak was posted at Salon and established a garrison in the midst of the Jungles of Nain and he cut down and cleared the most of the jungles. During 1838 he died and Darshan Singh again returned as Nazim with his brother Bakhtawar Singh, who held Salon till 1848. Later on he was succeeded by Nausoz Ali a weak and experienced person and during his tenure the Kanhpurias recovered all that they had lost during the time of predecessors. The Nazim found himself unable to interface with the powerful baron and could not enforce the payment by him the revenue of the Government, occassionally the fighting took place between the local taluqdars and the Nawab's officials because the chief of Baiswara never submitted without demur to the central authority of the estate.

The king of Awadh, Wajid Ali Shah, during 1856, was deposed of the territories including the area had been annexed by the East India Company. During that time the then district was created for this area having headquarters at Salon, was known as District of Salon and the area was placed in the Charge of the Captain Barrow. The passed procedure cf guilty and without opposition either from the people or the taluqdars and a years attention was devoted to the settlement of revenue and the arrangement of the details of the administration. However, the progress of this work was cut short by the outbreak of the freedom struggle during 1857.

Abruptly the British rule came to an end in this area and the inhabitants reverted to the shape of things which has prevailed on the eve of annexation. The Talukdars became again indepndent and resumed the control of their old estates. Every talukdar ensconced himself in his fort and prepared themselves to defend his property from others and tried to extend his possession.

The Kanhpuria Chief of the Tiloi and several of the Bais Talukdars also sent levies to help the freedom fighters. Immediately after departure of the British officials from the area the Kanhpurias of Nain descended upon the civil station near Salon and burnt down the Court-houses and destroyed all government records. The people of the area were thoroughly hostile to the British and without exception practically sided with the freedom fighters.

During the early October 1857, large scale levies were sent from this area to participate war against British Rule. The Nazim of the Salon Block took with him. More than 2000 men leaving behind instructions to enlist three or four thousand more. Jagraj Singh the brother of Rana, took 700 men and two guns from Shankarpur. By the end of November 1857, the British were forced to evacuate the Residency at Lucknow and retreat from the city and the revolutionaries in spite of their poor resources, prepared themself to retain the hard won freedom. The struggle went unabated till the city finally fell to the British in March 1858.

The talukdars of the area were obliged to reconsider their lot with the revolutionaries or to espouse the cause of the British. The proclamation issued by the British government soon after worked upon the fears of the talukdars for the future and continued presence of the revolutionary government activity. The revolutionary influences tended to prolong their antagonism and increased greatly the difficulties of the British Government. Later on it had been announced that the proprietary titles of the Talukdars were to be recognized and they were put in possession of the estates they held at the time of annexation. Talukdars were invited to receive their estates with a title from the British Government. After some time Queen's proclamation and Amnesty were made public and all persons who surrendered before the 1st January, 1859 were pardoned, barring of course those involved in the killing of Europeans.

The Civil Disobedience Movement was launched in 1930 and the area was in the forefront of this movement, Jawaharlal Nehru addressed four meetings in this area in February 1930. He visited the area again on March 29 to organize and revitalize the Satyagarh Movement of which the Salt Satyagarh was an integral part. Such was the enthusiasm of the people for this movement that countless volunteers were enlisted everywhere. The local Raja's led a large procession which culminated in a meeting addressed among others by Rafi Ahmad Kidwai and Ganesh Shankar Vidyarthi on April 1, batches of Volunteers poured into this area and paraded in a procession through the main town of the area. Rafi Ahmad Kidwai was appointed dictator of the Salt Satyagarh in the District and addressed a meetings in a large number of villages explaining the aims and objects of the movement. This area had the honour of starting the Salt-Satyagarh in Uttar Pradesh. On April 6, Rafi Ahmad Kidwai accompanied by volunteers to manufacture the salt in contravention of the Salt law but was forced to turn back as the owner of the premises would not permit them to make the salt on his land but on 8 April, the salt was manufactured publicly. Motilal Nehru playing the leading role, Jawaharlal Nehru's daughter Indira Gandhi then 12 years of age was also present. The manufactured salt was auctioned by Motilal Nehru. Numerous arrests were made, but this did not deter the people from manufacturing salt in very corner of the region. The Press Ordinance was issued and Government retaliated, the Prevention of Intimidation Ordinance, and the Unlawful Instigation Ordinance and declared the All India Congress Committee unlawful.

Rafi Ahmad Kidwai was deeply touched by the miserable plight of peasants who lived on the verge of starvation and consequently he organized a no-rent campaign in the area with the slogan "No land revenue to a government which is deaf to all grievances." For a month he occupied the waiting room of the Railway Station posing as transient passenger through buying and returning tickets to conceal his identity and hoodwink the police. This movement attracted thousands of persons and alarmed the government served an order on Rafi Ahmad Kidwai curtailing his movements and restricting his activities. Rafi Ahmad Kidwai ignored the order and he was arrested in June 1930 and convicted and sentenced him to six months imprisonment and a fine of Rs. 50 or in default of payment of fine to undergo six-week's labour. After the release of Rafi Ahmad Kidwai again threw himself whole heartedly in the peasant struggle that was being waged in the area. The tenants were rejected from their land holdings who were unable to pay the land rent, and the houses of the peasants and their properties were attached auctioned and they were made to suffer civil imprisonment. As a last resort the tenants refused to pay their rent till some settlement with the Government had been worked out. The Government adopted truthless measures as its endeavours to suppress the movement. The offices of the Congress were banned, sealed and locked, papers, accounts and flags were sealed, brutal lathi charges and firing on unarmed processionists resorted to and section 144 Cr.P.C. was promulgated throughout the area, when Rafi Ahmad Kidwai was addressing a gathering of peasants at Daulatpur Panwari, the police broke up the meeting forcibly and opened fire. But instead of achieving their purpose these repressive measures succeeded merely in adding fuel to the fire. Tens of thousands of additional persons became formal members of the Congress. Congress offices were established all over the District and Congress volunteers extremely toured the rural areas dissuading tenants from payment rent.

Suspension of Civil Obedience movement during 1931 and the signing of the Gandhi Irwin Pact gave the government some respite. The ban on the Congress was lifted and the release of political prisoners started.

Pandit Jawaharlal Nehru visited the District on June 25 and then again on the 6 December, 1931 and addressed largely attended people's meetings. Pandit Nehru exhorted the peasants to unite and urged them to withhold humiliation and repression. The people of the District enthusiastically responded to the call of the individual Satyagrah and large numbers of people courted arrest.

9. RELIGIOUS IMPORTANCE

Religious, archaeological and historical importance can be obtained from the museum established in the Sultanpur City which has been the District Headquarters of Amethi, Gauriganj and Musafirkhana Tehsils of the present newly reconstituted District, named as Amethi including Salon and Tiloi tehsils of the Rae Bareli district. Religiously the district has very importance, among the people of the District. Various religious places make it clear that the District has very much importance in the ancient historical period.

Important religious place of the District is Tikar Mafi Ashram located in the Amethi Tehsil of the District which is religiously very famous especially for the religious education and training.

Very famous Mazar located in the Jais Town has very importance among the minorities and the Jais is also known place for the famous noted Hindi Poet Malik Mohammad Jaisi.

There are other religious places which are very famous at local level especially the Nand Mahar.

Religiously very famous place Nand Mahar is known on the name of the Nand Baba which is a great worship place for the Yadav's Devi-Devata located in the Musafirkhana Tehsil of the District. According to the local people Lord Krishna's father Nand Baba stayed at Nand Mahar while going to Ayodhya. A great fair is being organized on the Kartik Poornima day every year in which the Yadav's worship their Kul Devi-Devatas and several devotees from various other districts also participate in this religious fair for worship.

There is no archaeological existence in the district except the forts of the Rajas and Maharajas. The Taluqdari system was observed in the district but it has no archaeological importance in this area. Religiously and historically the area has much importance especially in the case of Rajas and Kingship tenure of rulership. The local Rajas including Rajputs and Bhars were worshipping their family Devis and Devatas for their betterment.

10. DRAINAGE SYSTEM

The southern part of the district is drained by the Sai river and their tributaries on the basis of the geology, soil, topography and climate the district can be devided into Sai Catchment and Jais plain and riverine tracts of the Gomati river and lake tract.

The important river in the southern part of the district is Sai river. Sai river flows through the southern part of the district dividing it into almost equal halves. River Sai leaves Hardoi District and flows along the border of Unnao and Lucknow districts in the extreme north-west corner and flows southwards from some distance and turns eastwards till the railway line. It again bends southwards and passes through the western border and then turns east along the southern borders. From here its course is generally south-east, full of loops and bends and leaves. The Sai river flows through the district for about 100 km. The other rivers of the district are small and some drainage channels which are mostly seasonal during the rainy reason.

There are large number of marshes and irregular inland water bodies in the district. Most of the water bodies do not serve any useful purpose and these ponds cannot be developed for fish farming.

The Gomati river forms the northern border of the district Amethi. The district suffered from the problems of water logging during heavy monsoon rains both as a result of flooding by the natural drains and the undulated topography which causes the formation of several major and minor depressions. The existing drainage channels had been constructed mainly for linking the numerous ponds and tanks with each other. Several of these small drains are yet incomplete and do not connect to the ultimate natural drains. The natural drains are frequently choked by silting, leading to flooding of their Sai river southern part and Gomati river in the northern parts of the district while other rivers of the district are Majhuai, Mangar, Chamraura, Pili and Kunwar. The Amethi block is a part of the integrated drainage system of the river Chamraura, although different nalas join the stream. In general the drainage system of the area is dendritic and the general characteristic features throughout the plains are that the rivers meet at acute angles and several nalas from parallel or sub-parallel lines joins the main stream. There are several small streams. The river Chamraura flows from north-west to reach south direction. There are several tanks, *jhils* and lakes. There are several number of shallow *jhils* or swamps but none of them are of any considerable size or importance. The ground water potential and aquifers with primary water hydrology including various sources of irrigation, viz. canal irrigation, tubewell irrigation and well irrigation system. The distribution of drainage system in the district Amethi is depicted in **Map 1.10**.

11. INDICATORS

The innovation of natural infrastructure and developing the village information system has been carried out through adopting various independent indicators responsible for the dependent indicators of the village development of the Amethi Block:

A. Land Use

1. Total geographical area of the village.
2. Land not available for cultivation.
3. Barren and uncultivable land.
4. Cultivable land.
5. Cultivable Waste Land.
6. Permanent Pastures and grazing land.
7. Other uncultivable land.
8. Land under miscellaneous trees.
9. Fallow land excluding current fallow land.
10. Current fallow land.

B. Waste Lands

1. Total wastelands of the village.
2. Usar land.
3. Water logged land.
4. Banjar land.
5. Kankarili land.
6. Old Fallow land.
7. Fallow land.
8. Other types of waste lands.

C. Land Capability

1. Avval land.
2. Doyam land.
3. Soyam land.
4. Other types of land.

D. Agro-Afforestation

1. Total Agro-afforestation area of the village panchayat.
2. Agro-forestry land.
3. Social-forestry land.
4. Horticulture land.
5. Natural vegetation of the village.

12. CLIMATE

The District Amethi and Amethi Block experiences a rhythm of seasons with reversal prevailing winds. The reversal winds are regularly felt twice during the year. From November to May winds blow from the west and area of the continental character, whereas from June to October the reversal winds are of the oceanic origin and blow from the east. The complete reversal of wind's direction on a gigantic scale between summer and winter is responsible for the monsoon. The climatic conditions of the District Amethi has been depicted in **Map 1.8** in which annual rainfall zone has been shown.

The District Amethi experiences typical tropical climate with distinctive summer witness and rainy seasons. The mean daily maximum temperature rises to 40.6°C and mean daily minimum temperature dropped to 8.6°C on an average. The annual average rainfall is 1005 mm. The occurrence of dust-storm and thunder-storm in the hot-season is the special phenomena of the local climate in the District Amethi of Uttar Pradesh. The relative humidity rises to 85 per cent during south-west monsoon and decreases to 30 per cent during the summer season. The climate of the district is normal and suitable for agricultural-horticultural and Agro-Forestry and Social Forestry programmes.

I. ***Temperature***: The climatic temperature changes are due to the outcome of the monsoon currents. It tends to set up surface pressure differences, resulting in a low pressure over the land and high pressure over the adjoining area. May and June are the hottest months of the year. The average mean temperature during the months are 32.3°C and 31.07°C respectively. The extreme maximum temperature in the month of June is 45°C, which is due to the sun's rays falling almost vertically. Conversely, the extreme minimum temperature occurring in the month of December is 4.4°C, since the sun is low during these months owning to the earth's position during winter solstice.

II. ***Moisture and Humidity***: The most important factor which determines the potentiality of rainfall is the humidity level in the atmosphere. The humidity is high during the month of June, July, August and September. The winds are dry during the months of March, April and May. The temperature during these months are also high and the pressure is subsequently low and the relative humidity is comparatively low in the district.

III. ***Pressure of Winds***: The pressure of wind is 985 millibars during the month of June,

July and August. It rises to 1004 millibars during the month of December. A shallow cell of high pressure with sea-ward gradient develops over the district in December and January. In summer the pressure gradient is towards the land. The shallow depression visits the district in a cyclonic form ascend from the Bay of Bengal, but the high pressure wind cells in December and January reaches here from west causing rough weather and little rain.

IV. ***Rainfall***: The annual rainfall in the District Amethi is received within a short period of four months, i.e. mid-June to October. Rainfall during the other months is scanty and sporadic. The annual rainfall of the district is 1005 mm but it is unevenly distributed over the year.

(i) **Annual variability of rainfall**: The economic development of the farmers depend largely on the uniform distribution of annual rainfall. The annual rainfall variability is at 12 per cent or makes more, the area is susceptible to famine. In this regard District Amethi stands very often in the range of super-abundance or deficiency of rainfall. There are some area of the district which experiences drought conditions, while other areas are liable to face serious flood or water logging due to heavy rain.

(ii) **Seasons and Monthly Variability of Rainfall**: Seasonal and monthly variability of rainfall is rather more significant than the annual variability from the point of view of agricultural activities. If the rainfall during the months of June is insignificant or excessive, it can delay the normal sowing of Kharif crops and thus reduces the agricultural productivity. On the other hand, if the rainfall is heavy and alternate with long intervals floods and droughts are experienced simultaneously. The state of abnormality frequently causes havoc and the whole cropping pattern is affected seriously. Early or late retreat of the summer during September and October affects maturing of the Kharif crops while on the other land heavy rainfall may cause postponement or restrictions of sowing of Rabi crops.

Thus the weather elements together and considering the variations in their incidence the year is divided into three distinct seasons—

(a) The Seasons of Rain—Mid-June to October.

(b) The Cold Weather Seasons—November to February.

(c) The Hot Weather Seasons—March to Mid-June.

Thus the climate of the Amethi of the District Amethi is mild and healthy. Westerly winds prevail from October to June, gradually increasing in the strength as the hot season approaches. The northern part of the district receives more rainfall than the southern part.

13. SOURCE OF DATA

This study is based on the micro-level or grassroots level or village level innovation of available natural infrastructure of the village and to develop village information system due attention has been given for the collection of the data which can be used for propounding a theory. The data base varies from one theme to the another theme according to the nature of the problem. The source of data and collection of the available information base for this study are as follows:

(i) **Revenue Records**: Land Revenue Records from the Amethi Tehsil, Village Khasra

Maps of the Village Panchayats. Concerned Lekhpal of the all 44 village Panchayats. Khasra Register etc. from Muhafis-Khana of the Tehsil Amethi.

(ii) **National Atlas for Thematic Maps Organization (NATMO):** Maps of District Sultanpur and the District Rae Bareli, Uttar Pradesh.

(iii) **Survey of India Toposheets**: Toposheet no. 63 f/10, 11, 12, 14, 15, 16 and 63 J/3, 4, 7, 8, 11, 12 and 63k/5, 9 and other relevant toposheets of the Survey of India.

(iv) Annual Report of the Department of Forest, Government of Uttar Pradesh.

(v) Primary Census handbook of District Sultanpur and Rae Bareli, 2001 and 2011.

(vi) **Field Survey:** Most of the information on natural infrastructure has been innovation through direct field survey and visual observation on the ground has been depicted on the village Khasra map. Innovation of available natural infrastructure information system has been developed by village revenue officials, Village Panchayat representatives, villagers and members of the village level management committee of the respective Village Panchayat.

(vii) Resources Analysis for District-Sultanpur, Water-Technolgoy Centre Indian Agricultural Research Institute, PUSA, New Delhi.

(viii) Resource Analysis for District Rae Bareli Water-Technolgoy Centre Indian Agricultural Research Institute, PUSA, New Delhi.

(ix) **Action Fieldwork:** Action fieldwork has been carried out for collection of the qualitative data related to soil, land use, wastelands and the area under agro-afforestation and land capability at Village Panchayat level in Amethi Block.

14. CARTOGRAPHIC TECHNIQUES

Requisite cartographic techniques has been adopted to represent and depict the data in a graphic form. Choroplething techniques has been applied to show the distribution of land use and other indicators. Geographic Information System (GIS) has been adopted for the appropriate analysis of innovated natural infrastructure Village Panchayat level information system for appropriate analysis of the collected information on natural infrastructures at village level in Amethi. Various other techniques have also been adopted in a graphic form to analyse the data.

15. CHOICE OF THE TIME PERIOD

In this study the choice of the time period is very much matter of convenience, as the analysis on innovation of the natural infrastructure information system is based on the field observations, fieldwork and field survey and recently available land revenue records at village panchayat level in Amethi. The time period for the study has been made from 2001 to 2011 as the base year for the observation of temporal variations and decadal gap has been decided. Thus the changes in the land revenue record system has been noted over the period 2001 to 2011. The fieldwork, field observations and field survey work has been carried out during the recent year.

16. SOILS

There are older alluvial, saline and saline-alkaline soils found in the District Amethi. The soils of the area are transported from the sedimentary rocks of the Himalayas and product of the decomposition of these rocks have been deposited over the tract through water action. Therefore, the soils of the district fall under the alluvial group of the Indo-

Gangetic plain. The scientific data regarding the soils of the district are not appropriately available. The Soils of the District Amethi has been depicted in **Map 1.9**.

The brief account of the soils and land classifications based on the texture of the soils has been, however, prepared by the revenue authorities to assess the land revenue. The main factors governing the system of land classification are texture, colour, availability of water and the level of the land on the basis of the texture the soils of the district has been subdivided into the following groups:

(i) **Bangar Soil:** The upland plain known as Bangar soils vary from sandy loam to stiff clay; depending on the topography and drainage system of the area. Bangar soils arise on the basis of the principle that when a river grows older, its silt deposits subsequently occupy a lower position along its basin and the terraces and the flood plain are gradually left free by the spreading action of water during the old age of river. Such raised uplands, which are called Bangar, are again eroded by the newer channels but they always remain above the flood plain and Bangar lands are never affected by the flood.

(ii) **Loamy Soil**: Loamy soil is locally called as 'Domat Soil'. The loamy soil is extensively distributed over the *banjar* or upland plain. The patches of the loamy soils are found in the District Amethi. The patches of Domat soils are very well drained. The sand and clay are present in the soil has been almost equal in proportions, therefore, it can easily be recognized. Depending on the fertility, the Domat soil has been further subdivided into—

- Domat Avval
- Domat Doyam
- Domat Soyam

Depending upon the percentage of the sand the Colour varies from yellow to brown. The sub-soil is brownish and yellow, some times mixed together and some times with alternating layers. The water retaining capacity also varies according to the texture of the soil. Loamy soil is being preferred by the cultivators, since it can be used for double and multiple cropping.

(iii) **Clayey Loamy**: The Clayey loamy soil is locally known as Matiyar Soil and it has grey or yellowish grey colour at the surface. The sub-soil is in the lower horizons and it is dark grey in colour. In comparison to the Domat Soil the percentage of clay in it is high. Its water retention capacity is also very high. It has been observed that almost in every village Matiyar soil is found to occupy low level area, which results in the accumulation of rain water. Consequenly the soil responds well to paddy crop. Sometimes clay loam soil characterized by the pressure of calcareous concentration which are leached from the surface and get accumulated at various depths in the form of Kankar.

(iv) **Clayey Soil**: The clayey soil is locally known as "Dhankhar Soil" and it is grey to dark grey in colour. Dhankar soil as its name indicates that it is largely used for growing Dhan, i.e. paddy crop during Kharif cropping season. It has compact and cloudy structure and becomes sticky when wet and very hard when dry. It is mostly used for raising single crop of the late paddy while during other parts of the year it is almost left idle. The villagers call such fields with the clayey soil as "Jadahan Khet" (exclusively for paddy cropping). In fact ploughing and sowing are almost impossible when this land is dry. Clayey soil is found only in small patches. It mostly lies in the vicinities of the *Tals*, *Jhils*, and occasionally near low lying areas.

(v) **Khadar Soil**: Khadar Soil is mainly confined to the flood plains of rivers and *Jhils*. Khadar soil is generally light, but it is rich in the Calcareousness. Sand and clay are deposited at alternate layers on the beds. The soil in the vicinity of rivers has more sand in comparison to soil lying away from river beds. In the different areas of the district soil is loamy and makes cultivation non-profitable. Khadar soils are found relatively in the low-lying areas which are subject to occasional flooding and therefore, gets freshly deposited silt on the bed which enhances the fertility status of the soil and allows the growing of Rabi crops in the Amethi, District Amethi.

(vi) **Saline and Alkaline Soils:** The soils which has excessive proportions of soluble salts or exchangeable sodium, due to which crop production is seriously hampered are locally known as "**Reh** or **Usar**". The usar soil has high *p*H value of the soil. There is deficiency of calcium, nitrogen and micro-nutrients like zinc, extremely low water in take rate and hydraulic conductivity resulting in poor moisture transmission characteristics. The presence of calcium carbonate nodules called Kankar, often occurring at almost one metre depth and poor physical condition leads to the hindrance in the root penetration. Soils of this type are observed in almost all villages of the Amethi Block of the Amethi District of Uttar Pradesh.

17. NATURAL VEGETATION

The newly created District Amethi has an almost inhabited human occupance for over three millennia and centuries of plough and pastoral culture has reduced the natural vegetation except in the pockets of low-lying areas and some river banks, with a moderate and healthy climate and fertile soil of the area. Natural vegetation, especially flora presents no peculiarities. The patches of Dhak (*Butea frondosa*) are found here and there. In general the district is well wooded and contains magnificent groves of mango, jamun (*Eugenia Jambolana*) and Mahua (*Bassia Latitolia*) trees.

18. CONCLUSION

The sustainable economic development has been realized and commendable advances have been made in the diverse field of rural development programmes, would in reality be incomplete, if such rural development programmes fail to translate into better and dignified life for the majority of people who live in the remote rural area because the benefit of development must reach to the downtrodden deprived rural resourceless multitudes of the society.

To obtain the equal and sustained development proper planning and programme implementation is necessary. The planning is an organized process by which the people achieve its development goals or the planning is a means to achieve the development. The policy makers and planners should not only be technocrat to provide projections but they should study and innovate the available resources/infrastructure at village level before the formulation of village development planning.

The development is pursued with genuine concern for human values. The development means the evolution of the people to acquire better perception and best control over its natural and human infrastructure at National-State-District-Block-Village level perspectives.

The existing process of development is neither true, nor good, nor beautiful but it does not mean that everything is wrong. Really the development lies in the development of Science and Technology and its application by the common people.

The formulation of the planning process and the implementation of integrated rural development programmes are an information intensive task. The available natural resources/infrastructural information system at grassroots level, i.e. village level-block level-district level are not properly arranged while such available infrastructural information needs to be integrated for the development planning process.

The innovation of available natural, social, economic and institutional infrastructure at village level and developing village information system at grassroots level has become a challenging task for the policy makers and planners to formulate need based village development planning and also to assess the progress of on going implementation of various rural development programmes being implemented at village level/grassroots level. The people's participatory planning process and implementation of various rural development programmes through the peoples involvement may be an asset for successful implementation of the programmes for realizing the aspiration of the people for better and dignified quality of life.

The development is after all a multifaceted phenomenon, which can neither be understood nor promoted unless it is seen from the interdisciplinary and holistic perspectives. Moreover, the development is basically a human enterprise.

There are no such study carried out for the innovation of infrastructure and developing village information system at grassroots level some studies has been carried out to study the planning and development studies at macro and meso level while at the grassroots level insignificant studies have been carried out. Natural Resources Data Management System (NRDMS) programmes have been evolved to develop spatial data management technologies for integrated rural development planning and its demonstration at local level in the different districts of the country.

It has been felt that an integrated and faster development is required to bridge the gap between rural to urban and rurban. An appropriate faster development of rural areas are required to fulfil the increasing demand of ever growing cattle and human population. The formulation and implementation of rural development programmes at village level is an intensive, innovative informative and infrastructure (Natural and Human) oriented activity. Participatory planning and development should be initiated and sustained through the participation and involvement of the rural people at grass-roots level. The real meaning of democracy is important to the people and self-government is the main soul of the rural development and to establish village panchayat for self- governance, i.e. for the people, by the people and with the people, to ensure participatory planning and successful implementation of various rural development programmes for self-reliance.

In view of the above an attempt has been made in this study to innovate the available natural resources/infrastructure at village level in Amethi. To study the physiographic, soil, climate, land use, waste lands, land capability and agro-afforestation available resources at village level in Amethi Block. To develop village information system of the available natural infrastructure resources at village panchayat level in Amethi Block. To analyze and develop the geo-referencing of the available natural infrastructure village information system based on the village khasra map for all 44 village panchayats of Amethi Block, District Amethi.

The newly created District Amethi of Uttar Pradesh has been the study area while Amethi Block has been selected to study the innovation of available natural and human infrastructure at village panchayat level. There are five Tehsils, viz.—Salon, Tiloi, Gauriganj, Musafirkhana and Amethi of the district while there are 44 village panchayats in the Amethi Block which have been taken to study the innovation of available natural and human infrastructure and developing the village information system at grassroots level, i.e. village panchayat level in Amethi.

The topography of the district is irregular in shape but fairly compact tract of undulating land. In general the slope is from north-west to south-east. The land of the district is fairly flat and gently undulating, the highest points being the crown of the watersheds.

Geologically the district forms the part of the Gangetic plain which is the recent origin according to the geological chronology and reveals ordinary gangetic alluvium containing the nodules of calcareousness or Kankar pan. Geologically there is no noteworthy feature in the physiognomy of the district. It is characterized by unrelieved monotony of alluvial deposits which is completely shroud the old land surface to a depth of several thousand metres. The Kankarili and ravine land is found in the district. Older alluvial soils, saline, alkaline and saline-alkaline soils are found in the different parts of the district.

The Gazetteer of the district published in 1903 A.D. throws some light on the historical background of the area. It has been observed that the main land owning families of the past were the Rajputs of the different clans. These Rajputs were owning 76.16 per cent land of the total area of the district. The Rajkumars alone held over one-fourth of the land of the district while their kinsmen Bachhgotis and Rajwars owned 11.4 per cent and 3.4 per cent land respectively. The Rajkumars were main proprietors of the land.

Religiously, historically and archaeologically the District Amethi had an important key role in the overall development of the area. There are Hindus and Muslims religious important places in district. There are no archaeological existence in the district except the forts of Rajas and Maharajas.

The drainage system of the district plays an important role in the integrated development of the district. The southern part of the district is drained by the Sai River while northern part of the district is drained by the Gomati River. There are large number of marshes and irregular inland water bodies in the district. Most of the water-bodies do not serve any useful purpose and the ponds cannot be developed for fish farming.

The various categories of land use and different types of wastelands, land capability and area under agro-afforesation has been taken to innovate and analyse the natural resources/infrastructure and developing village information system at village panchayat level in Amethi Block of the District Amethi.

The agro-climatic conditions of the district, viz. Kharif-Rabi-Jaid cropping seasons are appropriate farming seasons for the agricultural practices in the district. The annual rainfall in the district is received for a short period of four months. Rainfall during the other months is scanty and sporadic. The seasonal and monthly variability of rainfall is rather more significant than the annual variability from the point of view of the agricultural practices.

The available land revenue records field observation, fieldwork and direct field survey has been the basic source of the collecting the available natural infrastructure information and development of the village information system at village panchayat level in Amethi Block of the District Amethi.

There are older alluvial soils, saline, alkaline and saline-alkaline soils, found in district. The bangar soil, loamy soil, clayey loam, clayey soil, khadar soil and saline and alkaline soils found in the district. Depending upon the fertility status of the soil—Avval, Doyam and Soyam land capabilities has been classified in the district by the revenue authorities for the fixation of revenue charges.

The district has an almost inhabited human occupance for over three millennia and centuries of plough and pastoral culture has reduced the natural vegetation except in some pockets of low-lying areas and river banks with a moderate and suitable climatic conditions and fertile soil of the area, except the patches of Dhak, Mahua, Jamun, Mango and other trees including Dhak etc.

Thus, it has been observed that the available natural resources or infrastructure of the area is well enriched except the saline-alkaline soils, water logged land and ravine land. The latest scientific and technological application for agricultural, horticultural, social forestry and agro-forestry programmes through people's participation and involvement may be an asset to meet the ever-increasing demand of growing cattle and human population for fuel wood, fodder, fibre, fruits, food grains, fisheries (F^6) and vegetables for integrated sustainable development of the Amethi.

2
Climate

1. INTRODUCTION

Amethi District experienced semi-arid climate characterized by hot summer and cold winter and rainy seasons. The hot summer months are May and June and severe cold conditions prevail during the months of December and January. By the end of June the monsoon starts and lasts till the end of September month. The rain gauge stations have been installed in the each tehsil headquarters of the district. The detailed analysis of recorded rainfall data have been obtained from each tehsil headquarters. There is no meteorological observatory in the district and the informations related to the climatic parameters other than the rainfall data are extrapolated from the data obtained through the meteorological observatories located at Lucknow City. In fact there is no such data of the climatic conditions of the Amethi Block and district. Some of the analysis done by Water Technology Centre of the Indian Agricultural Research Institute (IARI), New Delhi has been taken as base to analyse the climatic conditions of the Amethi block, to analyse availability of climatic natural infrastructure in district by agricultural, agro-forestry and social-forestry activities of the area.

2. TEMPERATURE

During the months of May and June the extreme mean minimum temperature of 41.1°C and 38.9°C are recorded. During the months of April the temperature is also quite not having mean temperature of 38.0°C. The extreme maximum temperature even recorded has been 45.6°C. The mean minimum temperature during the months of December and January has been 7.7°C and 7.9°C respectively. The lowest minimum ever recorded temperature has been 2.2°C. and 7.9°C. The minimum and maximum temperature of the district varies from one season to the another season.

3. RAINFALL

The recorded rainfall daily data of different tehsils of the district has been obtained from the Indian Meteorological Department. The normal average annual rainfall in the district has been 984.5 mm. It has been observed that the variation in the annual average rainfall has been high. The co-efficient of variation which is a unit free variation has been high for the average annual rainfall has been only 30 per cent in the district. The annual rainfall zone and the climatic conditions of the district Amethi has been depicted in **Map 2.8**.

The least annual normal rainfall of 885.7 mm and the highest of 1100.2 mm has been recorded in the district Amethi. The average annual rainfall maximum per cent of co-efficient of variations is as high as 60.4 per cent, while in other parts of the district the co-efficient of annual average rainfall variation has been just about 50 per cent. The annual rainfall is received on an average in about 46 days. More than 94 per cent of the annual rainfall is received during the period from June to September. It has been observed that July and August are the rainiest months during this period 61 per cent of the annual rainfall is received in about 26 days. About 4 per cent of the annual rainfall is received during the months of October and in two days on the average. The November and December are the months of very little rainy months in the district. It has been recorded that 10 to 15 mm of rain in one to two days has been observed in the rest of the non-rainy months of the year in the district Amethi.

4. ONSET OF EFFECTIVE MONSOON

The Water Technology Centre (WTC) of the Indian Agricultural Research Institute (IARI), New Delhi analysed daily rainfall data of the rain gauge station located at Tehsil headquarters, using a computer based forecasting technologies. The records of the daily rainfall data revealed that rainfall, phenomenon showers of high intensity common in the different parts of the district. The phenomenon of rainfall cannot be considered as effective monsoon for agricultural purposes particularly because they are followed by long dry spells which may affect the germination of seeds adversely, if the crop sowing are carried out immediately often such rains. To overcome such circumstances it is necessary to identify the effective monsoon seasonal rain from pre-monsoon rain showers in the various parts of the district.

From the agricultural points of view effective rainfall should be high enough to charge the soil profile with enough moisture after meeting the evaporation losses. An awareness regarding the texture of the soil, its moisture retaining capacity and the crops to be grown will be helpful in determining the amount of rain required for initiation of agricultural activities and thereby define the effective monsoon in the appropriate terms. The very limited rain gauge stations available rainfall data represent large areas of the district. The characteristics of the soil and cropping pattern varies greatly in such larger areas of the district. Therefore, the definition of effective monsoon could be taken only as a broad indicator in deciding the agricultural operations at particular location. The effective monsoon is defined as that monsoon which brings enough moisture in the soil even after meeting the evaporation losses, with the view of support agricultural operation. From the records of the daily rainfall data, a spell of seven days are identified through satisfying the following criteria:

(i) In the first rainy days spell of seven days is not less than 'e' mm where the 'e' means average daily evaporation.
(ii) The total rainfall during the seven days spell is at least equal (Se+10) mm.
(iii) Out of the seven days at least four days are rainy days with a minimum of 2.5 mm rainfall everyday.

The occurrence of such spell is called onset of effective monsoon (OEM). As such there is no meteorological observatory in the Amethi district. The daily evaporation value of Lucknow Meteorological observatory which is 6.7 mm has been used for analysis by the water Technology Centre of the Indian Agricultural Research Institute, New Delhi.

It has been observed that the effective monsoon in the area commence during the beginning of July and ends towards the end of September. The earliest probable date of the rainfall commencement with 68 per cent probability has been during the third week of June and the latest date with the same probability has been third week of July. Slightly lower probability of 50 per cent, interval time between earliest and latest rainfall commencement probable dates of onset effective monsoon rainfall varies, from July 1 to July 8, from one place to another in the district Amethi. The effective monsoon rainfall can be expected with a probability of 68 per cent during the last week of June in the different parts of the district Amethi and also at Amethi.

5. WET SPELLS

It has been recorded that the district Amethi experienced three distinct wet spells:

(i) **First Wet Spell** begins with the onset of effective monsoon in the first week of July and last about four weeks when about 450 mm rainfall which has been about 45 per cent of the annual average rainfall received in the district. The first wet spell at Amethi begins two days earlier and it is shorter by two days and yields a rainfall of 433.4 mm. It has been observed that the first wet spell lasts nine days longer in the area where it has a duration of 25 days in the other parts of the district Amethi.

(ii) **Second Wet Spell** has been recorded from the third week of August to the beginning of September when about 234 mm of rain has been received and the second wet spell has average duration of 17 days. The second wet spell at Amethi begins during the second week of August and lasts for 18 to 19 days yield about 29.6 mm rainfall. The second wet spell rainfall varies from one part of the district to the another part of the district.

(iii) **Third Wet Spell** has an average duration of nine days in the district. The third wet spell begins in the third week of September at Amethi and the second week of the September month. It has been observed that the area received 116 mm rainfall during the third wet spell of 8 to 11 days.

The third wet spell rainfall is helpful for the agricultural practices in the area.

6. CRITICAL DRY SPELL

It has been recorded that there are three distinct dry spells having duration exceeding 10 days in the district Amethi. A dry spell of duration exceeding ten days has been called as critical dry spell. Various distinct dry spells in the district are as follows:

(i) **First Critical Dry Spell** begins by the end of July and lasts for about 17 days. The duration of first critical dry spell varies from 16 days to 20 days in the district. The dry spell starts from July 29 and lasts for 16 to 18 days. The shorter first critical dry spell begins on July 26 at Amethi and other parts of the district Amethi of Uttar Pradesh.

(ii) **Second Critical Dry Spell** begins from August 22 which has been the normal date of occurrence and the duration of second critical dry spell is of about 18 days in Amethi. The second critical dry spell for 17 days starts on August 18 at Amethi while the dry spell starts on August 20 and 24 and lasts for 19 to 20 days in the other parts of the district Amethi.

(iii) Third Critical Dry Spell has an average duration of 24 days and starts during the first week of September. The critical dry spell has been as long as 41 days and begins as early as on August 27, while the duration of the third critical dry spell begins at Amethi for 20 days from September 9. The third critical dry spell begins two to four days later and it is shorter by three to six days in the different areas of the district. It has been observed that if the beginning of critical dry spell starts earlier the longer critical dry spell lasts in the various areas of the district Amethi.

6. SPECIFIC DURATION OF RAINFALL

The average daily rainfall records of various rain gauge stations of the district Amethi, recorded minimum, maximum and normal rainfall of the specific duration and the probabilities of the rainfall normal values being equal or exceeded as well as the expected minimum precipitation (rainfall) with their return period has been calculated using regression analysis by the Water Technology Centre of the Indian Agricultural Research Institute, New Delhi. The term probability has been used to mean the probability of being equalled or exceeded in any year's rainfall.

It has been observed that over 772 mm of rainfall accounting to 80 per cent of the annual rainfall has been received during the 26th and 37th weeks corresponding to the period July to mid-September at Amethi. Weekly highest normal rainfall is 90.0 mm which has been recorded during the 28th week, during the month of July and this rainfall has a probability 43.3 per cent. The second highest weekly normal rainfall of 89.7 mm occurs after the months of July. During this week the highest weekly maximum of 572.5 mm rainfall has been observed. The probability of average normal rainfall for this week has been lower at the 26.6 per cent rainfall. High normal rainfall of 76.8 mm has been registered during the 32nd week of the year at Amethi. The second highest weekly maximum rainfall of 39.9 mm has been observed during the 34th week against the normal rainfall of 65.3 mm whose probability has been found low at 23.3 per cent rainfall in the area. It has been observed fortnightly that the minimum, maximum and normal rainfall and the probability of normal rainfall at Amethi, the received rainfall during the seventh fortnights from the 13th to 19th fortnights has been 825 mm which has been over 85 per cent of the annual average rainfall. Fortnightly highest average normal rainfall of 151.6 mm having a probability of 40 per cent occurred during the 14th fortnight. During this fortnight the first highest fortnightly rainfall of 575 mm has been recorded in Amethi and the adjoining areas of the district. During the 16th and 17th weeks normal rainfall of 127.2 mm and 126.6 mm rainfall have been a normal probability of 43.3 per cent rainfall occurrence in the district Amethi.

It has been observed that minimum, maximum and normal monthly rainfall and the probability of the normal monthly rainfall in the Amethi during July to August are the months when the highest normal rainfall of 322.8 mm and 284.9 mm occur having probability of 43.3 per cent and 533 per cent in Amethi. The normal rainfall with that of September contribute to 81 per cent of the annual rainfall of the district. In addition the month of July having the highest normal rainfall has been recorded in the area. During the month of July highest monthly maximum rainfall of 983.5 mm as well as minimum rainfall of 63.4 mm have been recorded in Amethi. The recorded maximum precipitation (rainfall) of the varying duration from one day to 10 days and expected maximum precipitation (rainfall) during the same duration has been estimated with different rainfall periods at Amethi and it has been found that the highest one day maximum rainfall has been recorded 220.4 mm and the highest three days maximum of 361.4 mm rainfall was observed having maximum rainfall of 431.5 mm, 484.6 mm and 596.9 mm rainfall for the four, five and ten days respectively. The expected maximum precipitation from one to ten days duration with various return period of rainfall indicates that most of the rainfall values are in conformity with the figures observed confirming the reliability of estimation procedures of the normal rainfall. It has been observed that two days precipitation having five years return period of rainfall has been 184.3 mm which makes it clear that a precipitation at this level may occur once in five years, not only in the Amethi but the adjoining areas of the district Amethi.

It has been observed in the Salon Tehsil that almost the rainfall received during the 12 weeks from 26th to 37th weeks of the year has been 751.2 mm which has been about 79 per cent of the annual average rainfall. During the 32nd and 33rd weeks of the year about 82 mm rainfall having probabilities of 43 per cent and 40 per cent has been recorded respectively. The highest maximum rainfall of 428 mm has been observed during the 36th week of the year. Fairly high maximum rainfall has been observed during 38th to 41st weeks of the year. The uncommon rainfall has been recorded having maximum of 140 years in the 44th weeks which has been confirmed by the probabilities of the normal rainfall which ranges between 10 to 33 per cent rainfall in the Salon Tehsil of the District Amethi. In the Salon Tehsil during the six fortnights from the 14th to 19th fortnight about 743 mm rainfall has been received which has been about 78 per cent of the annual average rainfall. The first four fortnights out of these six fortnights share the same probability of the 40 per cent of annual average rainfall while the fifth and sixth fortnights have lesser probabilities of annual average rainfall. The last out of these fortnights have the lowest normal value of 86 mm annual average rainfall at Salon. It has been observed that highest maximum average rainfall of 600.2 mm has been recorded during this fortnight. During the 16th fortnight a minimum rainfall of 55.3 mm has been observed in Salon Tehsil of the District Amethi.

It has been recorded that from July to September, three months, have a normal rainfall of the 773 mm which has been about 81 per cent of the annual average rainfall in Salon. It has been observed that the August has highest normal rainfall of 316.2 mm which has a probability of 46.6 per cent. During July the probability of normal rainfall has been slightly lower at 43.3 per cent to the annual average rainfall. It has been observed that July and August have a minimum rainfall of 82.2 mm and 86.1 mm average rainfall. During the month of July highest maximum rainfall of 678.7 mm has been recorded in the Salon.

The highest one day maximum rainfall of 312.5 mm has been recorded. It has been observed that the maximum rainfall of 329.5 mm, 345.0 mm and 358.0 mm rainfall has been recorded for 2nd day, 3rd day and 4th day period respectively. In certain years maximum of 496.0 mm and 600.2 mm rainfall for 4th day and 10th day has been recorded respectively in the Salon Tehsil of the District Amethi.

Maximum rainfall of 159.2 mm for one day has been recorded in Salon, while a rainfall of maximum magnitude has occurred very rarely in Salon but closer values of average maximum rainfall has been observed more frequently. It has been recorded that two days maximum rainfall having a value of 329 mm can be expected only once in year. The maximum rainfall values of 231.9 mm and 265.7 mm for intervals of 3 days and 4 days has been expected once in several years. In general it has been observed that average annual rainfall has been found suitable for agricultural activities in Salon Tehsil of District Amethi.

Weekly observation of minimum, maximum and normal rainfall and their probabilities pertaining to the Musafirkhana rain gauge station has been analyzed by the Water Technology Centre of the Indian Agricultural Research Institute (IARI), New Delhi. The rainfall data and analysis of this study has been based on the Research Analysis for Integrated Development, for Sultanpur District and Rae Bareli District by Water Technology Centre, Indian Agricultural Research Institute (IARI), New Delhi.

The beginning with the 27th week provides a total rainfall of 642 mm which has been about 73 per cent of the annual average rainfall. The highest normal rainfall of 79 mm has been recorded during the 28th week and it has a probability of 40 per cent average annual rainfall. Highest weekly maximum rainfall of 585.7 mm has been observed in the 29th week of the year. The 28th preceding week has the same probability because during the week of second highest normal rainfall of 76.2 mm has been recorded at Musafirkhana. The 31st week has only a normal rainfall of 44.1 mm has the highest probability of 53 per cent average rainfall. It has been recorded that a maximum rainfall of 125.9 mm has been observed during the 42nd week against the normal rainfall of 4.7 mm while the probability of this normal rainfall has been only six per cent. Fortnightly observed minimum, maximum and normal rainfall at Musafirkhana indicates that the six fortnights beginning with the 14th and ending with the 19th fortnight provided 666 mm rainfall which has been about 75 per cent of the annual average rainfall. The highest normal rainfall of 141.0 mm during the 15th fortnight has been recorded with a probability of 36 per cent to the average annual rainfall. The highest recorded fortnightly average annual rainfall of 590.8 mm has been observed at Musafirkhana Tehsil of District Amethi. The normal average rainfall of 137.4 mm during the 14th fortnight has been recorded having a probability of 43 per cent to the annual average rainfall. The normal rainfall of 102.4 mm during the 17th fortnight has been recorded which has a one to one chance of being equalled or exceeded average rainfall of Musafirkhana Tehsil. It has been observed during the 21st fortnight that a maximum rainfall of 194.0 mm against the normal rainfall of 25.0 mm has been recorded which has a probability of only 26 per cent to the average annual rainfall at Musafirkhana Tehsil of the District Amethi of Uttar Pradesh.

Month-wise minimum, maximum and normal rainfall and their probabilities have been recorded from the rain gauge station at Musafirkhana. It has been observed that 80 per cent of the annual rainfall during the month of July to September has been recorded. During these three months July has been one of the heavy rainfall month having rainfall of 293.1 mm and the minimum rainfall during the month of July has been recorded 62.5 mm at Musafirkhana. The highest maximum rainfall of 1008.2 mm has been recorded during the month of July, while probability of the normal rainfall during the month of July has been 43 per cent. The month of August has a probability of the normal rainfall of 46 per cent against the normal rainfall of 223.3 mm. It has been observed that a maximum rainfall of 526.9 mm during the month of September has been recorded at Musafirkhana against the normal rainfall of 188.0 mm having a probability 50 per cent of average rainfall at the Musafirkhana Tehsil of the district.

It has been observed that the maximum precipitation for the duration of one day to ten days over several years especially three decades at the Musafirkhana rain gauge station and the expected precipitation for certain intervals has been recorded and the highest among the maximum has been recorded to occur during these years. One day highest maximum rainfall has been recorded 233 mm rainfall. It has been observed that maximum highest rainfall has been an unusually high value while occurrence of such highest rainfall at Musafirkhana has been very rarely recorded from several decades.

The average annual rainfall normal recorded at Salon, Amethi and Musafirkhana Tehsil of the District Amethi has been appropriate and suitable for agricultural activities except during the few years of experience regarding occurrence of maximum rainfall which caused flood and water logging and very low rainfall or dry spell in a critical situation lead towards dryness and created heavy loss to the agricultural practices in the District Amethi.

Thus it can be said that the average rainfall is more helpful for the farmers for agricultural cropping while very high or maximum rainfall or very low or minimum rainfall in the area leads towards flood and water logging and critical dry conditions respectively which leads towards very low agricultural productivity of the Amethi.

7. RELATIVE HUMIDITY

It has been observed that during the month of April and May atmosphere of the district is dry. The mean relative humidity in the morning during April and May are 33 per cent and 37 per cent respectively. The mean minimum relative humidity during the month of April has been 18 per cent and during the month of May it has been 20 per cent in the various parts of the District Amethi.

8. CLOUDINESS

It has been recorded that the mean cloud amount varies during the months of monsoon from 4 to 6 oktas in the morning and 3 to 6 oktas in the evening. Approximately 90 per cent of the clouds during the monsoon months are low level clouds. During the withdrawal of monsoon the low level cloud reduces to 3 per cent of the total cloud amount in the district. During the month of October and November the sky is normally clear in the morning having only 0.6 to 0.9 oktas of cloud cover and 0.6 to 1.3 oktas during evening. There is slight increase in the cloud cover during the month of January and February. The proportion of cloud cover varies about one to two oktas in the morning time as well as in the evening time. The slight increase in the cloud cover may occur due to the fog, mists or due to western disturbance in Amethi of the District Amethi.

9. WIND SPEED

There is quite moderate mean wind speed in the Amethi District. During the month of June about 12 kmph wind speed has been recorded in Amethi. During the monsoon wind speed ranges from 8 to 10 kmph while during the winter the mean wind speed varies from 4 to 6 kmph. In the Amethi, predominantly easterly winds prevail during the monsoon months. During the winter season westerly winds are dominant in the area. During the April and May months the western winds change their directions to eastern direction in the Amethi of the District Amethi of Uttar Pradesh.

10. HAIL

Hail storms are very rare in the Amethi. Even if at all hail storm occurs only during the months of March, April and May. Stray cases has been recorded during the month of October in the Amethi district.

11. DUST STORM

There is occurrence of dust storm during the pre-monsoon season from April to month of June. The occurrence of the dust storm has been recorded on an average about one or two days in a month in the district Amethi.

12. EVAPORATION

During winter minimum evaporation has been recorded about 2.2 to 4.0 mm per day. The rate of evaporation during summer has been found high. The evaporation in the district Amethi from 9 to 11 mm per day. The mean daily evaporation during the monsoon season has been recorded about 4 to 6 mm per day. The annual rate of evaporation in Amethi has been observed that 6 to 7 mm per day. It has been observed that the rate of evaporation is found very high during the summer season comparatively than the winter season in the District Amethi of Uttar Pradesh.

13. POTENTIAL EVAPOTRANSPIRATION (PET)

It has been recorded that the mean potential Evapotranspiration attains a maximum of 185 mm during the very hot summer months of May and June. During the monsoon months potential Evapotranspiration has been found much less and it ranges between 116 to 126 mm. During the winter potential Evapotranspiration varies between 47 to 73 mm. The annual potential Evapotranspiration has been observed 1379 mm in Amethi.

14. HOURS OF SUNSHINE

The hours of sunshine has been received on the average for 8 hours during the winter season. It has been recorded that during summer season 10 to 11 hours sunshine has been observed. The sunshine during the monsoon season has been recorded from 5 to 6 hours per day. The sunshine in the District Amethi has been found very appropriate and suitable for agricultural practices especially for Kharif, Rabi and Jaid cropping season in Amethi of the district Amethi of Uttar Pradesh.

15. CONCLUSION

The District Amethi experiences semi arid climate characterized by hot summer, cold winter and rainy season. The hot summer months are May and June, while severe cold conditions prevail during the months of December and January. By the end of June monsoon starts and lasts till the months of September.

During the month of May and June the extreme mean minimum temperature of 44.1°C and 38.9°C has been observed. In the month of April the temperature is also quite hot having mean temperature of 38.1 °C. The extreme maximum temperature has been recorded as 45.6°C. The mean minimum temperature during the month of December and January has been 7.7°C and 7.9°C respectively. The lowest minimum temperature has been recorded as 2.2°C and 7.9°C.

The least annual normal rainfall of 885.7 mm and the highest of 1100.2 mm has been recorded in the district. The average annual rainfall maximum per cent of co-efficient of variations is as high as 60.4 per cent. From the agricultural points of view effective rainfall should be high enough to charge the soil profile with enough moisture after meeting the evaporation losses. It has been observed that the effective monsoon in the district commencing with the beginning of July and ends with the end of September. The district experiences wet spell as well as critical dry spell in a different characteristics.

It has been recorded that the minimum, maximum and normal rainfall of the specific duration and the probabilities of the normal rainfall values being equal or exceeded as well as the excepted minimum precipitation (rainfall) with their return period has been analyzed. It has been observed that 80 per cent of the annual rainfall during the month of July and September has been recorded.

During the month of April and May atmosphere of the district is dry. The mean relative humidity in the morning during April and May has been recorded 33 per cent and 37 per cent respectively.

The mean cloud amount varies during the months of monsoon from 4 to 6 oktas in the morning and 3 to 6 oktas in the evening. Approximately 90 per cent of the clouds during the monsoon months are low level clouds. There is a slight increase in the cloud cover during the month of January and February.

There is quite moderate mean wind speed in the district. During the month of June about 12 kmph wind speed has been recorded in Amethi. During the monsoon season wind speed ranges from 8 to 10 kmph while during the winter season the mean wind speed varies from 4 to 6 kmph.

Hail storms are very rare in Amethi. Even if at all hail storm occurs, it is only during the months of March, April and May.

It has been observed that there is occurrence of dust storm during the pre-monsoon season from April to the month of June.

The minimum evaporation during the winter has been recorded about 2.2 to 4.0 mm per day. The rate of evaporation during the summer season has been found high. The evaporation status of the district ranges from 9 to 11 mm per day in Amethi.

The mean Potential Evapotranspiration has been recorded as 185 mm during the hot summer months of May and June. During the monsoon months Potential Evapotranspiration has been found much less and it ranges between 116 to 126 mm. During the winter evapotranspiration varies between 47 to 73 mm. The annual Potential Evapotranspiration has been observed as 1379 mm in Amethi.

The sunshine has been received on an average of 8 hours during the winter season. It has been observed that during the summer season 10-11 hours sunshine has been recorded. The sunshine during the monsoon season has been recorded from 5 to 6 hours per day. The recorded sunshine during the different season has been very appropriate and suitable for agricultural practices especially for the Kharif, Rabi and Jaid agricultural cropping season in Amethi of the district Amethi.

It has been observed that the climatic conditions of Semi Arid Zone in Amethi of the district Amethi has been an appropriate and suitable for agricultural practices in the district provided that the occurrence of climatic situations are normal because the most of the farmers of the area are engaged in the traditional agriculture farming.

The occurrence of the rainfall during the rain season and very hot summer during May and June and very cold season during December and January and the moderate hot cold weather in the area are suitable and feasible for the agricultural practices in the area and the agricultural farmers are also immuned with such agro-climatic conditions. The mild weather and semi-arid agro-climatic conditions of the district has forced the farmers for adoption of single, double and multiple-agriculture cropping in a traditional farming practices with a medium agricultural productivity and also the agro-forestry and social-forestry programmes to produce fuel wood, fodder, fibre, fruits, food grains, fisheries (F^6) and vegetable to meet the increasing demand of growing cattle and human population.

In spite of the above if there is occurrence of very high rainfall or dry critical spell, spell and very high evapotranspiration in the climatic conditions may adversely affect the agricultural practices, but in view of these worst climatic conditions farmers of the district are not having any techno-scientific know how to combat the worst agro-climatic conditions like dry land agriculture farming or immediate farming after the flood or utilization of water logged lands in the district.

The techno-scientific awareness and latest knowledge like Israel the farmers of the area can also develop agricultural agro-forestry and social-forestry practices in the district to meet the increasing demand of growing cattle and human population.

Thus the agro-climatic conditions of the Amethi and the District Amethi are suitable for moderate agricultural practices. The agro-climatic conditions available in the area as natural infrastructure are feasible, suitable, viable and acceptable to the cultivators single, double cropping and also for the multiple cropping as per the farmers awareness. The techno-scientific knowledge and awareness of the farmers and exploitation of existing seasonally agro-climatic conditions the farmers may develop the agricultural-agro-forestry and social-forestry programmes which can fulfil the increasing demand of growing population for fuel wood, fodder, fibre, food grains, fruits, fisheries (F^6) and wood, vegetable and it can also generate employment and improve the socio-economic conditions of the deprived rural multitudes surviving in the remote rural areas of the Amethi of the district Amethi.

3

Soils

1. INTRODUCTION

The soils of the District Amethi are formed through the alluvial deposits carried out by the river drainage system comprising Gomati, Sai and Ganga rivers and their tributaries.

The Gomati river forms the northern boundery of the district while river Sai flows from the south-east direction of the district. The physiographic variations are very prominent in the District Amethi. The textural variations in the soil profiles of the district are apparent. The horizontal differentiates in the soil profiles are not prominent in the Amethi. The physiographic variations of the soil in the District Amethi has contributed to the development of soils of different physical make up and chemical characteristics of the soil. The developed soil in the Gomati River watershed and its adjacent areas have predominantly coarse fragments while the further areas from the river catchments has increasingly finer fragments of the soil. Such variations in the soil has discernible in the soils of different areas of the District Amethi. The soil structure of the district can be divided into the different physiographic tracts—

i. Ravines of the River Gomati.
ii. Flat lands of River Gomati.
iii. Sai river uplands.
iv. Sai river low lands.
v. Sai river flat lands.

(i) **Ravines of the River Gomati** occur along the both sides of the river which has undulated land surface, sandy soils and intercepted with shallow ravines.

(ii) **Flat lands of River Gomati** occur adjacent to it and the flat land presents an even land escape having restricted drainage system and the soil is domat suitable for agricultural practices.

(iii) **Sai river uplands** occur on the both sides of the river Sai, having coarse textured soils and gradual slope from north to south and occurrence of ravine land, having undulated topography.

(iv) **Sai river low lands** area has been developed in the depressions within the elevated regions and the area has restricted drainage system and the area remained waterlogged from a prolonged time.

(v) **Sai river flat land** occurs in the southern part of the district and the area has flat topography. The physiographic variations of the district has contributed to the development of the soils of the varying physical make up and chemical characteristics of the soils of the District Amethi.

The younger alluvial and older alluvial tracts and saline and alkaline soils of the District Amethi have been depicted in **Map 1.9**.

2. SOIL SAMPLE SURVEYS

The soil sample surveys have been carried out in the different areas of the some blocks of the District Amethi. The soil sample surveys of the district have been carried out by the soil survey organization of the Sharda Sahayak Command Area Development Authority and Department of Agriculture, Government of Uttar Pradesh. Some important soil series have been established in the different areas of the district. Soil sample surveys have been carried out representing well drained productive soils and problematic soils suffering from drainage impedance and sodic land or alkaline-saline soils. The developed information to soil sample surveys has been found immensely valuable for land use planning and for the conservation of available land resources of the district. The complete appraisal of soil productivity capabilities and potentials, natural hazards, and to divide management planning strategies for enhancing the soil feasibility status and soil productivity to increase the agricultural productivity. It has been an imperative task to observe comprehensively and quantitative analysis and physical and chemical characterization of the soils in the different blocks of the District Amethi. The special soil sample survey was carried out through undertaking traverses in the various parts of the district which was based on preliminary examinations and visual observations of the angered soil samples of the different soil types and based on the soil problem area of the district. Various soil samples has been collected from different locations in the district having depth from 0-30 cm layers and the soil has also been collected from the various layers of the soil profile. Mechanical composition of the collected soil sample has been carried out, especially the proportionate distribution of clay, sand and silt water retentive capacity, *p*H value of the soil, electrical conductivity, cation exchange capacity and exchangeable sodium percentage through the application of standard techniques. The fertility status evaluation of the soil has been carried out to analyze the macronutrients (N.P.K.) and micronutrients (Zinc, Copper, Iron, Manganese etc.). The analysis of the fertility status of the soil through sampling techniques has been carried out at different areas of the district, with the view to maintain the fertility status of the soil to increase the agricultural productivity of the area and to improve the socio-economic conditions of the deprived rural multitudes of the Amethi.

3. CHARACTERISTICS OF THE SOIL

There are two main characteristics of the soil, i.e. chemical and physical characteristics. Therefore, it is necessary to analyze the physical and chemical characteristics of the soils of the Amethi through soil sample survey carried out by the water technology centre of the Indian Agricultural Research Institute (IARI), New Delhi and Sharda Sahayak Command Area Development Authority and Department of Agriculture, Government of Uttar Pradesh.

A. Physical Characteristics of the Soil

The physical characteristics of the soil has been analyzed through mechanical composition, water retentivity, soil depth and soil salinity and alkalinity to assess the requirement of the soil to increase the agricultural cropping fertility status of the soil.

I. **Mechanical Composition of the Soil:** The mechanical composition of the soil consists of clay, sand and silt proportions in the soil. The soil samples has been obtained up to the depth of 0-30 cm in the various parts of the district. It has been observed that the area distribution of various soil textures varies from one area to another area in the district. The distribution of the textural groups on surface soil of the District Amethi has been depicted in **Map 3.13**. It has been noticed that loam soils has been excessively found in the Jagdishpur, Shukul Bazar, Jamo, Shahgarh, Deeh, Amethi, Bhadar and Bhnetua blocks of the district. The clay laom soil has been found in the Gauriganj block of the district while silt clay laom soil has been observed in the Singhpur, Tiloi and Bahadurpur blocks of the District Amethi.

The soils of the Jagdishpur, Jamon, Amethi, Bhnetua, Gauriganj, Sangrampur, Shahgarh blocks are having medium to fine textured soil. It has been recorded that medium textured loam soils are predominantly found in the Jamo (80%), Bhadar (66%), Sangrampur (77%), Jagdishpur (76%), Shahgarh (65%) and Amethi (61%) while fine textured soils (silty clay loam and clay loam) are predominantly seen in the Chhatoh (14%), Gauriganj (82%), Tiloi (83%), Bhnetua (61%), Singhpur (86%), and Bhadar (17%) are found under coarse textured sandy loam soils. The larger areas adjacent to the Gomati river especially Shukul Bazar (37%) and Musafirkhana (23%) are under coarse textured soils. The other areas of district are predominantly under the medium textured soils. It has been observed that the loam soils are found in the Bahadurpur (97%), Chhatoh (86%). It has been observed that the fine textured silt loam to clay loam and the both textural groups are found in Salon Block of the district having 35 per cent to 47 per cent area under loam. There is occurrence of silt loam soils under loamy sand to sandy loam and 12 per cent under loam soils. The silt loam soils occur in 20 per cent to 24 per cent in the Singhpur Block while the other parts of the District Amethi are found under the loam to clay loam soils. The clay content variations are discernible within a broad textural group particularly in the case of medium and fine textured soils. The lower level variations in the clay contents and discernible textural group of clay loam in which clay content within a textural group has been responsible for variations in the associated soil characteristics especially water retentivity, cation exchange capacity and exchangeable sodium proportion in the different parts of the District Amethi of Uttar Pradesh.

The problematic soils and texturally variant soils which has limited soil fertility status due to excess of kankar granules in the different sub-soil layers soil testing has been carried out. Soil testing has been carried out for the coarse textured soils, medium textured soils and fine textured soils. The textural variations with depth have been shown in the soil profiles. There has been a tendency for clay content to increase with depth in most of the soil profiles. The limitations of the agricultural productivity have been incidental to the sodicity, impeded drainage and excess of the kankar granules are associated with the fine textured soils, in most of the areas of the District Amethi.

II. **Water Retentivity:** The soil water retentivity testing has been ensured on the basis of saturated soil samples to 0-3 bar and 15 bar suctions corresponding to the field capacity and permanent wilting point and moisture content. The moisture content by weight has been converted to volumetric moisture content by multiplying it with an average bulk density, water storage at field capacity and plant water availability and the coarse textured soils, lowest profile water storage analysis has been carried out by the Water Technology Centre, Indian Agricultural Research

Institute, New Delhi. The soil of the area tends to finer in texture especially for the water storage and increase of the water to higher values, viz. 33.2 to 34.2 and 24.2 to 24.8 cmm^{-1} and classifying medium textured soils and fine texture soils at the different soil profiles of the area.

III. **Soil Depth**: Generally the depth of the soils are very deep in the Amethi. Most of the soil profiles of the area has excess of iron and lime granules which has imposed depth limitations for the crop rooting. The profile of the soil characteristics and potential of plant rooting depth in the different soil profiles in different areas of the District Amethi. Varies from one area to another based on the depth of the soil. It has been observed that in the barrier free soil, the potential rooting of the plants and in the depth it could be 130 cm to 150 cm depth or more deeper in the different soil profiles of the district.

B. Chemical Characteristics of the Soil

The chemical characteristics of the soil has been analyzed based on the cation exchange capacity of the soil and salinity and alkalinity status of the soil.

I. **Cation Exchange Capacity of the Soil**: The cation exchange capacity of the soil samples has been carried out through sodium acetate (N) for saturation and ammonium acetate (N. *p*H 7.8) solution for the displacement of the observed sodium adopting the procedure. It has been recorded that the cation exchange capacity has been found lowest in the coarse textured soils in the loamy soil to the sandy loamy soils. The cation exchange capacity has been increased in the finer textured soils varying from 16 to 23 ma/100 g soils in the silty clay loam and clay loam soils. Generally the cation exchange capacity of the soil corresponds according to the clay content in the soil. It has been observed that the clay content in soil has been found more in the low lying area, *Tals*, *Jhils* etc. The *tals* and *jhils* are found excessively in the Amethi.

II. **Salinity and Alkalinity of the Soil**: The salinity status of the soils in the different areas of Amethi has been evaluated in terms of electrical conductivity (EC_2). The electrical conductivity has been determined in 1:2 ratio soil water extract using Elico make EC Meter.
The alkalinity of the soil has been evaluated in terms of *p*H value of the soil and percentage of exchangeable sodium. The *p*H value of the soil water suspension through using glass electrode. The exchangeable sodium of the soil has been determined through using normal ammonium acetate for replacing the exchangeable cations.

Generally, the soil samples has been taken from 0-30 cm depth from the surface layers. Moderately high *p*H value of the soil has been found ranging 8.6 *p*H to 9.8 *p*H value of the soil. It has been observed that the most of the alkaline soils of the Amethi has low EC_2 values (0.2 to 0.8 mm ho/km) in some portions of the district. The alkaline soils contain sufficient soluble salts which has been observed by the relatively higher EC_2 values (1.2 to 3.8 mm ho/cm). The saline affected soils *p*H values of the soil varies from 8.5 to 9.5 and > 9.5 *p*H value of the soils of the moderately high and high sodic affected soils of the District Amethi. The saline soils of the area has been found in most of the parts of the district. The salt affected soil based on the *p*H value of the soil which varies from less than 8.5 *p*H value of the soil and more than 9.5 *p*H value of the soil of the District Amethi has been depicted in **Map 3.12**. It has been noticed that more than 9.5 *p*H value of the soil has been found in the Singhpur, Tiloi, Chhatoh, Sangrampur, Salon, Bhadar and Bhnetua blocks of the district, while *p*H value of the soil ranging from 8.5 to 9.5 has been recoreded partially in the Singhpur, Shukul Bazar, Jagdishpur, Shahgarh, Gauriganj, Chhatoh, Bhnetua blocks of the district. It has been observed that less than 8.5 *p*H value of the soil has been found in the Shukul Bazar, Musafirkhana, Jamo, Bahadurpur, Deeh, Salon, Sangrampur and Amethi blocks of the District Amethi.

4. PHYSICAL AND CHEMICAL CHARACTERISTICS OF THE SOIL

The physical and chemical characteristics of the soil has been carried out in the village Ramdaipur of the Amethi Block by Water Technology Centre of the Indian Agricultural Research Institute (IARI), New Delhi. The physical and chemical characteristics of the soil have been analysed from 0-30 cm depth, 30-50 cm depth, 50-90 cm depth, 90-130 cm depth and 130-150 cm depth soil samples in the village Ramdaipur of the Amethi Block of the district in which textural class, moisture percentage, *p*H value of the soil etc. have been analyzed.

Fig. 3.2: Physical and Chemical Characteristics of the Soil (Village-Ramdaipur, Amethi Block)

Depth in cm.	*Texture Class*	*Moisture %age (g/g) Soil*		*pH*	*EC_2 mm ho/cm*	*ESP*	*CEC meg/ 100 g soil*
		Field Capacity	*Wilting Point*				
0-30	Silty Clay Loam	26.4	6.7	9.8	1.2	17.4	29.4
30-50	Silt Loam	23.3	6.5	9.2	0.8	16.6	22.8
50-90	Silt Loam	24.5	6.7	8.8	1.1	12.3	20.7
90-130	Silty Clay Loam	28.2	7.1	9.0	0.9	15.8	26.2
130-150	Silty Clay	30.2	7.8	9.0	0.7	16.2	32.5

i. Textural Class

Textural class of the soil has been assessed based on the different depth of the soil. The silty clay loam textural class soil has been found at the depth of 0.30 cm while silty loam soil has been seen from 30-50 cm depth and 50-90 cm depth soil. Silty clay loam soil texture has been found up to the depth of 90-130 cm and silty clay soil has been observed up to the depth of 130-150 cm soil.

ii. Moisture of the Soil

The moisture of the soil has been evaluated through field capacity and wilting point in the different depths of the soil layers. It has been observed that up to the depth 0-30 cm soil the moisture field capacity has been found 26.4 per cent and up to the depth of 30-50 cm and 50-90 cm moisture field capacity has been seen 23.3 per cent and 24.5 per cent respectively. The moisture field capacity up to the depth of 90-130 cm has been found 28.2 per cent and 30.2 per cent moisture percentage in the field capacity upto the depth of 130-150 cm has been recorded in the Ramdaipur village of the Amethi Block of the District Amethi.

The wilting point at different layers of the soil varies from 6.7 point at the depth of

0.30 cm and 50-90 cm depth soil profile and 50-90 cm depth soil profile and 6.5 wilting point has been recorded at the depth of 30-50 cm layers of the soil while 7.1 wilting point has been found at the depth of 90-130 cm and 7.8 point wilting has been seen up to the depth of 130-150 cm depth of the soil profile.

iii. *p*H Value of the Soil

The *p*H value of the soil varies according to the depth of the soil. The high *p*H value of the soil 9.8 has been found at the depth of 0-30 cm and 30-50 cm depth *p*H value of the soil has been 9.2 *p*H. The *p*H value of the soil 8.8 has been found at the depth of 50-90 cm of the soil layers and 9.0 *p*H value of the soil has been seen at the depth of 90-130 cm and 130-150 cm layers of the soil in the soil profile at the village Ramdaipur of the Amethi Block of the District Amethi.

iv. Electrical Conductivity (EC_2)

The salinity status of soil has been evaluated in terms of electrical conductivity (EC_2) which has been determined in the 1:2 ratio-of the soil water. Electrical Conductivity of the Soil (EC_2)/mmho/cm has been found 1.2 at the depth of 0-30 cm soil layers 0.8 at the depth of 30-50 cm depth, 1.1 EC_2 at the depth of 50-90 cm depth and 0.9 and 0.7 EC_2 has been found up to the depth of 90-130 cm and 130-150 cm depth of soil layers respectively. Higher electrical conductivity has been recorded at the top layer (0-30 cm depth) of the soil profile in the village Ramdaipur of the Amethi Block of the District Amethi.

v. Exchangeable Sodium Percentage

Exchangeable Sodium Percentage (ESP) has been evaluated through applying standard techniques. The exchangeable sodium percentage (ESP) has been calculated through the following relationship:

$$ESP = \frac{\text{Ex. Na (me/100g soil)}}{\text{CEG (me/100g soil)}} \times 100$$

The high percentage of exchangeable sodium having value of 17.4 per cent has been recorded at the depth of 0-30 cm soil layers and high *p*H value having 9-8 *p*H value of the soil has been recorded. It has been found that higher *p*H value of the soil and higher percentage of exchangeable sodium are consistent. The exchangeable sodium 16.6 per cent has been found at the depth of 30-50 cm and 12.3 per cent ESP has been recorded at the depth of 50-90 cm soil layers while 15.8 per cent exchangeable sodium and 16.2 per cent Exchangeable sodium has been found at the depth of 90-130 cm and 130-150 cm of soil layers in the selected village Ramdaipur of the Amethi Block.

vi. Cation Exchange Capacity

The Cation Exchange Capacity (CEC) has been evaluated meg/100 g soil samples taken in the different depths of the soil layer. The Cation Exchange Capacity has been found 29.4 per cent (meg/100g soil) at the depth of 0-30 cm and the highest CEC 32.5 has been observed at the depth of 130-150 cm of the soil layer. The lowest 20.7 per cent

CEC has been found at the depth of 50-90 cm and followed by 22.8 per cent at the depth of 30-50 cm and 26.2 per cent Cation Exchange Capacity has been observed at the depth of 90-130 cm of the soil layer in the village Ramdaipur of the Amethi Block.

Thus the high *p*H of the soil and Exchange Sodium Percentage (ESP) in the different soil samples indicate the higher level soil sodicity in the Amethi. The High *p*H value of the Soil and Exchange Sodium Percentage (ESP) has been observed at the different layers of the soil depth.

5. SOIL PROFILE

It has already been mentioned that detailed soil surveys have been carried out by the Soil Survey Organizations of the Sharda Sahayak Command Area Development Authority and the Department of Agriculture, Government of Uttar Pradesh. The collected soil samples at various layers of the soil profiles have been tested and analyzed by the Water Technology Centre of the Indian Agricultural Research Institute (IARI), New Delhi. The soil testing and analysis has been carried out to determine the mechanical composition of the soil especially the distribution of Sand, Silt, Clay and Water retentivity at field capacity and permanent wilting point of the soil and the *p*H value of the soil, electrical conductivity in 1:2 soil-water extract (EC_2), Cation Exchange Ratio (CEC) and the Exchange Sodium Percentage (ESP) has been tested and analyzed through using standard scientific techniques. The evaluation of the fertility status of the soil has been carried out for the macro nutrients, viz. Nitrogen, Phosphorus and nutrient index of the soil and the values of the micro-nutrients of the soils have been analyzed for the zinc, copper, iron and manganese contents through adopting the standard procedure.

The soil sample survey has been carried out in the different parts of the district. The soil samples have been collected at village Ramdaipur of Amethi Block, Bhnetua and Purabdwara village of Bhnetua Block and Darpipur village of the Gauriganj Block and Madhopur and Paksarwana villages of Salon Block, village Nasirabad of the Chhatoh Block and Banbheria village of the Singhpur Block of the district in which the soil analysis at different layers of the soil has been carried out.

It has been observed that the fine textured soil with hard kankar layers below 110 cm depth and poor drainage water system having low permeability of the water and high value of the *p*H in the soil has been found in the soil profile of the village Banbheria of the Singhpur Block.

It has been recorded that in the village Paksarwan and village Madhopur of the Salon Block, the medium texture soil of medium water drain ability and low salinity and alkalinity problem has been found in the Madhopur Pathak village while fine texture soil underlaid with kankar layers below 90 cm depth and poor drainage system having low permeability of the water and high *p*H value of the soil has been recorded in the village Paksarwan of the Salon Block.

It has been found that there is fine textured soil having low drain ability and the soil has very low salinity and alkalinity problem in the Nasirabad village of the Chhatoh Block.

It has also been found that the soil has deep and inadequately drained and soil having medium to low permeability of the water and the soil has salinity and alkalinity problem in the Bhnetua village and deep soil underlaid with hard layer below 150 cm depth and inadequate water drainage system and the soil has low permeability of the water and the salinity and alkalinity problem of the soil in the profile of the soil sample in the village Purabdwara of the Bhnetua Block has been recorded. The low permeability of the water has been due to the hard kankar pan and occurrence of the calcareousness in the soil at the deph of 150 cm which is hindrance in plant growth.

It has been observed that there is medium deep soil underlaid with hard layer kankar pan due to calcareousness in the soil below 150 cm depth of the soil profile and poor water drainage system having low permeability of the water and occurrence of the salinity and alkalinity has been found in the village Darpipur of the Gauriganj Block of the District Amethi.

The mechanical composition and chemical properties of the top 30 cm soil layer has been analyzed in the different sampling locations at Amethi, Raipur, Phulwari, Ramdaipur and Trilokpur villages of the Amethi Block. It has been observed that 57.6 per cent sand, 17.3 per cent silt and 24.1 per cent clay in the loamy soil has been found in the soil texture at Amethi and 54.2 per cent sand, 20.8 per cent silt and 24.7 per cent clay in the loamy soil has been observed in the village Raipur Phulwari and 20.1 per cent sand, 46.3 per cent silt and 32.4 per cent clay having silt clay loam textured soil has been observed in the village Raidaipur of the Amethi Block, while 31.2 per cent sand, 37.0 per cent silt and 30.6 per cent clay having silty clay loam soil has been recorded in the village Trilokpur of the Amethi Block. The *p*H value of the soil has been recorded 8.6 in Amethi and Raipur Phulwari village and 9.8 *p*H in Ramdaipur and 9.4 *p*H value of the soil has been found in the Trilokpur village of the Amethi Block.

Fig. 3.3: Mechanical Composition and Chemical Properties of 0-30 cm Soil Layer (Amethi Block)

Villages/ Sampling Location	*Mechanical Composition*			*Textural Class*	*pH*	*EC_2 mm ho/cm*	*ESP*	*CED meg/ 100g soil*
	Sand	*Silt*	*Clay*					
Amethi	57.6	17.3	24.1	Loam	8.6	1.7	7.2	22.3
Raipur Phulwari	54.2	20.8	24.7	Loam	8.6	2.3	12.7	21.6
Ramdaipur	20.1	46.3	32.4	Silty Clay Loam	9.8	2.9	17.4	29.4
Trilokpur	31.2	37.0	30.6	Silty Cla Loam	9.4	4.3	16.6	26.8

The Electrical Conductivity (EC_2) in 1.2 soil-water extract has been found 1.7 in Amethi, 2.3 in Raipur Phulwari, 2.9 in Ramdaipur and the highest 4.3 in village Trilokpur of the Amethi Block. The exchangeable Sodium percentage has been found 7.2 in Amethi, 12.7 in Raipur Phulwari, 17.4 in Ramdaipur and 16.6 in Trilokpur villages of the Amethi Block. The cation exchange capacity (mev/100g. soil) has been found 22.3 in Amethi, 21.6 in Raipur Phulwari, 29.4 in Ramdaipur and 26.8 in Trilokpur villages of the Amethi Block.

The representative soil profile of the Ramdaipur village of Amethi Block has been analysed from 0.30 cm depth up to 130-150 cm depth profile at different layers in the village Ramdaipur of the Amethi Block. The detailed analysis of the soil structure including physical and chemical characteristics of the soil or mechanical composition and chemical properties of the soil in different layers of the soils has been analyzed by the Water Technology Centre of the Indian Agricultural Research Institute, New Delhi.

I. **0.30 cm depth** : Silty clay loam soil, colour light brown (7.5 YR 6/4 dry) angulary blocky; sticky when wet and very hard on drying; low water permeability, roots sparse, no mottling; iron and lime granules not apparent and light effervescence.

II. **30-50 cm depth** : Soil texture silt loam, yellowish brown in colour (10 YR 5/4 dry); angular blocky; sticky when wet and very hard on drying; low water permeability; roots few, few mottles, iron and lime granules not apparent, light effervescence.

III. **50-90 cm depth** : Silty loam soil, yellowish brown colour (10 YR 5/4 dry); angular blocky; sticky when wet and very hard on drying; low permeability; no roots few mottles; fine granules of iron and lime; moderate effervescence.

IV. **90-130 cm depth** : Soil texture silty clay loam, colour grey brown (2.5 YR 5/2 dry) angular blocky, sticky when wet and very hard on drying; very low water permeability no roots, several mottles, granules of iron and lime moderage effervescence.

V. **130-150 cm depth**: Soil texture silty clay, colour grey (2.5 YR 6/2 dry), angular blocky, sticky when wet and very hard on drying; very low permeability, no roots, several mottles, granules of iron and lime; moderate effervescence.

It has been observed that the soil is deep and there is inadequate drainage system. The water permeability of the soil has been found low and there is occurrence of the salinity and alkalinity in the soil of the area.

5. FERTILITY STATUS OF THE SOIL

The fertility status of the soil has been evaluated for the macronutrient (nitrogen, phosphorous and potash) and micro nutrients (zinc, copper, iron and manganese) contents of the soil to determine the fertility status of the soil to grow the crops and plants to meet the increasing demand of growing cattle and human population.

I. Macronutrients of the Soil

The macronutrient status of the soil in different parts of the district has been assessed in terms of nutrient index through using the values achieved from the District Soil Testing Laboratory, Sultanpur. The Nutrition Index (NI) has been calculated as per details—% sample VL × 1+% samples L × 2+% samples.

$$\text{Nutrition Index} = \frac{M \times 3 + \%\,\text{sample MH} \times 4 + \%\,\text{sample H} \times 5}{100}$$

VL = Very Low
L = Low
M = Medium
MH = Medium High
H = High text values of given nutrients.

The value of the Nutrition Index has been categorized based on the nutrient level of the soil.

Nutrition Range	Nutrition Level
< 1.8	Very Poor
1.8-2.6	Poor
2.7-3.4	Average
3.5-4.2	High
> 4.2	Very High

The Nutrition Index values of the soil for Nitrogen, Phosphorous and Potash (NPK) for the soils of the different soil profiles of the various parts of the district has been analyzed as per the fertility status of the soil.

Fig. 3.4: Macro Nutrients and Micro Nutrients Status of the Soil (Amethi)

Villages sampling Location	*Macro Nutrients*			*Micro Nutrients*			
	Nitrogen	*Phosphorus*	*Potash*	*Zinc ppm*	*Copper ppm*	*Iron ppm*	*Manganese ppm*
Amethi	1.8	1.6	3.1	0.50	1.36	12.65	45.05
Raipur Phulwari	1.6	1.3	3.3	0.62	2.45	13.20	50.20
Ramdaipur	1.8	1.6	3.0	0.46	1.85	18.45	46.25
Trilokpur	1.9	1.5	2.9	0.35	0.88	16.40	35.20

i. **Nitrogen**: It has been observed that the Nitrogen Status of the Soils of the District Amethi and Amethi Block is very poor. The nutrient index indicates that the values of nutrients in the soil varies within a narrow range of 1.5 to 2.1. Comparatively between 1.8 to 2.1 nutrient value of the soil. It has been found that the Amethi Block has relatively poor nitrogen status, i.e. 1.6 to 1.9 nutrient value. The nitrogen status of the soil varies within the block from one village to the another village. The macro nutrients found 1.8 in Amethi and Ramdaipur, while in the village Raipur Phulwari nitrogen index has been found 1.6 and 1.9 nutrient value in the Trilokpur villages of the Amethi Block of the District Amethi.

ii. **Phosphorus**: There is poor status of the phosphorus in the soil of the District Amethi. The phosphorus value of the soil in the nutrition index ranges from 1.4 to 2.6 phosphorus value. Relatively very poor phosphorus status (ranging from 1.4 to 7) in comparison to nitrogen status has been recorded in Amethi. The soils within the Amethi Block shows a larger variations in the phosphorus status of the soil. It has been observed that there are 1.6 phosphorus value in the Amethi and Ramdaipur villages and 1.5 phosphorus value in the soil is found in the village Trilokpur. The lowest macro nutrient phosphorus value of 1.3 has been found in the village Raipur Phulwari of the Amethi Block.

iii. **Potash**: The content of the Potash in the soils of the District Amethi has been unlikely medium to high level. The nutrition index values of the potash varies from 2.7 to 3.7. The soils of the Amethi has comparatively lower potash content with the nutrient index values less than 3.0. The variations in the Potash status in the soils of the Amethi is not very large within the block because the potash content in the soil varies from one village to another. Very low 2.9 nutrient index value of potash content in the soil has been found in the village Trilokpur, Ramdaipur 3.0, Amethi 3.1 and 3.3 potash nutrient index of the soil in the village Raipur Phulwari of the Amethi Block. Thus the potash content in the soil has been found medium to high in most of the villages of the Amethi Block of the District Amethi.

II. Micronutrients

The zinc, copper, iron and manganese contents in the soil have been analyzed by the Water Technology Centre of the Indian Agricultural Research Institute, New Delhi based on the samples collected from different locations in the District Amethi. The critical limits for the micro nutrients has been established by the Indian Council of Agricultural Research (ICAR), under All India Coordinated Micro Nutrient Research Project.

Fig. 3.5: Micronutrients of Soil

	Micronutrients	*Critical Limit (ppm)*
1.	Zinc	0.6
2.	Copper	0.2
3.	Iron	2.4 to 4.5
4.	Manganese	1.0

The micronutrients in the soil of the Amethi have been found below the lower limits within which the soils are regarded as deficient in the soil micro nutrients. The distribution of the various micronutrients from one village to another village of the Amethi block of the District Amethi is as follows:

(i) **Zinc**: The content of the zinc in the soil varies from 0.3 to 1.25 ppm in the District Amethi. The critical limit of the zinc in the district has been observed 0.6 ppm. There are some soils in the district which are having deficient zinc content in the soil. The deficient zinc content in the soil has been found in most of the villages of the Amethi Block where the average zinc content in the soils has been found about 0.4 ppm. The soils within the Amethi Block are very considerably deficient in the zinc content. It has been observed that the zinc content in the soils of the district has been closely related with the *p*H value of soil. The relationship of the zinc with the *p*H value of the soils of the district has also been proved and determined statistically through the regression analysis. The micronutrient zinc content in the soil has been found 0.50 ppm at Amethi, 0.46 ppm at Ramdaipur, 0.35 ppm at Trilokpur and 0.62 ppm in the village Raipur Phulwari of the Amethi Block.

(ii) **Copper**: There are high copper content in the soils of District Amethi. The copper content in the soil varies from 0.65 ppm to 4.5 ppm. The very high copper content 2.45 ppm has been found at the village Raipur Phulwari and medium copper content 1.85 ppm has been seen in the village Ramdaipur and low copper content value 1.36 ppm at Amethi and very low copper content 0.88 ppm has been observed in the village Trilokpur of the Amethi Block.

(iii) **Iron**: The content of the iron in the soils of the Amethi varies from 7 ppm to 36 ppm in the District Amethi. The critical limit of the iron content in the soil varies from 2.4 ppm to 4.5 ppm. It has been observed that the soils of the Amethi of the District Amethi have been found sufficiently rich in the iron content. The soil of the various villages of the Amethi Block has been found quite rich in the iron content of the soil.

(iv) **Manganese:** The content of the manganese in the soil samples of the Amethi Block ranges from 10 to 55 ppm. It has been observed that the critical limit of the manganese has been 1.0 ppm. There are sufficient content of the manganese in the soils of Amethi. The soils of the area are having sufficient manganese content in the soil which has been rich in the micronutrient contents of the soil.

It has been observed critically that the highest manganese content of 50.20 ppm has been seen in the Raipur Phulwari village, medium 46.25 ppm in the Ramdaipur village followed by 45.05 ppm manganese content in the soils at the Amethi and very low 35.20 ppm content in the soil has been observed in the village Trilokpur of the Amethi Block.

The soil fertility status of the district indicates that the coarse texture soils and medium to fine textured soils are having impeded drainage water system and salinity and alkalinity problem of the soil are having poor nitrogen and phosphorus status in the soil. It has also been recorded that the medium textured soils of the district are not adversely affected by the salinity and alkalinity problem of the soil which has relatively better content of the nitrogen and phosphorus nutrient content in the soils of the Amethi. It has been observed that based on the nutrient index value of the soils of the various villages of the Amethi block is having higher macronutrient content in the soils with respect to the fertility status of the soil in the selected areas of the district.

6. LAND CAPABILITY

The land of the area has been classified into different land capability classes through applying different criteria. The four types of the land capability classes are found in the Amethi Block of the District Amethi.

- **The first class land capability** classification occupied maximum of 75 per cent areas of the different villages of the Amethi Block.
- **The second class land capability** classification has been found approximately 25 to 35 per cent of the area of the district. In this classification the lands are having limitations to the agricultural crop productivity due to moderate drainage system impedance and soil sodicity and poor fertility status of the soil. The areas having coarse texture soil adjoining to the river Gomati suffered from moderate drainage of the soil and flood/natural hazards due to which the fertility status of the soil has been found comparatively lower than the other areas of the district Amethi.
- **Third class land capability** classification of the land occupies mostly about 30 per cent area of the Amethi Block. Such lands have limitations to the agricultural crop productivity due to marked water drainage impedance, high alkalinity and salinity in the soil, limitations to the crop rooting at certain depth and hard Kankar layers in the sub-soil. The areas adjacent to the Gomati river are subject to severe erosion and occasional inundation with flood water in Shukul Bazar and Jagdishpur, Musafirkhana Blocks of the District Amethi.
- **Fourth class land capability** classification of the land occupies about 10 per cent land of the area. Such types of the land have limitations to the crop productivity and such areas are the low lying land and these lands arise from the pronounced marshiness, very high salinity and alkalinity, extreme root zone restriction due to soil concretion and hard kankar in the surface of the sub-soils of the District Amethi.

7. LAND IRRIGABILITY

The classification of land Irrigability has been ensured on the basis of the soil and land conditions of the area. The soils of the different areas have been grouped into soil Irrigability classes as per their limitations for the sustainable irrigation of the land. Thereafter the land classes according to the Irrigability of the land has been determined. The Irrigability classes of the soil has been defined in terms of the degree of soil limitation. The criteria for classifying the soils into Irrigability classes has been ensured for different agro-climatic conditions.

Fig. 3.6: Land Capability Classification

Sl. No.	*Sub-Class*	*Land Capability Class*	*Characteristics*	*Land Limitations*
I.	Class-I	Land capability has no sub-class.	Low level land with deep depth, well drained soils of satisfactory texture and structure of the soil not affected by harmful factors like Kankar granules, erosion wetness and salinity and alkalinity of the soil.	No productivity limitations suitable for intensive agricultural production for the climatically adopted crops.
II.	Class-II	Sub-Class designated as erosion hazard 11e, soil wetness 11w limitations 11s.	Gently to moderately sloping land with deep soils of fairly satisfactory texture and soil structure affected by the moderate erosion, wetness and salinity and alkalinity of the soil.	Productivity limitations due to moderate erosion drainage impedance and wetness, drought-ness and salinity and alkalinity, crop cultivation adopted to soil limitations.
III.	Class-III	Sub-class designated as erosion hazard 111e; soil wetness 111w and root zone limitations 111s.	Gently to strongly sloping land with moderately deep soils having very slowly permeable sub-stratum; affected by marked erosion, wetness, presence of the shallow water table and the salinity and alkalinity of the soil.	Productivity limitations due to severe erosion marked drainage impedance and wetness poor water holding capacity and high salinity and alkalinity cultivation with precautions against permanent land damama and careful selection of crops adopted to soil limitation.
IV.	Class-IV	Sub-class designated as erosion hazard ive; soil wetness ivw and root zone limitation iv5.	Strongly sloping to steep land, deep to shallow sandy or concretionary soil with low moisture holding capacity, affected by severe erosion, pronounced marshiness droughtness and alkalinity.	Productivity limitations due to very severe erosion pronounced marshiness/drought ness and very severe root zone limitations occasional cultivation of crops in rotation with grasses.

Source: Soil Survey Manual, All India Soil and Land Use Survey Organization, IARI, New Delhi (1970).

The land irrigability classes of the district Amethi has been depicted in **Map 3.14**. It has been noticed that land irrigability classes of different category has been noticed significantly distributed in most of the blocks of the district Amethi.

The land Irrigability classes has been defined in terms of soil conditions identified as soil Irrigability classes and conditions of the land including topographic conditions such as availability of outlets, need for sub-surface, water drainage system and depth of the water table. The criteria for the land Irrigability classification has been presented based on the Soil Survey Manual, All India Soil and Land Use Survey Organization, IARI, New Delhi.

Land Irrigability Class I has an occurrence of maximum of 25 to 30 per cent area in Amethi Block. The land irrigating class I has an occurrence of predominantly deep soils of loam to clay loam texture with high water detentivity and the area is free from salinity and alkalinity problem of the soil, proper water drainage system and erosion problem and sub-surface depth restriction to crop rooting and the Class I land irrigability has water table wells below the potential crop root zone in the sub-surface soil.

Fig. 3.7: Land Irrigability Classification

Sl. No.	Soil Properties	Soil Irrigability Classes				Non-Irritable Class
		A	B	C	D	
1.	Effective soil depth (useful to crops)	90 cm	45-90 cm	22.5 to 45 cm	75-22.5 cm	< 7.5 cm
2.	Texture of surface 30 cm soil	Sandy loam to clay loam	Loamy sand Clay	Sand Clay	Sand Clay	Any texture
3.	Soil permeability (of least permeable layer)	5.0-50 mm/hr	1-3.5 mm/hr 50-130 mm/hr	0.3-1.3 mm/hr 130-250 mm/hr	Less than 0.3 mm greater than 2.50 mm/hr	Any texture
4.	Available water holding capacity to depth of 90 cm	12 cm or more	9-12 cm	6-9 cm	2-6 cm	< 2 cm
5.	Coarse fragments, Gravels and stones (775 mm)	< 5%	15-15%	5-25%	35-65%	> 69%
6.	Gravel and kankar (725 mm)	< 15%	15-35%	35-55%	55-70%	> 70%
7.	Salinity	< 4	4-8	8-16	12-16	> 16
8.	EC_x 10^3 saturation extract	mmhos/cm	mmhos/cm	mmhos/cm	mmhos/cm	mmhos/cm
9.	Salt affected (visual 6% of area of affected.	< 20%	–	20-50%	–	> 50%
10.	Severity of Alkali Problems	ESP < 15%	–	ESP > 15%	–	–

Source: Soil Survey Manual, All India Soil and Land Use Survey Organization, of the Indian Agricultural Research Institute (IARI), New Delhi (1970).

Fig. 3.8: Specification of the Land Irrigability Classes

Land Characteristics	Land Irrigability Classes			
	Class-I	Class-II	Class-III	Class-IV
Soil Irrigability Classes	A	A to B	A to C	A to D
Topography				
Slope Surface Grading	< 1% No. restriction	1-3% Moderate restriction	3-5% Moderate to Severe restriction	5-10% Severe restriction
Drainage				
Outlets	Suitable outlet available	Suitable outlet available	Suitable outlet available	No. drainage outlets available
Sub-surface	No sub-surface drainage needed	No sub-surface drainage needed	Surface drainage needed	No natural outlets available cost of pumping
Depth of Water Table	More than 5 Metres	300 to 5.0 metres	1.5 to 3.0 metre	1.5 metres and below

Note: Land Irrigability Classes V and VI are non-irrigable land, irrigability Class V however could be irrigable after reclamation of the land.

Source: Soil Survey Manual, All India Soil and Land Use Survey Organization, (IARI), New Delhi (1970).

Land Irrigability Class II occupies more than 47 per cent area in the Amethi Block. Irrigability Class II land areas has deep coarse textured soil from low to medium water retentivity and slightly to moderate susceptole to erosion and medium to fine textured soil and it has moderate water drainage impedance salinity and alkalinity problems of the soils and restrictions to the surface grading of the soil in the various villages of Amethi.

Land Irrigability Class III occupies more than 30-34 per cent area in Amethi Block. Land irrigability has coarse textured soils of low water retentivity and moderate to severely susceptible to erosion and occasionally inundated by flood flows and fine textured soil with marked water drainage impedance salinity and alkalinity problem of the soil, and sub-surface soil depth limitations to the crop rooting and moderate to severe restrictions to the surface grading of the soil. The Land Irrigability Class III would give high pay off under irrigated agriculture when erosion severity and droughtness in the coarse textured soils are minimized and ameliorative measures has been adopted to reclaim alkalinity and salinity problem of the soils and drainage impedance in finer texture of the soils in the various villages of the Amethi Block.

Land Irrigability Class IV occupies about 15 to 22 per cent area in the Amethi Block. The land irrigability has permeable drought soils. The land is severely susceptible to erosion and regularly inundated through flood flows and it has fine textured soils with severe water drainage impedance, water logging, salinity and alkalinity problems of the soil and the soil depth limitations to crop rooting. Agriculturally productive irrigation would be possible through adequate ameliorative measures are taken against adverse soil problem in the various villages of Amethi Block, District Amethi of Uttar Pradesh.

8. RECLAMATION OF USAR LAND

"Usar Land" in general term includes soils having an excess of soluble salts, an excess of exchangeable sodium or both. The usar soils having high exchangeable sodium percentage (ESP > 15) are called sodic soils or usar land which has higher *p*H value of the soil more than 8.5 higher enrichment of the soils clay with exchangeable sodium. Excessive salt affected soils are agriculturally unproductive. The main reclamation procedure of the usar land is to replace the exchangeable sodium with calcium—

$$2\ Na + (Clay) + \text{Counter affidavit} ++ = \text{Counter affidavit} ++ (Clay) + Na^{+}$$

The required efficient exchange reactions are sufficiency of moisture and removal of reaction product through leaching and flushing of the usar soils.

The usar land reclamation conventional measues and organic and inorganic amendments to provide calcium is gypsum (Counter affidavit SO_4, $2H_2O$). The another material applied for the reclamation of usar land is pyrite. The main ingredients of the soil is iron, pyrites, responsible to bring sulphur in the usar soils. The chemical reactions of the soil in the presence of water is as follows—

$$2\ Fe\ S_2 + 7O_2 + 2H_2O = 2\ FeSO_4 + 2H_2SO_4$$

$$FeSO_4 + 2H_2O = Fe(OH)_2 + H_2SO_4$$

The chemical equations make it clear that the water and oxygen are main ingredients needed for the oxidation process in which sulphur in pyrite is converted into sulphuric acid, to ensure the oxygen and water availability in the soil while pyrite requires to the surface dressing of the soil. The sulphuric acid reacts with native insoluble calcium carbonate to form soluble calcium sulphate. The calcium through base exchange reactions converts the sodium clay in the calcium clay.

The organic and inorganic amendments required for the reclamation of usar land based on the severity of the usarization and to re-maintain the fertility status of the soil. The gypsum and pyrite requirements of the soil varies based on the *p*H value of the soil. Proportionately higher quantities of the gypsum and pyrite is required to reclaim the excessively affected usar soils having advanced stage of soil deterioration in terms of the fertility status of the soil. There are certain suggestions and points which should be considered before application of the gypsum and pyrite as an amendment for the reclamation of usar soils of any area.

Fig. 3.9: Pyrite (22% Sulphur) and Gypsum (Tonnes/Ha)— Requirements for Usar Land Reclamation

pH value of the soil	*Coarse Textured Soil*		*Medium Textured Soil*		*Fine Textured Soil*	
	Pyrite	*Gypsum*	*Pyrite*	*Gypsum*	*Pyrite*	*Gypsum*
8.6	–	–	–	–	0.42	0.49
8.7	–	–	–	–	0.85	1.00
8.8	–	–	0.42	0.49	2.13	2.57
8.9	0.42	0.49	1.53	1.80	3.40	4.00
9.0	1.20	1.41	2.55	3.00	4.68	5.50
9.1	1.20	2.00	3.40	4.00	5.53	6.50
9.2	2.55	3.00	4.42	5.20	6.63	7.80
9.3	3.20	3.77	5.10	6.00	7.45	9.00
9.4	3.60	4.24	5.98	7.04	8.60	10.12
9.5	4.25	5.55	6.80	8.00	9.35	11.00
9.6	4.90	5.76	7.65	9.00	10.25	12.06
9.8	5.53	6.50	8.50	10.00	11.56	13.60
9.9	6.80	8.00	10.20	12.00	12.00	14.12
10.0	7.23	8.50	11.05	13.00	12.75	15.00
10.1	7.65	9.00	11.50	13.53	13.25	15.60
10.2	8.07	9.50	12.00	14.12	14.00	16.47

Note: The pyrite quantity need to be multiplied by 1.47 and 0.743 for pyrite having 15 per cent sulphur and 3 per cent sulphur respectively.

Source: Reclaiming Alkali Soils, Central Soil Salinity Research Institute (CSSRI) Karnal, *ICAR Bulletin No. 2* (1973).

(i) Proper preparation of field and pre-amendment treatment of water irrigation to bring the soil near to field capacity and moisture content of the soil.

(ii) Properly crushed material should be spread uniformly and thinly as possible in the most soil surface to ensure maximum air availability to the soil to facilitate oxidation during the application of pyrite.

(iii) Requisite time should be given for the application of pyrite to oxidize on the surface of the soil. The pyrite application should be ensured during summer to provide advantage of higher temperature for the excessive effective oxidation in the soil.

(iv) The small plotting of the land should be ensured and filled with the good quality irrigation water to facilitate leaching and flushing of the soluble salts in the soil.

(v) Preferably suitable high yielding variety of rice crop should be transplanted in the reclaimed usar land. Proportionate number of seedling per hole should be increased and the distance between row to row should be decreased to compensate poor survival of the plant seedlings and to maintain appropriate plants. Requisite

Fig. 3.10: Tolerant Crops on Usar Land

Field Crops			*Vegetables*			*Trees*		*Grasses*
Highly tolerant	*Medium tolerant*	*Sensitive*	*Highly tolerant*	*Medium tolerant*	*Sensitive*	*Highly tolerant*	*Medium tolerant*	*Highly to medium tolerant*
Dhaincha	Castor	Pulses	Spinach	Cabbage	Beans	*Babool*	Eucalyptus	Blue Panic
Rice (IR.8).	cotton	Pigeon pea	Kale	Cauli	Peas	*(Acacia-*	hybrid	*(Panicum*
Jaya. Pusa	Sorghum	Sunnhemp	Asparagus	flower	Musk	*Nilotica)*	Zizyphus	*Antidotale)*
2-21, CSRI	Maize	Gram Pea	Sugar	Radish	Melon	*Bilayati*	Species	Para Grass
(Damodar)	Mustard	Linseed	beets	Carrot	Okra	*Babool*	Eucalyptus	*(Bracharia*
Padma	(Raya	Sesamum	(Romons	Garlic	tomato	*(Prosphis*	tetaricornis	*munica)*
CSR-5	RH7811	Lentil	Kaya-06)	Onion	Potato	*Juliffora)*	Neem	Rhodes
TKN6 X IRS	RLM1988			Turnip	Brinjal	*Siris*	*(Azadira*	Grass
Wheat	Varuna)				Chilies	*(Albizia*	*Chtaindica)*	*(Chloris*
(Kalyan					Cucumber	*Pebbeck)*	Arjun	*gayana)*
sona							*(Terminalia*	Karnal
S-308,							*Arjuna)*	Grass
HD 1553							Bakain	*(Diplachne*
HD 1981							*(Azadir*	*fusca)*
HD 1982							*achita*	
HD 2236							*exotica)*	
HD 2009								
HD 916 WG								
157 WG								
357 C-306								
Karahia 65								
Barley								
Vijoy, Retna								
Ambar								
DL68,								
DL120								
Oats								
Barseem								
Sugarcane								
CO975								
CO-453,								
CO1148								
B-37172								

Source: Irrigation Water Quality and Crop Management in the Union Territory of Delhi, Indian Agricultural Research Institute (IARI), New Delhi, *Research Bulletion No.* 9 (1976).

application of organic and inorganic amendments including fertilizer with Nitrogen, Phosphorus, Potash and Zinc to maintain the fertility status of the soil.

(vi) Generally, the rice-wheat crop rotation has been recommended. The cultivation of rice-barseem during the first year of usar land reclamation has been helpful in maintaining the fertility status of the soil. The cultivation of barseem required higher doses of irrigation water which would be helpful in the faster reclamation of usar land.

(vii) The cultivation of Dhaincha (*Sesbania acyleata*) as green manuring of the soil should be carried out. The cultivation of Dhaincha should be ploughed under 6 to 8 weeks after sowing the dhaincha crop, rice cropping should be transplanted as per requirements of the soil to maintain its fertility status.

It has been observed that proper application of gypsum and pyrite and cultivation of paddy in the usar land in the several villages of the Amethi has shown pronounced effect. Comparatively the application of gypsum and pyrite has indicated slightly better performance in the paddy cultivation than wheat cultivation. The pyrite application also supplies the micronutrients in the soil to maintain the fertility status of the soil.

The use of chemical amendments, i.e. inorganic and organic amendments of the Farm Yard Manure (FYM), organic waste and plant residues in the reclamation of usar land requires to be promoted. The adoption of salt resistant crop cultivation should be encouraged in the reclaimed usar land. The highly affected usar land should be diverted for the plantation of forest trees and shrubs. The agricultural crops, trees, grasses and vegetables having salt tolerant capacity should be encouraged in the salt affected land or usar soil of any area.

The reclamation of usar land planning in the various villages of the Amethi Block has usually resulted through improper poor quality management of irrigation coupled with very poor water drainage system due to the lack of properly designed drainage system. Therefore, it has been imperative to improve the management of proper irrigation and to develop appropriately designed water drainage system to obtain the reclamation of usar soils on the basis of long term sustenance with the view to improve agro climatic conditions of the area to meet the increasing demand of the growing cattle and human population for fuel wood, fodder, fibre, fruits, fisheries, foodgrains (F^6) and wood vegetables etc.

9. CONCLUSION

It has been observed that the soils of the district are formed through the alluvial deposits carried out by the river drainage system. The soil structure of the district has been divided into ravines of the Gomati river, flat lands of Gomati, Sai river uplands, Sai river low lands and Sai river flat lands.

The younger alluvial and older alluvial tracts and saline and alkaline soils having hard kankar layers are found in the district Amethi.

The soil sample surveys has been carried out to assess the fertility status of the soil by the Soil Survey Organisation of the Sharda Sahayak Command Area Development Authority, Department of Agriculture, Government of Uttar Pradesh and the collected soil samples has been tested and analyzed by the Water Technology Centre of the Indian Agricultural Research Institute, New Delhi.

The physical and chemical soil characteristics of the soil has been analyzed. The physical characteristics of the soil has been explained through mechanical composition of the soil water retentivity, soil depth and salinity and alkalinity problems of the soils.

The chemical characteristics of the soil has been analyzed based on the cation exchange capacity of the soil and salinity and alkalinity problems of the soils.

The physical and chemical characteristic of the soil has been carried out at Amethi, Raipur Phulwari, Ramdaipur and Trilokpur villages of the Amethi Block. The textural classes of the soil has been found silt clay loam, silt loam, silty clay loam and silty clay in the different villages of the Amethi. The field capacity, wilting point of the soil, *p*H value of the soil, Electrical Conductivity Exchangeable Sodium percentage and cation exchange capacity of the soil has been analyzed, which can be maintained and managed to increase the fertility status of the soil. The textural class of the soil and moisture of the soil, *p*H value of the soil and Electrical Conductivity value of the soil, Exchangeable sodium percentage and cation exchange capacity of the soil analysis indicates the higher level of soil sodicity in the Amethi. The high *p*H value of the soil and Exchangeable Sodium Percentage has been found at different layers of the depth of the soil.

The soil samples have been collected, tested and analyzed at different depths, layers of the soil profile in the various villages of the Amethi. The macronutrient and micronutrients value of the soil have also been analyzed.

The mechanical composition and chemical properties of the top 30 cm soil has been analyzed in the selected villages of the Amethi. The mechanical composition of the soil including sand, silt and clay percentage has been explained and the textural classes and *p*H value of the soil and other physical and chemical properties of the soil for the selected at 0.30 cm depth, 30-50 cm depth 50-90 cm depth, and 130-150 cm depth soil texture and other properties of the soil have been explained to maintain the fertility status of the soil. The macronutrient and micronutrient values of the soil have been analyzed through the nutrition index of the soil. The nutrition index of the soil ranges from 1.8 (very poor) to 4.2 (very high) value of the soil has been found in the various villages of the Amethi.

The macronutrients (Nitrogen, Phosphorus and Potash) and micronutrients (Zinc, Copper, Iron and Manganese) content of the soil in the various sample locations at Amethi, Raipur Phulwari, Ramdaipur and Trilokpur village has been analyzed. It has been observed that higher value of the macronutrient and micronutrient contents at different layers of the soils have been found at Amethi. The land capability classification has been carried out and four different types of land capability classes have been explained based on soil texture and content of the soil and fertility classes of the soil.

Land irrigability classes have been classified in Amethi through A, B, C, D irrigability classes of the soil based on the soil properties. Agricultural production is possible through adopting adequate ameliorative measures to maintain the fertility status of the soil.

The reclamation of the usar land has also been carried out in different villages of Amethi. Proper bunding, levelling, leaching, flushing, good quality water irrigation management, flushing, proper application of organic and inorganic amendments and cultivation of salt tolerant high yielding variety agricultural crops, trees, grasses, vegetables and multi-cropping through techno-scientific management of the reclamation of usar land to increase the fertility status of the soil and to enrich the macronutrient and micronutrient content of the soil for cultivation of the crop, with the view to conserve the environmental conditions of the area and to utilize the available natural resources of the Amethi, and to maintain the eco-system on the one hand and on the other to meet the growing demand of growing cattle and human population of the area for fuel wood, fodder, fibre, fruits, fisheries, food grains (F^6) and wood, vegetables through which socio-economic conditions of the deprived resourceless rural poors can be improved for sustainable eco-friendly integrated development of the Amethi and adjoining areas of the other districts.

4

Water Resources

1. INTRODUCTION

The water is the basic source which affects the survival of plants and affects the agricultural crop production in any area. The district Amethi is located in the mid of the Gangetic plain and Gomati river. The district is endowed with an abundant water supply of this vital natural resource. It has been recorded that the average annual rainfall in the district has been about 1000 mm of which approximately 93 per cent of the rainfall occurs during the period from June to October. Thus the district Amethi falls under the medium rainfall regions of the country. The deep alluvial soils of the Gangetic plain in the district offer ample opportunity for the moisture of the soil, water storage and recharge of the under ground water. It has been observed that there is abundant ground water which could be harnessed for facilitating the irrigation system for the agricultural crops through installation of shallow and deep tube-wells. The canal irrigation is another source of irrigation system and the area is irrigationally served by the various branches of canal irrigation system of Sharda Sahayak Command Area Irrigation Development Programmes of the Department of Agirculture, Government of Uttar Pradesh.

It is known fact that the district Amethi lies in the heart land of the Sharada Sahayak Canal Command Area Development Programme. The major natural drainage system of the district are the tributaries to the Gomati river and Sai river, Chamraura and Kunwar all of which are the tributaries of the Ganga River. The river Sai flows from the south-western border of the district while the Gomati river flows from the south-eastern border of the district Amethi. The Sai river leaves the Salon Block on the south-east corner of the district. The Gomati river, a major tributary of the river Ganga, flows across the district entering along the northern border of the district in the Shukul Bazar Block and Musafirkhana Block of the district. The ravine areas of the district are undulating and there are large and small numerous water sheds in the district.

2. RIVER SYSTEMS

The district Amethi has an abundance of surface water resources. The Gomati river has been untapped for irrigation practically. Similarly, the Sai river is also practically untapped for irrigation except for a few minor lift irrigation systems and large canal network provides the irrigation water to most of the areas of the district. An attempt has been made to describe briefly the origin and coarse of the river Gomati and Sai rivers which are well known for natural water resources of the district.

I. Gomati River

Approximately 3 km east of the Pilibhit town in the Pilibhhit district the Gomati river rises near Mainkot at an elevation of the 20 metres at North latitude 28°34′ and east longitude 80°7′. The Gomati river drains in the areas lying between the river Sharda and Ram Ganga in upper reaches and the Ghaghara river at the lower reaches. The Gomati river receives its first tributary the Gachai after flowing 24 km and later on the Gomati river receives another two tributaries, namely Jomkai and Barua while flowing through the District Shahjahanpur. There are two more tributaries, namely Chaha and Sarayan, which join the Gomati river in Kheri-Lakhimpur and Lucknow districts. The Gomati river flows to the south-east direction and passes through Lucknow city and district and flows through the Barabanki district, Shukul Bazar and Musafirkhana Blocks of the district Amethi, Sultanpur and Jaunpur districts before its confluence with the Ganga River. The total length of the Gomati river is about 940 km and drains covering an area of 30,433 sq. km from its origin to the confluence with the river Ganga.

II. Sai River

The river Sai is the most important tributary of the Gomati river which joins on its right bank. The Sai river rises in the Hardoi district and traverses in the district Amethi, Rae Bareli, Pratapgarh and Jaunpur districts before joining the Gomati river at about 60 km up stream of the river confluence of the later with the river Ganga. The Sai river has a drainage area of about 11,115 sq. km starting from its origin up to the confluence with the river Gomati and later on with the river Ganga.

Most of the areas of the district are low-lying and there are numerous lakes, ponds, water logged land and marshy land. Chamraura river is a small tributary in the Amethi. The natural water resources of the district are sufficient provided that appropriate planning system is developed to tap the available water resources of the area.

In spite of the available natural water resources of the district, Sharda Sahayak Irrigation Project has been planned and augmented to irrigate to the lower reaches of the Sharda Canal Command Area by diverting the water flows of the Ghaghara and Sharda rivers with the view to increase the annual irrigation intensity up to 96 per cent to the total cultivable land especially 60 per cent in the Kharif and 36 per cent in Rabi cropping season. Lower Sharada Barrage was constructed across the Sharada river about 160 km down stream of Banbasa where the headworks of the Sharada Canal system is located. A 260 km long cumecs take off has been constructed to augment the lower reaches of the area. The feeder channel across the river Gomati and Sai river is on its way. The Gomati-Sai Doab area has been earmarked covering the district Amethi covering 28 per cent irrigation of the total planned area under irrigation through Sharda Sahayak Command Irrigation Project.

The Singhpur distributary enters in the district near Sidhnagar village of the Singhpur Block. The design discharge of the Singhpur distributary at its take off point has been 8.80 cumecs, while there are 5 minors and distributaries that take off from Singhpur distributary, each with a culturable command area of more than 500 hectares, in addition to it there are 35 smaller minors with total culturable command area of 8887 hectares.

The Inhauna distributary enters near Nanla village in the Singhpur Block of the District. The distributary runs through Singhpur Block for a length of 7.26 km. The design discharge capacity of Inhauna distributary as its off-take point is 18.46 cumecs. There are two minors that take off from Inhauna distributary each with culuturable command area more than 500 hectares. In addition to the above there are 6 smaller minors with a total culturable command area of 1618 hectares. The total culturable command area of Inhauna distributary system is 4226 hectares.

In addition to the extensive development of the available natural surface water resources in the district Amethi, the ground water resources contributes substantially to the water irrigation system in the area, the existing ground water potential and the feasibility to increase the exploitation of available ground water resources has to be planned scientifically to fulfil the irrigational requirement of the district. The ground water and tanks has been the major source of irrigation, prior to introduction of canal irrigation system in the district. The extensive development of the available surface water resources in the district the ground water resources contributed to the irrigation up to a limited water irrigated 20 per cent of the net cultivated area of the district and 29.6 per cent of the net irrigated area. It has been observed that with the adventure of canal irrigation and irrigation from wells and tank irrigation has been practically reduced to insignificance. The extensive irrigation canal network and to older alluvial soils of the district, it has been anticipated that the ground water resources may provide more support for agricultural irrigation in the years to come. The existing available ground water potential and the feasibility for the increased exploitations of ground water resources has to be considered attentively for appropriate exploitation of the available natural ground water resources to meet the increasing demand of irrigation facilities of the cultivators of the district Amethi.

3. DRAINAGE

The Sai river, Gomati river, Kunwar, Chamraura are the main natural drains in the Amethi district. The district suffers from the problem of water logging during the heavy rains in the monsoon season. The excessive rain has been responsible for the problems of water logging, flooding by the natural drains and the undulated topography has caused the formation of several major and minor watersheds. Approximately more than 40 per cent area of the district has been found flood-prone area. The soil survey classification of the district indicates that approximately 50 per cent of total geographical area of the district has been identified and classes as poor to very poorly drainage system of the area.

Although there is extensive irrigation canal network functioning in the district, there is no major problem of the water logging in the district, resulting from the rise of sub-surface water table, except in the few pockets of the district. It has been the occurrence due to large scale development of ground water for irrigation purpose. Thus the problem of drainage system has been essentially the removal of excess surface water during the monsoon season. The occurrence of heavy rains has caused large scale water logging, affecting, practically mostly all the villages of the district. Approximately, 22 per cent of the total geographical area of the district has been covered by the existing drainage networks. The existing drainage networks have been constructed with the objectives of linking numerous ponds and tanks with each other. The most of the smaller linking drains are yet incomplete and do not connect to the ultimate natural drains of the area. Natural drains, namely Gomati, Sai, Kunwar and Chamraura rivers are frequently chocked by silting leading to flooding and water logging of their embankments. There are several blocks in the district which do not have any drainage network.

In view of the inadequacy of the existing network of drains, in controlling the problems of flooding and water logging, the Department of Irrigation, Government of

Uttar Pradesh, has planned for the remodelling and cleaning of the existing drains and constructions of several new drains in the different blocks of the district Amethi. A plan has also been formulated for cleaning up the large natural drains including Chamraura, Sai, Kunwar and other local level silted natural drains. Most of the drainage channels has also been constructed for linking the numerous ponds, tanks, lakes with each other. There are several small drains which are incomplete and do not connect with the ultimate natural drains as per the slope and direction of the water flow to the low-lying areas. The Chamraura and Sai rivers are frequently chocked through the deposit of the silt, which leads to flood along their embankments. The agricultural field drains are quite different from the main or intermediate drains. The drainage system for the cultivated land is required to be designed to remove the surface water faster enough to prevent drainage, overflow, for the standing crops in the field and to maintain the sub-surface water table at such level which is conducive to the crop growth. The most important factors influencing the drainage design system are the drainage requirements of the crops and topography of the cultivated field, soil characteristics, underground status of the water able, rainfall, system of irrigation water. The plan formulation for the drainage design parameters should be based on crop losses due to water logging and higher cost of drainage for the removal of the accumulated water at the earliest. The planning for the cultivated field drainage system is based on the field conditions and it requires need based planning through the field investigation at grassroots level.

There are variant approaches for the design of drainage system which has been a conventional practice because the slope up to 0.5 per cent on the flat field, constant drainage design and development has been assumed. The drainage rate has been about 10 mm/per day (10 cusec/sq. mile). The design and development of drains system adopting this approach could not prove very effective and useful to drain out the excessive accumulated water in the cultivated field especially in the low-lying areas. An appropriate design and development of drainage system required to be developed critically on the basis of the information related to the excess water flow and drain out requirements of the cultivated crops in the field.

The tolerance of the crops to the surface poundings varies in their proportionate tolerance to the water logging situations. The immediate removal of the excess water within the minimum time is must to prevent the damage of the crops in the field. It has been observed that most of the crops indicates the sign of deterioration of crops if the water logging situations prevail beyond two to three days. Especially for the maize and vegetable crops the removal of excess water period must be within a day. The wheat crop can tolerate the surface ponding maximum up to three days. There are other important food grain crops which can tolerate thc water ponding in the crop field within one to five days. The general variety of paddy crop cannot tolerate the total submergence of water for more than 5-7 days, through grown conditions of the water submergence in the field crops.

The drainage system for the crop field must maintain the sub-surface water table below the root zone of the crops to control the damage of the crops. The design and development of sub-surface drainage system in field which are threatened by the rising water table closer to the surface of the field. It has been observed that the water table levels during the post-monsoon season has reached up to two metres of the surface in the low-lying area in the district. The design and development of the agricultural field drainage is mostly to be based on the surface drainage system and requirements of the cultivated crops.

The surface drainage is required to remove the excess rain water and irrigation water from the cropping field. The development and drainage design and construction should be carried out based on the two or three connective days of the rainfall and resulting runoff for this time at the specified level of the risk of the crops.

4. GROUND WATER

The ground water infrastructure one of the most important sources of irrigation because more than 55 per cent irrigated areas of the district Amethi are being irrigated through the ground water source of irrigation. The largest alluvial tracts of the Indo-Gangetic plains, in which the district Amethi lies, has provided an ideal geo-hydrological situation for the integrated development of ground water resources. The individual farmers and the government agencies have given due recognition and considerable emphasis for the development of ground water resources. Numerous open wells, tube-wells, dug wells, bore-wells have been constructed and installed which are uniformly speed allover the district at village level. Those areas which are not covered under the Canal irrigation system, as per the policy of the State Government, deep tubewells have been installed in most of the villages of the District Amethi. The cultivators are being encouraged to install shallow tube-wells and dug-cum-bore-wells to control the water logging and augment canal water irrigation system. Appropriate progressive development of ground water resources through private tube-wells, government tube-wells and other sources of irrigation have been implemented in the various villages of the district. The geo-hydrological conditions and the analysis of the ground water table fluctuations and the assessment of the available ground water resources and the identification of the scope for the further development of the ground water resources of the district have been analyzed in detail by the government agencies.

i. Ground Water Status

The movement and occurrence of the ground water resources of the district is dependent on the sub-surface formation of the water ground and hydrological characteristics of the area. The district Amethi lies in the large scale alluvial tract of the Indo-Gangetic plains, the ground water occurs in the deposits of the loose unconsolidated sediments in the unconfined conditions. The lithology of the soil and sub-surface soil comprise sandy clay of varying thickness of sand and gravel of the tectonic to the recent age. The aquifer zone extends to depths more than 100 m below the ground surface. The transmissivity values lie in the range of 1000 to 2000 m^2/day. The specific yield of the functions in the ground water has been in the range of 10 to 18 per cent in the district Amethi.

It has been recorded that the average annual rainfall of the district has been 1000 mm which has also been the basic source of ground water recharge. The canals and wells another source of irrigation has also been another source of ground water recharge in the district. The data to the depth of ground water has been observed and recorded for the both pre-monsoon and post-monsoon ground level in the district Amethi. The average depth of the water table has been 5.3 m below the ground level before the monsoon and rises to 3.9 m below ground level at the end of the monsoon in the study area. It has also been observed that in spite of the extensive development of the ground water the water table has reached upto within two metres of the ground surface in south to eastern parts of the district due to low-lying areas and accumulation of excess rain water and irrigation water of the adjoining areas.

ii. Ground Water Tab. ⁻luctuations

The ground water table of the district fluctuates and varies from one village to the another and also based on the occurrence of the rain fall in the district. The rain fall has been principal source of the ground water recharge. In general the annual rainfall occurs during the monsoon season, which affects the water table during the rainy season are most significant. The average rise of the ground water table during the monsoon season has been 2.35 m but declined by nearly half 1.3 m during the another situations and it varies from the location to the another observation points. The average rise of the ground water table for a decade at the different locations of the wells ranges from 1 m to 2.3 m leading to an average annual rate of 1.83 m rise of ground water table all over the district, during the period when rainfall exceeded the normal by more than 60 per cent. The rise in the ground water table reflected in the water table rise by rarely 2.9 m during that particular year in comparison to the average value of 1.83 m for analysis period in the district Amethi.

The seasonal fluctuations of the water table reflect primarily the effect of monsoon rainfall on the ground water, the net effect of the both recharge and draft of the ground water during a year. The water table rose by as much as 2.1 m because of the unusually high rainfall in the district. In the subsequent years there has been once again marginal rainfall in the district which has lead towards a marginal fall in the average elevation of the water table. Thus, the ground water table fluctuations and its ups and downs depend on the average rainfall of the area.

iii. Ground Water Draft

The extraction of the ground water in the district Amethi has been carried out through tube-wells, dug-wells, and dug-cum-bore-wells. Very nominal dug-wells are operated by the Persian wheels in the district. The average annual draft of the ground water depends on the number and yield of wells operational in each year and the capacities of the water pumps in the area.

The Central Ground Water Board (CGWB) established the following norms based on the sample survey of the minor irrigation units in the area:

• Shallow tube-wells/Electric Motor Driven Pump sets	2.28 ha.m./year
• Shallow tube-wells/Diesel Engine operated Pump sets	3.32 ha.m./year
• State Government Tube-well	10.85 h.m/year

Thus Ground Water Estimation Committee of the Ministry of Irrigation, Government of India, recommended the following norms for estimating the draft for various ground water tables. The norms for estimating the draft provides the irrigation potential of each type of well.

Ground water Structure Type	*Command Areas*
• Dug well	1.0 ha
• Dug Well with Persian wheel	1.5 ha
• Dug-cum-bore well	1.5 ha
• Pump set on dug-cum-bore-well	5 ha
• Private tube wells	5 ha

It has been assumed that approximately 60 cm of water is required for irrigation of the one hectare of the land. The irrigation potential may be expressed in terms of the corresponding volume of ground water extracted for each type of ground water structure in the respective types of wells of any area in the district.

The obtained values are comparable to the estimated ground water draft by the Central Ground Water Draft by the Central Ground Water Board, Government of India. Therefore, the following norms can be adopted to analyse and estimate the ground water draft structure of any area including the study area of the district Amethi.

- State Tube-wells — 10.85 ha.m/year
- Private Tube-wells — 3.32 ha.m/year
- Dug well — 3.32 ha.m./year
- Dug well with Persian wheal — 0.9 ha.m./year
- Dug-cum-bore-well — 0.9 ha.m/year

The annual draft of the ground water structure ranges from 660 million cubic metres to 1260 million cubic metres annually during different years. It has been observed that the ground water draft has directly affected the ground water level in the district by lowering the water levels in the area about 1.00 m during the respective period in the district Amethi.

iv. Ground Water Recharge

The principal source of the ground water recharge has been the rainfall seepage from Canal irrigation distribution system and irrigation return water flow. The Ground Water Estimation Committee, Ministry of Irrigation, Government of India, has recommended specific norms for the estimation of ground water recharge from each of the above sources of irrigation prevailing the different parts of the district:

1.	Recharge from rainfall	=	20 per cent of normal rainfall
2.	Recharge due to seepage from unlined canals	=	1.8 to 2.5 cumecs/million square metres of wetted area.
3.	Return seepage from irrigation fields under canal irrigation	=	35 per cent of water delivered at outlet.
4.	Return flow from irrigation underground water irrigation	=	30 per cent of water delivered at outlet.

The ground water recharge through return seepage includes losses in the field irrigation channels of the district.

(i) **Recharge from Rainfall**—The average annual rainfall of the district has been recorded 1000 mm. The ground water recharge corresponding to the average annual rainfall as per the above norms has been 174 mm or 772 million cubic metre in volume units has been analyzed for the district Amethi.

(ii) **Recharge from Canal Seepage**—The irrigation water seepage losses from canal systems has been estimated based on a rate of 2.5 m^3/sec./million sq. m. of the wetted area of the canal branches of the district. The designed discharge capacity of 33 cumecs has been recorded for the canal branch in the district. The wetted perimeter and seepage losses have been estimated for different reaches between

41st and 110th km, about 56 cumecs or about 10 per cent of the available water in the canal system has been lost as seepage and added to the ground water reservoir recharge in the district. The large canal distributaries with discharge capacity of more than 3 cumecs (100 cusecs) are approximately considered as branch canal irrigation system. The wetted perimeter and corresponding seepage losses for the Amethi canal distributary has been estimated to be about 13 per cent of the design discharge capacity of 8.6 cumecs from canal water irrigation system.

The seepage losses from smaller canal distributaries have been assumed at the rate of 7 per cent as per the standard norms. The total recharge from the canal irrigation network system has been estimated as follows:

1.	Recharge from the branch canals (10 per cent of design discharge)	=	16 cumecs.
2.	Recharge from the large distributaries (13 per cent of design discharge)	=	9 cumecs
3.	Recharge from other distributaries of canal branches (7 per cent of water let into distributaries)	=	9 cumecs

The total recharge from the canal irrigation system has been 34 cumecs in the district, if the canal system has been running throughout the year at its design discharge capacity, when the actual water discharges has been much lower than the corresponding design discharge values. The seepage losses would not be reduced to the same proportion for the reduced level of water discharge from the canal irrigation system. Thus the net water recharge in the district Amethi through canal irrigation system upto the outlet has been 514 million cu.m.

Return seepage from the canal irrigated fields has been estimated as 35 per cent of the water delivered at the outlet. The water recharge from the return flow from irrigated areas to the ground water has been 430 million cubic metre.

The total average annual recharge from the various sources has been estimated as follows:

1.	Recharge from rain fall	=	922 MCM.
2.	Recharge from canal irrigation system and return flow from canal irrigated area	=	1430 MCM
3.	Recharge from return flow from the ground water irrigated areas	=	439 MCM

Thus there is sufficient ground water recharge from rain fall, seepage from canal distribution system of irrigation and irrigated return flows constitutes principal sources of ground water recharge in the district Amethi of Uttar Pradesh.

The estimated seepage losses in the canal irrigation system (514 million cu.m) and return seepage from the canal irrigated areas (321 million cu.m.) for the branch canal irrigation system. There are Inhauna, Singhpur, Subeha distributaries, which do not originate from the Sultanpur and Jaunpur Canal branch system, contributes effectively in the ground water discharge of the district. Thus the total average annual recharge from rainfall, (772 m cu.m.) + canal irrigation (900 m.cu.m) + ground water irrigation

(360 m cu.m.) = 2032 million cu.m. for the last decades has been estimated in the district Amethi of Uttar Pradesh.

v. Ground Water Balance

The net change in the ground water table indicates a marginal fall in the ground water table of the district. The total, ground water inflow or outflow is not known but it can be estimated from the above equations. The change in the depth of 0.1 m corresponds to a net drop in the ground water storage of 55.3 MCM. The ground water balance in the aquifer has been as follows:

Change in the ground water storage = (Total recharge) – (Total draft) – (Ground water inflow/outflow from the basin)

In view of the above equations the ground water out flow may be estimated as 1045 million cu.m. in the district Amethi.

vi. Ground Water Potential

The ground water natural resources can be apportioned into static and dynamic components of the water table of the district.

The static water resources are the amount of water table available in the permeable portion of the aquifer below the zone of water level fluctuations of the area.

The dynamic water resources are amount of the water available within the zone of water table fluctuations. The ground water resource usable has been essentially the dynamic component which has been the result of the periodic water recharge through the rainfall, seepage from canal irrigation water and ground water irrigation return flow in the ground water of the district.

The water table fluctuations approach can be used for estimating the ground water potential. The maximum recorded depth/thickness of the zone of water fluctuations has been determined through the hydrographs of the observations of wells. The water fluctuations have been taken with the difference between the lowest pre-monsoon and highest post-monsoon under ground water level. The depth of the zone of water fluctuations varies in the range of 3.9 to 10.0 m. The specific field of aquifer in the zone of water table fluctuations for the Indo-Gangetic alluvial plains can be assumed to be uniform at 12 per cent in the district Amethi.

The gross recharge = (Thickness of water table fluctuation zone) × Specific Yield × Area of the district

The average annual rainfall for a decade has been information related to the rainfall in the district. The average estimated recharge potential for the normal rainfall has been estimated up to 4066 MCM. The estimated mean gross recharge includes the effects of the ground water recharge components and the discharge from the aquifer during the monsoon season. The outflow from the ground water reservoir during the non-monsoon period and the effluent of the river etc. has not been included. The average ground water outflow has been about 1150 MCM as per the occurrence of the pre-monsoon and post-monsoon period in the district. The Ground Water Estimation Committee of the Government of India recommended that the net recoverable recharge has been 70 per cent of the gross recharge.

The existing status of ground water exploitation upto 57 per cent indicates that about 43 per cent increase in the ground level water development. In the canal irrigated areas, ground water development through shallow tube-wells and dug-wells has to be promoted in the such areas where water table fluctuations has been recorded comparatively very low. The areas which are not facilitated by the canal irrigation network, further identification of ground level water development through the deep tube-wells has been recommended. The analysis of the ground water recharge on the basis of quantum of irrigation activity through the various canal irrigation network in the district. It would be an appropriate step, if the canal irrigation system could be extended promoted at larger extent. The water seepage from canal water irrigation system and return flow from the irrigated fields through the canal irrigation network are the major important contributors of the annual ground water recharge and it is an important task of the implementing non-government and government Institutions to maintain the tempo of the ground water level development at least at the prevailing rate of ground water recharge in the district Amethi.

5. UTILIZATION OF WATER RECOURSES

The district Amethi has adequate water resources and well knit distribution of irrigation water system. The water resources and infrastructure available in the form of irrigation facilities and ground water resources are properly exploited to meet the minimum requirements of the people of the district. The important features of the utilization of water resources are:

1. Canal water and ground water sources of irrigation.
2. Canal water irrigation is the main source of irrigation through the Sharda Sahayak Irrigation Programme.
3. Ground water is the major source of irrigation in the district even in the canal network irrigated areas and there are ample opportunities for the further development of the ground water resources of the district.

The subsequent changes in the cropping pattern from one cropping season to the another leads to varying demands of irrigation management. The reasons of the under-utilization of irrigation potential in the canal irrigated areas requires to be understood based on a comparative analysis of the changing irrigation water demands and supplies of the district. The demands of the irrigation water are governed by the dynamics of the cropping systems and irrigation application efficiency while the irrigation water supply is governed by the water conveyance efficiency of the water distribution system. The imbalance in the irrigation water supply and demand conditions may be corrected through judicious conjunctive use of surface water and ground water or increased efficiency of operation at the irrigation system level and at the field level or at the both levels of the irrigation system. The monthly canal water availability at the field level has been compared with the corresponding irrigation water requirements of the agricultural cropped areas of the district. The feasibility of increasing the irrigation intensity, conjunctive use of surface and ground water possible improvements in the cultivated field water management has been analyzed.

The canal water discharge at the field level and irrigation water availability has been determined by the water conveyance efficiency of the canal network of irrigation system. It has been governed by the seepage losses during the water conveyance in the different components of the irrigation network leading up to the irrigated land.

(i) **Seepage losses**: The seepage losses in the district has been estimated based on the estimated wetted perimeter and the losses amount to be approximately 10 per cent of the discharge capacity of the canal irrigation system. The seepage losses based on the water perimeter has been estimated about 13 per cent of the design discharge capacity has the Amethi Canal distributary. The other large distributaries of the canal irrigation system have their design discharge capacity varying from 5.7 to 14.2 cumecs. The gross water supply to the remaining canal distributaries and minors (less than 100 cusecs discharge capacity) has the standard value of 7 per cent has been adopted to estimate the seepage losses. The net seepage losses upto the outlet have been about 27 per cent of the water supplies available at the head of the canal distributary. The water losses based on the water loss rates of 0.75 cumecs and 1.52 cumecs per million sq. metres has been estimated for monsoon and non-monsoon seasons respectively. Comparative estimated water seepage losses have been about 27 per cent. *The proportionate increase in the seepage water losses have been noticed from lower flows in the non-monsoon season.* The seepage losses during Feburary-April are approximately double than the July-October when the available discharge in the former period has been about half of the later seepage losses. The estimated seepage water losses of 27 per cent has been for the canal water flows at the capacity discharge without considering changes in the seasonal water flows. Thus the actual water seepage losses in the district may be much higher than the 27 per cent water seepage loss. It has been estimated that about 30 per cent of the water delivered at the outlet and the seepage water losses in the unlined canal water courses has also been found to be 30 per cent seepage water losses in the district.

(ii) **Irrigation Water Availability**: It has been observed that the estimated seepage losses of the water in the different components of the irrigation water distribution system, i.e. 10 per cent in the canal branches network, 13 per cent in the large and network and 7 per cent in the small level irrigation distributaries and 30 per cent in the water course, has been the water conveyance efficiency of about 51 per cent in the study area. The efficiency is derived assuming that the irrigation water distribution system is operating at its design discharge capacity. It has been observed that when the operating level is low in the design capacity, the percentage of seepage losses in the larger components, specially the branches canals and large distributaries, will be more than the assumed design capacity, while sometimes these proportions may occur twice or more. The real conveyance efficiency may be below 51 per cent as described above. The variations of the water flow from design discharge values requires to be adjusted for every cropping season based on the water flows from the canal water irrigation system. It has been observed that water discharge availability in the canals did not exceed 8800 ha.m. per branch in the district. The average operating water discharge of the canal systems has been about a half of the canal water capacities. The real proportionate water losses may be more than the estimated losses at full flow water capacity of the canals. The assumed double rate of the water losses in the canal branches and large distributaries reduces the water conveyance efficiency upto 38 per cent in the canal irrigation network. The real efficiency value and the estimated water losses have been about 51 per cent for the particulars period in the district. An uniform conveyance efficiency of 45 per cent has been assumed in deriving the canal irrigation water availability at the cultivated field level in the district Amethi.

(iii) **Irrigation Water Requirement**: The water requirement for the irrigation varies from time to time and cropping pattern. There are the consumptive use of water by the various crops effective rainfall and water requirements for the particular purposes specially for the preparation of the land, leaching and flushing of the usar soils, green manuring and other agricultural practices in the field. The irrigation water requirements for the long term cropping pattern in the irrigation command area of the branch canals and the actual cropping pattern for the district has been evaluated and analyzed the important main Kharif crops are rice and sugarcane, wheat has been the major Rabi crop in the district.

The consumptive use of irrigation water for the crop or its potential evaporated is a function of the stage of crop growth and the evaporate demand of the atmosphere and climatic conditions of the district. The reference value of evaporation occurring from a free water surface may be estimated from the meteorological information. The consumptive use (CU) during any interval for the given crop is then given by the product –

$$CU = E_0 \times K_C$$

The K_C is the crop co-efficient which varies with the variety of crop and stage of the crop growth. The monthly values of $E_0 \times K_C$ and CU for the different crops have been analyzed by the Water Technology Centre of the Indian Agricultural Research Institute, New Delhi.

The different procedure of estimating effective rainfall has been applied for the paddy crop, because the paddy crop irrigation has been ensured by the check basin method and presence of standing water during the monsoon or paddy cropping season. The water percolation losses on account of significant water losses from the paddy field. There is considerable water storage space above the standing water has been about 10 cm. or more and a large fraction of the rainfall becomes effective during the paddy cropping season. In reality the effective rainfall is determined by continuous water balance approach system, but this approach is not always feasible broad guidelines based on empirical estimates are applied. Approximately more than 80 per cent monsoon season rainfall has been assumed more effective for the paddy crops but it does not account for the high intensity rainfall occurrence and a substantial proportion of rainfall and the run off without being effective for the growth of crops. The high intensity rainfall during the rainy season. 80 per cent of the first 100 mm rainfall and 70 per cent of the rainfall is excess of 100 mm rainfall has been considered to be more effective for the paddy cropping in the area.

The net irrigation water requirement includes water need for the land preparation, leaching and flushing for salinity and alkalinity control or unavoidable percolation water losses in the paddy crop field. The monthly water requirement for different crops has been assumed for the possible conditions in the area.

(a) The agricultural operation of crop productions for kharif-paddy are initiated during the first week of July and the cultivation operation is completed during the 10 days. The field operations of the Rabi crops are initiated during the middle of November.

(b) Water requirement for flooding and land preparation for the paddy crop fields.

• Saturation	—	60 mm
• Evaporation (4 mm/day)	—	40 mm
• Percolation losses (5 mm/day)	—	50 mm
• Flooding for transplantation	—	100 mm
Total	=	250 mm

(c) The land preparation for the other crops requirements has been neglected, because the rainfall during the June for Kharif crops and the residual moisture in the field for Rabi crops are sufficient for the cropping pattern.
(d) The application of irrigation efficiency is 85 per cent for the paddy and 65 per cent for the other crops.
(f) The average duration of the paddy crop is 135 days and no irrigation is envisaged 15 days before harvesting of paddy crop.

The monthly water requirements for different crops are computed by the Water Technology Centre of the Indian Agricultural Research Institute, New Delhi—

Net irrigation water requirement at the field (WR) = (Consumptive use – Effective Rainfall +Land Preparation+Percolation losses)

and

$$\text{Gross Irrigation Requirement (IRR)} = \frac{\text{Net Irrigation Requirement}}{\text{Water Application Efficiency}}$$

The monthly water requirements of the branch canal command area for each crop has been multiplied by the Corresponding area of the particular crop actually irrigated during the different months of the cropping season.

It has been estimated by the Sharada Sahayak Irrigation Project of the district has been designed and planned for irrigation intensity of 96 per cent in the cultivable field out of which 60 per cent in the Kharif and 36 per cent in the Rabi cropping season.

Kharif			Rabi		
(i) Paddy	—	33 %	Wheat	—	24%
(ii) Sugarcane	—	7 %	Potato	—	7 %
(iii) Other Kharif Crops	—	20 %	Gram and other crops	—	5 %
Total	—	96 %	Total	—	36 %

The long term agricultural cropping pattern envisaged for the cultivated field and irrigation water requirement has been assumed but in reality it varies from one crop to the other crop and for the different crop fields as per the agro-climatic conditions. The long term water requirements of the cropping pattern need to be assumed through applying reliable estimates of the annual average rainfall of the area. The mean monthly rainfall from June to October are highly variable and the analysis of the rainfall information indicates that due to exceedence of the mean monthly rainfall values has been 35 to 50 per cent in the district. To increase the reliability of the irrigation water requirements, rainfall values of higher certainty than the mean rainfall would be applied. The 70 per cent mean rainfall may be used as a reliable value at a higher certainty than the mean possibility of exceedence being the range of 70-80 per cent. The values of 70 per cent mean monthly rainfall for the monsoon season and zero for the non-monsoon season as reliable rainfall for the irrigation requirement of the area has been appropriately suggested for the district Amethi.

6. CONJUNCTIVE WATER USE

Analysis has been carried out through highlighting the features of the irrigation activities in the district—

(a) The areas irrigated by the canal network, canal water supplies for exceeded demands excluding the July month.
(b) The irrigation intensity of 96 per cent cannot be achieved from canal water supplies alone in the irrigated areas of the canal network.
(c) It is necessary to develop additional sources of irrigation water through increasing canal water availability and ground water extraction of the water. The diversification of the cropping pattern including paddy, pulses, oil seeds may reduce the demand of water during the Kharif season when the canal irrigation water is in the short supply as per the demand and supply of the irrigation water.

The existing situation of the irrigation, water demand and supply when restricted to the canal irrigated area, provides the wastage of water; the long term view, for the district indicates to a water scarcity conditions, unless there is a change in the cropping pattern. The benefits of the conjunctive water utilization of the water resources has to be identified. The conjunctive utilization of water in reality in the field, is yet to be incorporated into the water irrigation planning and management of canal network systems in the district.

The conjunctive water use based on the estimates of the water available in the different situations rainfall, canal water supplies, ground water and moisture storage at the various points of time—

(i) The district has been classified as a medium rainfall area. There is immense scope for introducing pulses and oil seeds in the larger areas because the crops are dependent on soil moisture storage out of the annual rainfall of the district.
(ii) The canal irrigation water supplies has been planned to meet the irrigation requirements of the gross cropped area with a view to achieve an irrigation intensity of 96 per cent.
(iii) The total ground water draft has been estimated potential of 2100 MCM and about 57 per cent of the ground water potential has been utilized.

The possible policies of the conjunctive water use are as follows:

(a) The excess pumping of ground water during July and October to augment the canal irrigation water supplies for the paddy crops.
(b) Reducing the ground floor draft from November to January as canal water flows are adequate during this cropping season.
(c) To increase the water pumping during February and March when canal water supply begins to recede but the wheat crop is sensitive to the water stress in the cultivated field.
(d) The changes in the cropping pattern and diversion of the land for social forestry agro-forestry and horticultural plantation in the cultivated field.

7. WATER MANAGEMENT

The irrigation water distribution system comprises water courses, cultivated field channel and management in the basic responsibility of the cultivators and various measures for a farm water management has been initiated.

1. Consolidation and rectangulation of the field.

2. Construction of drainage water to carry water from the canal outlet to the land holding cultivators.
3. On farm water conveyance through cultivated field channels.
4. Construction of drainage network for diverting water from water course to the field channels.
5. Proper leveling of the land to provide efficient appropriate distribution of irrigation water, soil moisture conservation and removal of the excess rainy water from the field.
6. Adoption of suitable irrigation techniques for efficient water application in the crop field.
7. Construction of field drains for the removal of excess surface/sub-surface water and linkage of the drains with the suitable outlets of the cultivated fields.

It has been observed that the fragmented land holdings with irregular field boundaries are the common features of irrigation system of the district. The scattered cultivated fields with irregular boundaries has increased at larger extent of the water courses to the distribution of water to the various cultivated fields in the district. The fields are often unsuitable for the modern techniques of water application such as border strips and furrows. The diversion of the cultivated fields existing into rectangular plots and consolidation of the land ownership on the basis of suitable exchanges between different owners of the field has been essential for increasing the water management efficiency at the level of cropping field in the district.

The existing scheme of rectangulation of the cropping field to maintain in the area of the field the same as before with minimum disturbance in the location of the land holdings of different cultivators. The alignment of the water course along the middle of the area by relocating the outlet so that the water course is straight and minimize the length. Keeping in view of the topography such layout can be planned to facilitate the adoption of an appropriate irrigation system like *wara bandi* and use of modern techniques of the irrigation water facilities. It is not easy to implement the rectangulation scheme of the field because the cultivators to adopt the changes in their cropping land. The rectangulation and consolidation of the field has to be attempted whenever new canal irrigated areas are developed for canal irrigation and mobilization and involvement of the cultivators adopt the latest techniques of rectangulation of the crop field for the uniform distribution of the water management in the cropping fields in the Amethi district of Uttar Pradesh.

The essential, efficient, effective prerequisite of the irrigation water management is the construction of water courses to convey the canal water to the fields of individual cultivators of the area. It is feasible and advantageous to line the water course to minimize water seepage losses to a larger extent of the land through appropriate alignment and control of the bed slope of the irrigation channels of the area. The lining construction may enable the equitable distribution of the irrigation water between the farmers sharing the release of the water from the outlets of the canal. The water seepage from the field channels are influenced by the soil permeability, wetted area of the channel and the distribution of irrigation water at the field level.

1. **Lining of the Length of Water Course:** The losses of the water through seepage in water course are function of the duration of water flow. The optimum length of the lined section of the water course has been estimated on the basis of the net water saving.

   The saving from the losses of the water seepage through construction of lining of length of drainage network from its head to end has been analyzed—

Length of lined section of channel	*Saving in seepage loss*
25%	65%
50%	80-85%
75%	95%

The 50 per cent lining of the upstream sections of a water course approximately 80 per cent seepage losses can be checked. The following suggestion may be helpful for the lining of the field channel—

1. The length of water course from the canal outlet up to the turnout lined.
2. The bends and curves are lined to control the erosion of the field soil.
3. Approximately 3 metres of the length from the upstream to the down stream of the structures of the turnouts/crossing are lined to control the soil erosion. The extension and training programmes to educate the cultivators are very important while carrying out construction of lining of the drainage irrigation networks.

2. **Alignment of field channels**: The contour map of the outlet of the area depicting contours at 30 cm intervals and indicating the field boundaries are important for planning layout of water course and field channels. The irrigation channels are aligned along the ridge line with the view to irrigate larger extent of the fields in the outlet and to minimize the number and length of canal branch water course. The deviations from the ridge line and the provision of fills and cuts are required to obtain suitable field channel alignments.
3. **Channel Section and Bed Slope**: It has been observed that the rectangular cross section has been found superior than trapezoidal section for the field channels and water course. The rectangular cross-section is easy to construct and maintain because it occupies the less land area. It discourages the animals from crossing the field channels. In the trapezoidal section breaches occurrence of the unequal settlement and removal of soil embankments have been observed in the field channels. The bed slope of the channel is field from the canal outlet. Attempt should be carried out to run the water course in partial cutting and bank filling maintaining the water surface elevation in the field channel at least 15 cm from the ground level to permit the gravity of the irrigation in the cultivated fields.
4. **Lining Materials**: The proper lining materials for the field channels and water course are carried out through local masonry including bricks, flat stones, stones, lateriate blocks and cement. The joints of bricks, stones, lateriate blocks, are constructed in cement or lime mortar and pointed. Plastering of the in side of the channel fields has been optional. It has been constructed to remove efficiency in the workmanship resulting leakage at the joints. The pre-fabricated non-reinforced concrete channels are suitable for lining small field channels.
5. **Field Channel Structures**: The field channel structures are constructed for water diversion, to control soil erosion, and crossing of the obstructions in the field channels. The diversion structures are constructed with checkgates of the values are provided in a masonry box. The efficient check gates comprising a pre-cost of R.C.C. frame and sheet metal shutter working in a watertight rubber seal strips. Low cost turnout can be made through using the commercially available cast iron river values commonly used in the underground pipelines. The structures for the erosion control comprises straight drop spillways and pipe drop spillways. The structures are provided with suitable gates at the inlet to control the water surface,

elevation in the upstream section of the field channels. The field channel structures are designed for specific structural and hydraulic requirements. Straight deep structure has been commonly used to check the soil erosion in the structure of farm irrigation management. The field channel structures are constructed to maintain the channel grades within permissible limits and allow the water from higher to lower fields. The chute spillways are most economically viable and the channel is brought down along the sloping side of an earthen embankments. The pipe drop spillways has been required to carry the field channel to a lower field across the earthen embankments of the irrigated field management. The field channel structure consists of an inlet tank and an outlet tank joined together at the bottom with an underground pipelines. The check gate, serving as a cross regulator is provided on the up stream side of the inlet tank. It has been observed that pipe culverts are oftenly used to carry out water course and field channels across the ridges and drains in the irrigated fields of the area.

The field drains has to be constructed at the lower ends of the crop fields to facilitate the escape of excess run off water reaching the field due to rains and excess irrigation water in the district. The construction of the field drains must join the outlet of the irrigation drain. The important problem has been faced in the efficient functioning of the field drains water logging due to weed growth and lack of field outlets to allow the water from the cropped fields into the drain field. The subsequent maintenance of the drains by removal of the weeds and clearing the ridges and bunds which restricts entry of water into the drain has been essential to ensure frequent and efficient surface drain network of the district.

Thus, the water management of the field channels, canal water, branches of the canals and distributaries and the control of water seepage and to check the soil erosion of the irrigated network system at grassroots, level has become an utmost top priority task of the not only farmers but irrigation implementing institution in an integrated manner with joint efforts of the people's, private water resources to grow crops and to increase the agricultural productivity and to promote optimum utilization of available water resources and to conserve the environmental conditions of the areas and to meet the increasing demand of ever growing cattle and human population of the study area.

8. CONCLUSION

Thus it is known fact that the water is the basic source which affects the survival of plants and also affects the agricultural crop production of the area. It has been observed that there is abundant ground water resources which could be harness for facilitating the irrigation purpose for the agricultural crops through installation of shallow and deep tube-wells. The canal irrigation is another source of irrigation system.

The major natural drainage systems of the district are the tributaries of the Gomati and Sai rivers, Chamraura and Kunwar rivers. The Gomati river flows from north-west border of the district covering Shukul Bazar, Jagdishpur and Musafirkhana blocks of the district. The Sai river flows from south-east portion of the district covering Salon Tehsil of the district Amethi. Most of the areas of the district are low-lying and there are numerous lakes, ponds, water logged land and marshy land. Chamraura and Kunwar rivers are small tributary. The natural resources of the district are abundant and sufficient provided that appropriate techno scientific planning systems are evolved to top the available natural water resources of the district.

The irrigation network of the district is another source of the drainage network to facilitate the irrigation of agricultural land. Sharada Sahayak Command Area Irrigation Project of the Government of Uttar Pradesh has constructed irrigation network especially Inhauna, Singhpur and Amethi distribution tributaries and other minor of the Sharada Sahayak Irrigation Project in the district.

The ground water and tanks were the major sources of irrigation, prior to introduction of canal irrigation system. The extensive development of the available surface water resources and the ground water resources contribute to the irrigation up to a limited extent. It has been observed that the ground water resources provide irrigation of 20 per cent of the net cultivated area of the district and 29.5 per cent of the net irrigated area. The existing ground water potential and the increased exploitation of ground water resources have been considered attentively for appropriate exploitation of the available natural ground water resources of the district.

The Gomati, Sai, Chamraura and Kunwar rivers are the main natural drainage system of the district. Due to excessive rain, low-lying areas, the water logging, flood etc. are major problems of district. It has been observed that about 40 per cent of the district has been found flood-prone area and about 50 per cent area of the district has been classified and identified as poor drainage system of the area. The extensive canal network, resulting from the rise of sub-surface water table except in few patches of the district. The occurrence of heavy rainfall caused large scale water logging. There are variant approaches for the design of drainage system having conventional practices because the slope up to 0.5 per cent on the flat field. Constant drainage design and development has been attempted. The tolerance of the crops of surface poundings varies in their proportionate tolerance to the water logging conditions of the district. The surface drainage requires the removal of excess rain water and irrigated water in the cropping land. The development and drainage design and construction should be carried out within the limited days to avoid the crop loss in the field.

The ground water infrastructure has been one of the most important source of irrigation because more than 55 per cent area of the district are irrigated through the ground water source of irrigation through open-wells, tube-wells, dug-wells, bore-wells which are uniformly distributed all over the district. Appropriate progressive development of ground water resources through private tube-wells, government tube-wells and other sources of irrigation has been implemented in the district.

The movement and occurrence of the ground water resources of the area is dependant on the sub-surface formation of the underground water and hydrological characteristics. The lithology of the soil and sub-surface soil comprises sandy clay of varying thickness of sand and gravel of the tectonic to the recent age. The acquifer zone extends to the depths greater than 100 m below the ground surface. The transmissivity values lies in the range of 1000 to 2000 m^2/day. The specific yield of the formations in the ground water has been in the range of 10 to 18 per cent in district.

The ground water fluctuations varies from one village to another based on the occurrence to the rainfall of the area. The rainfall has been principal source of ground recharge. The seasonal fluctuations of the water table reflect primarily the effect of the monsoon rain fall on the ground water, the annual ground water fluctuations represent the net effect of the recharge and draft of the ground water in the district.

The average annual draft of the ground water depends on the number and yield of wells operational in the area and the capacities of the water pumps. It has been observed that about 60 cm of water is required for irrigation of the one hectare land. The irrigation potential may be expressed in terms of the corresponding volume of the ground water extracted for each type of ground structure in the respective types of wells of any area in the district. The annual draft of the ground water structure ranges from 660 million cubic metres to 1260 million cubic metres in the district.

It has been recorded that ground water recharge from rainfall is 20 per cent, due to seepage from unlined canals 1.8 to 2.5 cumecs/million square metres of the wetted area and return seepage from irrigation fields under canal irrigation is 35 per cent of water delivered at outlet and the return flow from irrigation, underground water irrigation is 30 per cent of water delivered at the outlet in the district. It has also been recorded that there is 16 cumecs recharge from the large distributaries and 9 cumecs recharge from the other distributaries of the canal branches of the district.

The net change in the ground water table indicated a marginal rainfall in the ground water table of the district. The total ground water inflow or outflow is not known but can be estimated that the change in the depth 0.1 m corresponds to a net drop in the ground water storage of 55.3 MCM. The ground water outflow has been estimated as 1045 million cu.m. in the district.

The ground water natural resources has been aportioned into the static and dynamic water resources. The depth of the zone of water fluctuations varies between 3.9 to 10.0 m. The specific field of aquifer in the zone of water table fluctuations of the area has been assumed to be uniform at 12 per cent in the district. The estimated average recharge potential for the normal rainfall has been estimated up to 4066 MCM. The average ground water outflow has been about 1150 MCM as per the occurrence of pre-monsoon and post-monsoon seasons.

The seepage water losses based on the water perimeter have been estimated about 13 per cent of the design discharge capacity. The other large distributaries of the canal irrigation system have their design discharge capacity varying from 5.7 to 14.2 cumecs in the district. The gross water supply to the remaining canal distributaries and minors (less than 100 cusec discharge capacity) has the standard. Value of 7 per cent has been adopted to estimate the seepage losses. The water losses based on the water loss rates of 0.76 cumecs and 1.52 cumecs per million sq.metres have been estimated for the monsoon and non-monson seasons respectively. The comparative estimated water seepage losses have been about 27 per cent in the district varying from 5.7 to 14.2 cumecs.

It has been observed that the different irrigation water distribution systems, i.e. 10 per cent in the canal branches network, 13 per cent in the large canal network and 7 per cent in the small level irrigation distributaries and 30 per cent in the water course has been the water conveyance efficiency of about 51 per cent in the district. The water requirements for the irrigation varies from time to time and cropping pattern. The consumpting use of water by the various crops effective rainfall and water requirements for the specific purposes including land preparation, leaching and flushing of the saline/ alkaline soil, green manuring and other agricultural practices. The consumption use of irrigation water for crop or its potential evaporation is function of the stage of crop growth and the evaporative demand of the atmosphere and climatic conditions.

The conjunctive water use based on the estimates of the water available in the different situations—rainfall, canal water supplies, ground water and moisture storage at the various point of time. The irrigation water distribution system comprises water courses, cultivated field channels and field water application network. The water management is the basic responsibility of the farmers and various measures must be adopted for the appropriate and optimum utilization of available water resources of the district. The diversion of the cultivated field existing into rectangular plots and consolidation of the land ownership on the basis of suitable exchange between different owners of the field has been essential for increasing the water management efficiency at the level of the cropping field.

The water seepage from the field channels are influenced by the soil permeability, wetted area of the channel and distribution of the irrigation water at the cropping field level. The losses of the water through seepage in the water course are function of the duration of the water flow. The optimum length of the lined section of water course has been estimated on the basis of net water saving. The saving from the losses of the water seepage through construction of lining of length of drainage network from its need to end. The 50 per cent lining of the upstream section of water course, approximately 80 per cent water seepage losses can be controlled.

The contour map of the outlet of the area depicting contours at 30 cm intervals and indicating the field boundaries are important for planning layout of water course and field channels. The channels season and the bed slope of the channel must be fixed from the canal outlet. Attempt should be carried out to run the water course in partial cutting and bank filling and maintaining water surface elevation in the field channel at least 15 cm from the ground level to permit the gravity of the irrigation in the cultivated land.

The proper lining material for the field channels and water course must be carried out through local masonry work including bricks, flat stones, stones, laterite blocks, and cement. The joints of bricks, stones, laterite blocks are constructed in cement or lime mortar and pointed and the plastering of the inside of the field channels is optional. The field channel structures are constructed for the water diversion to control soil erosion and crossing or the obstructions in the field channels.

The field drains has been constructed at the lower ends of the crop fields to facilitate the escape of excess run off water reaching the field due to rains and excess irrigation water of the crop field. The construction of field drains must join the outlet of the irrigation drain. The subsequent green maintenance of the drains by removal of the weeds and clearing the ridges and bunds which restricts entry of water into the drain has been essential to ensure frequent and efficient surface drainage network of the district.

Thus, it is evident that the average annual rainfall in the district has been sufficient to carryout the agricultural practices, the surface water through river systems, lacks, ponds and accumulated water in the low-lying areas are enough to fulfil the water requirements of the area. The ground water resources are sufficiently available in the area which is utilized through dug wells, tube-wells and bore-wells to irrigate the cropping land. The drainage discharge capacity and water seepage losses have to be controlled to avoid the water loss and check the soil erosion of the area. Appropriate irrigation water management through techno-scientific interventions are necessary and proper practical demonstration to save water and management of available water resources of the area must be ensured and water resources should be stored and safeguarded to utilize during the drought period and also during the pre-monsoon and post-monsoon time. Since the water is life, therefore, it should be properly utilized, maintained as per the need of the cattle and human population, plants and agricultural cropping requirements of the farmers to meet the increasing demand of the growing population for the conservations of the available natural water resources and to improve the eco-system and environmental conditions of the area and to ensure sustainable Eco-Socio-Economic sustainable development of the Amethi.

5

Forests

1. INTRODUCTION

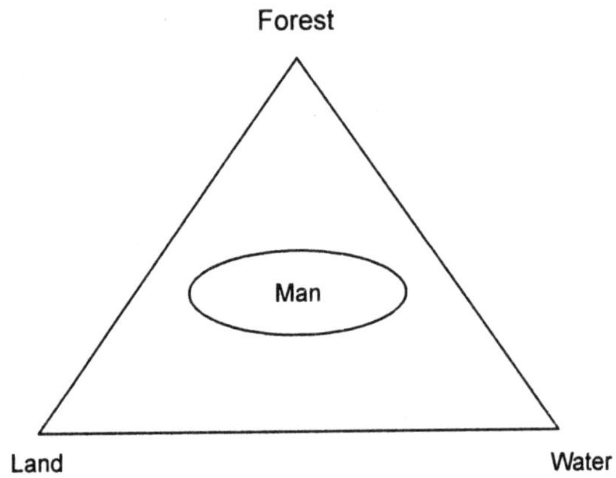

The forest is one of the important natural resources out land and water. The *Jal, Jangal, Jameen* (Water, Forest and Land) are one of the most important natural resources for the survival of not only for the human beings but also for the animals and wild life of an area. The natural resource forest plays an important role in maintaining the ecological balance of an area. The forest provides raw materials to the industries, fire wood, fuel wood, fibre, fruits etc. to the people, grazing land and fodder to the cattle.

The forest, land and water has been the basic source of survival for men and animals in universe. It has been recorded that the centuries ago the district Amethi had been covered with extensive forest, flora and fauna. The *jhils*, tanks and ponds were very famous for the birds.

Due to the ever increasing population, urbanization and growing need of the people, the forest cover of the area had been gradually destroyed and turned into cultivated lands and to develop for the settlements etc. In the beginning of the twentieth century the patches of thick forests were formed in the Pargana of Inhauna and extensive forests were spread along the banks of the river Sai and the patches of Sarpats (grass), forests etc. had been available in the catchments of the river Gomati. The patches of Dhak forests were found throughout the district. During the period of "grow more food" (अधिक अन्न उपजाओ) National campaign following the Second World War the forests were recklessly cut down. However, during the early fifty the extensive afforestation measures had been suggested and initiated by the Department of Forest, Government of India and State Government including the district authorities as well, and trees like Dhak, Khair, Babool, Sisham, Neem, Arjuna, Kanji, Seras, Eucalyptus, Mango, Mahua, Jamun etc. trees were planted.

Thus the tremendous growth of the population and the ever growing demand of fuel wood, fodder, fibre, food grain, fruits, fisheries (F^6) and vegetables and goods and services have imposed severe constraints on the natural resources especially on the forest and green cover. The over-exploitation of natural forest resources continues to affect the quality of life of the deprived resourceless rural poor people surviving in the remote rural areas who constitute more than 76 per cent of our population. Excessive and over-exploitation of our natural resources especially the forest cover for meeting the growing demand of ever increasing cattle and human population has led to forest and environmental degradation and rapid growth in the wastelands of the district Amethi of the Uttar Pradesh.

The massive deforestation has taken place in the reports which has been fairly well documented. The massive deforestation of the country shook the nation to know that 50 per cent of our forest cover and it was partly in response to this tremendous forest degradation and the associated ecological crisis.

The excess use of chemicals and insecticides and pesticides and industrial growth have had serious negative impact on the life of the common people. In the historic words of our Former Prime Minister Late Smt. Indira Gandhi ji, "*In his arrogance with his increasing knowledge and ability, man has ignored his dependence on the earth and has lost communion with it. He no longer puts his ear to the ground so that the earth can whisper its secret to him. The national song which inspired our freedom movement describes our land as one endowed with water forest, fruits, rich with greenness of growing plants, we must make this true not only of India but of all lands.*"

This statement was delivered by Late Smt. Indira Gandhi, the then Prime Minister of India, while participating in the World Environment Congress at Stokholm in the year 1966 and the Government of India established Ministry of Environment, Forest and Wildlife to protect our environment and to ensure 33 per cent forest cover on the land.

The major impact of deforestation and ecological degradation that has taken place has been largely on the rural poor because half of our nation depend for its fuel wood and fodder on the common lands. About 40 million hectares of our forest land with trees has been considered without tree cover and we have lost 26 to 52 per cent of our grazing lands. To achieve the ecological balance we need at least 33 per cent of our land under forest cover while at the moment we have less than 11 per cent. There is very serious ecological crisis. To maintain the ecological balance it is necessary to afforest 50 million hectares of land every year.

The former Prime Minister Late Rajiv Gandhi emphasized for massive afforestation and the target for afforestation on wastelands. Massive afforestation and people's movement. The implementing agencies like forest departments could not achieve the earmarked target because it was not people's movement.

The continued deforestation has brought us face to face with major ecological imbalances and environmental crisis. This trend must be halted. The present fire wood requirement of our country is more than 130 million tonnes are fetched from the forest.

In the forestry sector out of 75 million hectares of land 40 million hectares land is degraded and 30 million hectares land is without forest cover and 5 million hectares land is without shrub. The environmental and ecological crisis has emerged on account of excessive exploitation of our natural land resources leading towards the growth of land degradation. It is known fact that a village in Amethi Block is known as **Jangal Ram Nagar**, due to the excess forest cover Jangals of the villages have been given a name because the Raja Amethi lived in Ram Nagar and the **Bhupati Bhawan** was built in the Ram Nagar due to the extensive forest cover the **village panchayat** was named as Jangal Ram Nagar and the revenue village was also named as Jangal Ram Nagar covering 48 purwas or tolas or hamlet villages but the present time the Jangles of the villages have been deforested and very insignificant jungles are left in the village while the most of the jungle/forest land has been brought under cultivation, settlements and development programmes of the villages. The area under forest of the district has been 0.44 per cent to the total geographical area of the district which is very insignificant against earmarked 33 per cent forest cover to maintain the ecological balance of the district and to conserve the environmental conditions of the area and to restore the Eco-system of the villages of Amethi.

2. RESERVED FORESTS

The area under reserved forest of the state of Uttar Pradesh has been recorded about 17 per cent of the total geographical area of the district. However, it is much higher than that of neighbouring Sultanpur district and lower than the Rae Bareli district. There are reserved forests and protected forests in the district. The per capita forest area in the district is 0.0021 hectares against 0.0012 hectares in Sultanpur District and 0.0025 hectares in Rae Bareli District of Uttar Pradesh, as a whole the reserved forest is 0.0367 hectares. According to the forest statistics of Uttar Pradesh 7 cu.m. of timber and 425 cu.m. of the fuel wood were produced from reserved forest of Rae Bareli District while the production of timber was 26 cu.m. and fuel wood production was 258 cu.m. in the Sultanpur district. The two tehsils, namely Tiloi and Salon of Rae Bareli District and three Tehsils, viz. Amethi, Musafirkhana and Gauriganj of Sultanpur district have been taken together and new district has been created which is named as district Amethi. Therefore, according the status of reserved forest and its proportionate distribution in district must be considered and compared and estimated based on the area under the Reserved Forests in Rae Bareli and Sultanpur districts respectively. There are four ranges of the forest in the district Amethi—

1. Musafirkhana Forest Range.
2. Amethi Forest Range.
3. Salon North Forest Range.
4. Salon South Forest Range.

The forest departments of the district controls and maintains the road side afforestation and afforestation on village panchayat common land lying as waste lands in the district. There are 1258 hectares forest land in Musafirkhana Forest Range and 33 hectares reserved forest land controlled and maintained by the Amethi Forest Range in the district Amethi.

3. FLORA

The district Amethi used to be covered with extensive forest centuries ago with the increase of forest cover which has been gradually destroyed and turned into cultivated lands. During the beginning of the twentieth century patches of thick forest were formed in the Pargana of Musafirkhana and Salon and extensive forest cover had been spread along the banks of the Sai river and Gomati river. The Kadu Nala reserved forest had been protected and forest picnic spot was developed at the Kadu Nala Jungles which has been maintained till date by the Department of Forest, Government of Uttar Pradesh under the control of Forest Range Officer, Musafirkhana and the District Forest Officer. During the National campaign of the "Grow More Food" following the Second World War the forests were recklessly cut down and the forest land had been deforested with the view to increase the cultivated land of the district to fulfil the increasing demand of growing human population of the area. However, during the early fifty various afforestation programmes were launched and initiated by the Department of Forest and various varieties of trees, like—Dhak, Khari, Babool, Shisham, Neem, Arjun, Kanji, Siras, Eucalyptus, Mango, Jamun, Mahua and various other trees has been planted to prove the afforestation programme with the view to improve the Eco-system and environmental conditions of the area and to meet the increasing demand of fuel wood, fodder, fibre, fruits, wood etc. of the growing cattle and human population of the area.

4. FAUNA

The excessive deforestation and increasing land degradation has led towards negative impact on the fauna of the district Amethi. Due to the large scale deforestation the number and varieties of wild animals have greatly decreased during this century. The tiger and wild buffalows used to be found in the jungles along the banks of rivers. The larger carnivore has disappeared and even the wolf and hyena have become very rare in this area. The Indian Antelope has been protected species in the district Amethi. The jackals, porcupine, hare, fox, jungle cats, mongoose, and palm squirrel are some of the animals found in the district.

5. BIRDS

The main game birds found in the district are of several varieties of ducks, partridges and pigeons. There are some species of geese, coots, jacana, moorhen, snipe and sand piper are also met within the area. During the cold weather there are plenty of water fowl and snipe, quail being fairly common in the district. The pink headed-duck has become extinct in the area. The green pigeons are becoming rare due to filling/cutting of fruit and berry trees in the district. The pigeons and partridges have become rare due to indiscriminate netting. Besides these game birds there are innumerable other ordinary varieties of birds to the plains. The Khunna ka Tal in the Amethi is very famous for the bird sanctuary where during the cold weather season the national and international birds used to come and the birds also used to stay in the cold weather at the tals of Tiloi and Salon of the district Amethi.

6. AFFORESTATION

The afforestation has been an important sector of any policy, plan formulation for the integrated, sustainable development of an area. Apart from sustaining the ecological balance of the area afforestation helps to meet the various social requirements of the people. The afforestation approach has been distinguished into "Production" and "Social" Forestry. Production forestry aims at producing wood for industrial and household use. The major objectives of the Social Forestry includes the supply of fuel wood, small timber, fodder etc. to fulfil the needs of increasing cattle and human population of the area. The social forestry has been considered an asset to the rural economy, resource conservation and development of the area. The afforestation has been helpful in preventing soil erosion and sedimentation of reservoirs and increase soil moisture retentivity and long term improvement of micro-climate of the region. The development of pastures and grazing lands has also been important for the livestock development programmes of the district.

The large scale development of the commercial forestry has very limited scope in the district, while the emphasis has only been given to the development and promotion of social forestry in the Amethi. Large scale afforestation and tree planting prgrammes have been in the full swing from last two decades. The non-availability of land has been one of the major constraints for the development of afforestation programmes of trees have been launched along the roads, canal banks and on the waste lands and usar land. The Department of Forest is striving hard to bring additional land under tree plantation. The afforestation programme has been a part of the statewise social forestry project with the assistance of the World Bank, a social forestry department is functioning in the district, with the following objectives:

1. To meet the demand of fuel wood/fire wood and reduce the demand of gobar.
2. To ensure soil and water conservation.
3. To increase the availability of raw material for small scale and cottage industries.
4. To generate job opportunities for the rural poor.
5. To meet the demand of fodder, pasture and grazing and for the animals.
6. To meet the demand of timber wood.
7. To improve the eco-system and conserve the deteriorated environmental conditions.

The social forestry programmes have achieved the different types of land under plantation.

- Along National Highways.
- Along Roads.
- Along Rail Roads
- Small category forest.
- Village Panchayat and community land.
- Along with canal banks.
- Barren, usar, banjar or waste land's plantation.

It has been observed that there is very nominal plantation of the trees in the Amethi whereas Musafirkhana has the highest number of plantation along with the banks of river Gomati and Kadu Nala. Salon has also ample tree plantation followed by Tiloi and Gauriganj where nominal social forestry and tree plantation programmes have been implemented. The social forestry programmes in the district are not successful because there is no people's participation and peoples movement in the afforestation and need of the people has been ignored which has led to nominal success of the afforestation scheme in the district.

I. Tree Species Selection

The criteria for the suitability of tree species for the use under social forestry programmes has been very much different from those in the raising orchards or forestry plantations. The main important considerations which are applicable and adoptable for selecting the tree species in the social forestry programmes are as follows:

1. Multi-purpose use of the trees should be a prerequisite.
2. To ensure that the tree should yield one oı more of the principal produces like fuel wood, fodder, leaf, edible flower, fruits, fibre etc.

3. The tree yielding produce for the rural and cottage industries in comparison to the industrial use should be preferred under the social forestry plantation scheme.
4. Leguminous tree species are to be generally preferred against non-leguminous ones wherever possible under forestry project.
5. To ensure easy establishment of the afforestation programmes at a short time as possible.
6. Species which are not browsed by cattle in the areas where cattle grazing has been common may be preferred under the afforestation programmes.
7. The species which can withstand repeated pruning of their branches, species which pollard well should be preferred. It will ensure the availability of fire-wood on a sustainable system.
8. Species with good copping power, so that once planted many more rotations could be had under the social forestry scheme.
9. The choosen species must find acceptance of the local people.
10. The fuel wood, fodder, fibre, wood plant species should be planted and the pastures and grazing land should be carried out in multi-disciplinary approach under the afforestation prgrammes to meet the increasing demand of the growing cattle and human population.
11. To maintain the ecosystem and improve the environmental conditions of the area. While the formulation of the plantation programmes, a number of suitable species must be adopted as per the needs of the local people for whom the social-forestry plantation is being planted and raised. As per mutual agreement of the local people the final selection of the plant species should be carried out to ensure successful afforestation programmes.

II. Suitable Species for Social Forestry

There are several suitable tree species, to ensure the successful, sustainable development and the people's participatory social forestry plantation programmes—

(i) Mahua (Madhuca Longifolia)

The Mahua is considered to be one of the few most important tree species for plantation under the social-forestry programmes particularly on the community waste lands which are not very productive due to soil alkalinity/salinity or ravines and water logged land or banjar land. There are few trees whose even petals in addition to fruits and seed are edible among which Mahua is one of the such plant species. Mahua is rather a slow growing species and requires great care and sustained protection up to 4 to 5 years. Mahua starts yielding adequate flowers and fruits in 10 to 15 years.

The oil is being extracted from the seeds of the Mahua plants has a great demand in the pharmaceutical and soap industry and it is also edible. The Mahua seed cakes are used as bio-fertilizer enabling the development of the bio-fertilizer industry in conjunction with oil extraction industry. The seasonal branch wood of the Mahua plant has been suitable for cottage industry for making agricultural implements. The flowers of the Mahua tree can be fermented to produce excellent alcoholic drink for the export purpose. Mahua being a long rotation plant it can be planted in the first row or last row of the road and along the canal banks. The dead branches of the Mahua plant would also serve as a fuel wood and also for the other purposes as per its suitability. The bark of Mahua plant is used in the tanning industry. The flowers are rich in rector and serves as a source of honey. The reddish green wood is a good source of fuel energy.

Mahua tree plantation is preferably suited for planting along tanks, bunds, canal side, road side as last rows and inter-mixed planting in village wood lots. Network of bee keeping can be developed in such areas and highly quality honey extracted and marketed providing income generating employment in the rural area.

(ii) Jamun (Syzygium Cumini)

The Jamun species assumed the considerable importance in the social forestry scheme because the flowers of the Jamun yield very high quality honey. Fruits of the Jamun tree are highly nutritious. Jamun plant grows well in damp and marshy localities providing small timber for farms/agricultural implements and bark for tanning purpose. The Jamun plant's wood is a good raw material for making farm implements, boats and oars.

(iii) Mango (Mangifera Indica)

Very commonly known tree as Mango, produces delicious and nutritious fruits which is in great demand all over the country. Mango fruit is an important shade for avenue and the mango tree is planted by every household. The mango tree is of the great importance in the social forestry and horticultural programme. Mango plantation is ideally suited for road side plantation and village wood lots or the village bagh. Although the mango plants are grown mainly for fruits, the wood can be used in veneer, shoe heel, packaging cases and high grade charcoal industry. The mango co-operatives are growing and marketing would not only provide enormous employment opportunities but the mango will also yield high dividends for the people engaged in the mango tree plantations.

(iv) Kanji (Pangamia Pinnata)

The Kanji is a very important tree in the social forestry programme because it can be raised in the refractory sides which are semi-alkaline and semi-water logged land. The seeds of the Kanji plant yield brown-red oil used for relieving pain in arthritis. Kanji is useful antiseptic in treating human and animal skin diseases. The Kanji tree starts yielding fruits from four to five years.

The Kanji plants profusely set which contain 35 per cent fatty oil used in leather dressing, soap making and lubrication industry. The Kanji seed cakes are processed to develop the bio-fertilizer for the agricultural fields. The wood of the Kanji plant is used in the small scale village industries for making carts, farm implements and furniture. The large scale Kanji plantation would facilitate the development of oil extraction co-operatives providing suitable employment to the unemployed youths. Ancillary unit for the bio-fertilizer production and poultry feed from the seed cakes of the Kanji would develop in generating employment opportunities and increasing the income of the deprived rural poor multitudes.

(v) Eucalyptus

The eucalyptus is one of the very important species in the social forestry programmes. Eucalyptus has assumed great importance due to fast growth and high coppicing power. The eucalyptus is free from browsing damage and provides fire wood in place of cow

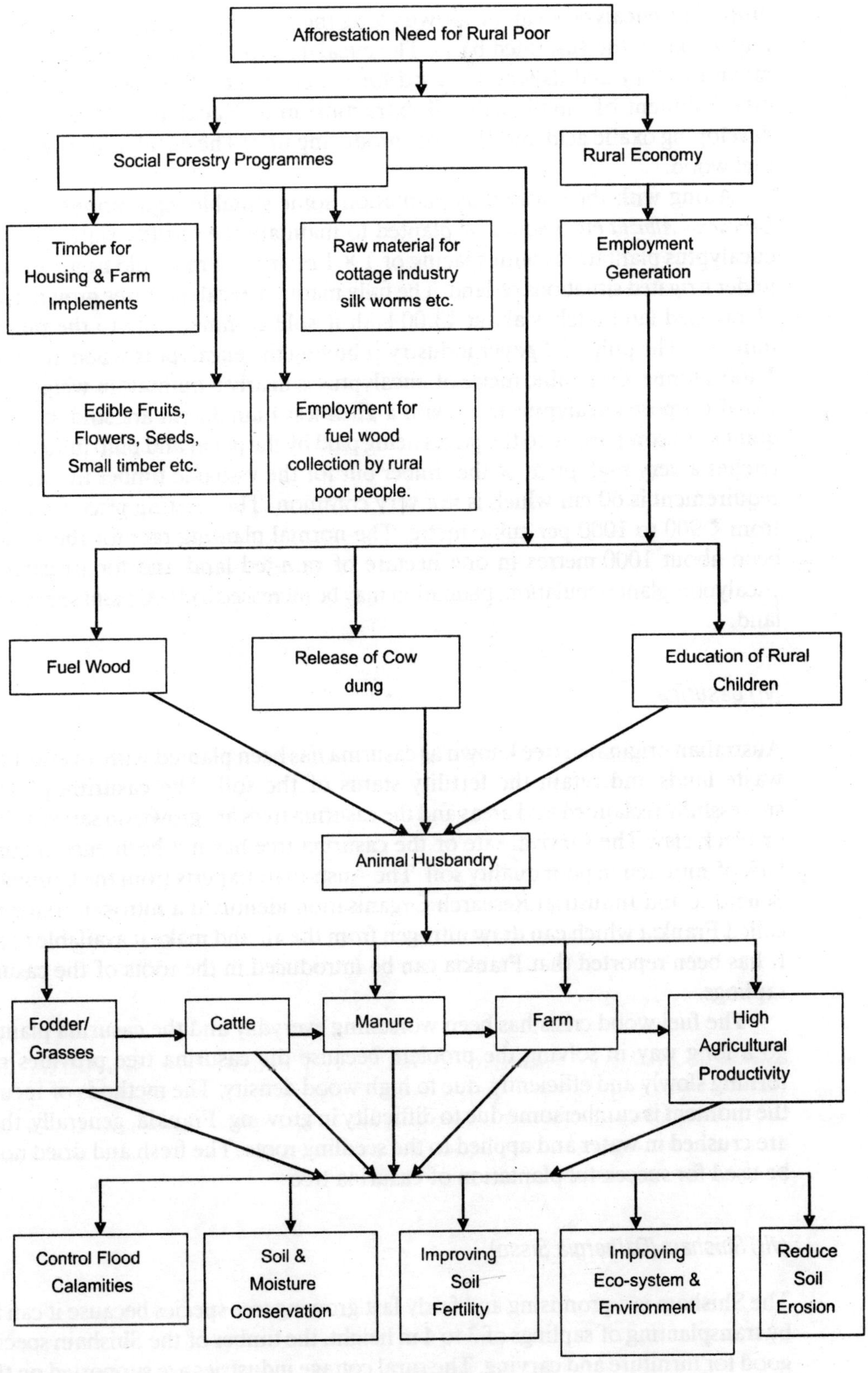
Afforestation Need for Rural Poor
Social Forestry Programmes
Rural Economy
Timber for Housing & Farm Implements
Raw material for cottage industry silk worms etc.
Employment Generation
Edible Fruits, Flowers, Seeds, Small timber etc.
Employment for fuel wood collection by rural poor people.
Fuel Wood
Release of Cow dung
Education of Rural Children
Animal Husbandry
Fodder/ Grasses
Cattle
Manure
Farm
High Agricultural Productivity
Control Flood Calamities
Soil & Moisture Conservation
Improving Soil Fertility
Improving Eco-system & Environment
Reduce Soil Erosion

Fig. 5.11

dung. The eucalyptus can be grown along the field boundaries with the view to obtain fuel wood on the sustained basis. The eucalyptus provides raw material for paper and rayon industry and its wood is used for match splints. The leaves can be used for the establishment of small scale oil extraction unit. Eucalyptus bark can be used for developing oxalic acid and charcoal producing unit. The eucalyptus is a good source of fuel wood.

Along with the eucalyptus plantation some suitable leguminous tree species like *Leucaena Albizia* etc. should be planted to maintain the fertility status of the soil. The eucalyptus plantations with spacing of 1 × 1 m or 1 × 2 m would be an ideal plantation under irrigated situations of land. The fully matured eucalyptus tree grown in one hectare of rain-fed land fetches about ₹3.00 lakh if sold commercially to the pulp and paper industry. The pulp and paper industry is buying the eucalyptus wood at approximately ₹ 400 a tonne. One cubic metre of eucalyptus is nearly 5 quintals in weight. For the fuel wood purpose eucalyptus trees with a girth less than 30 cm are sold at ₹ 20 to 25 per quintal. In comparison to the prices being paid by the paper and pulp industry eucalyptus fetches a very high price of the timber but for the valuable timber the minimum grith requirement is 60 cm which is not very common. The existing price for timber ranges from ₹ 900 to 1000 per cubic metre. The normal planting rate for the eucalyptus has been about 1000 metres in one hectare of rain-fed land and for irrigated areas the eucalyptus plant population, plantation may be increased to 1600 plant species per hectare land.

(vi) Casurina

Australian originated tree known as casurina has been planted with an effort to improve waste lands and retain the fertility status of the soil. The casurina plantation has successfully reclaimed arid areas and the casurina trees are grown on sands, salty marshes or black clay. The survival rate of the casurina tree has not been encouraging, due to lack of nitrogen in poor quality soil. The Australian Experts from the Common Wealth Scientific and Industrial Research Organisation identified a nitrogen fixing bacterium called Frankia which can draw nitrogen from the air and make it available to the plants. It has been reported that Frankia can be introduced in the roots of the casurina plant saplings.

The fuel wood crisis has been worsening everyday and the casurina plantation can go a long way in solving the problem because the casurina tree provides good fuel, furning slowly and efficiently, due to high wood density. The methods of inculcation at the moment is cumbersome due to difficulty in growing. Frankia, generally, the nodules are crushed in water and applied to the seedling roots. The fresh and dried nodules can be used for successful plantation of casurina tree.

(vii) Shisham (Dalbergia Sisso)

The Shisham is a promising and fairly fast growing tree species because it can be grown by transplanting of saplings of 3 to 4 m height. the timber of the Shisham species is very good for furniture and carving. The rural cottage industries are supported on the timber of Shisham. The carved species have great export market. The branches can be pottered at an interval of one of three years for fuel wood. The Shisham tree are planted along the boundaries of the field and the Shisham plantation is suited for well drained alluvial areas. The Shisham wood is extremely used for furniture, musical instruments, agricultural implements, cart, carriage and wheel works and rural industry and carving the small artisans. Shisham wood can be converted to charcoal and the leaves are used as fodder. The branch wood of the Shisham plant is extensively used for fuel wood by villagers.

The Shisham trees being a fast growing tree it can be planted closely (2×3 m) for fuel wood and a shade line by the side of road and canal, the Shisham plant is generally planted 10 m apart. The Shisham tree can be planted in the central rows of road, rail and canal plantation of the species is raised, for the support of small scale industry and Shisham plant can support various rural cottage industries and provide adequate self-employment avenues to the interested people and hence it can increase income of the deprived rural poor multitudes.

(viii) Arjun (Terminalia Arjun)

The Arjun tree species can be grown in the saline/alkaline area and if grown in a closer spacing of 1×2 m or so then from the 3rd and 4th year it is possible to rear insects which produce tassar cocoons. The leaves of the Arjun plant has been found suitable for feeding tassar silk worms and Arjun plants are useful in tassar silk industry. The bark, leaves, roots and fruits contain tannin and it is used in tanning industry.

Arjun plant has the salt tolerant capacity. The vast stretches of the usar could be planted by the Arjun plants to develop the tassar silk industry. Thus the Arjun plant would not only help in the reclamation of usar land and it will also generate gainful employment for the rural poor. The Arjun plants are recommended for planting in the middle rows of canal side and road side plantation in the four or five rows where the land is either alkaline or saline or semi-water logged lands. The plantation of Arjun trees are suitable for the groups of the village around these areas can rear tassar silk worms and make the tassar silk through co-operative systems. It has been recorded that one hectare Arjun plantation can support a family of landless labourer on permanent basis providing considerable income generation. The bark of the Arjun tree is an ideal raw material for tanning used in curing raw hide. The bark of Arjun tree can be removed periodically without damaging the tree. Thus the Arjun tree species can be used for establishing rural cottage industry of tanning raw hide with the view to generate employment and increase the income of deprived rural multitudes surviving in the remote rural areas of the district.

(xi) Jetropha

The Jetropha is a Australian tree species. The Australian local name is "Jerusalem Artichoke" which is being grown in the marshy land and water logged lands in Australia. The Australian people plant the Jerusalem Artichoke plants for the reclamation of the waste land and the water logged lands and marshy lands. The end product of the Jerusalem Artichoke has been used for manufacturing the Ethanol and other by-products.* The Jerusalem Artichoke plants has been planted in India but it has been renamed in India as 'Jetropha'. The Jetropha plants have been planted on the waste lands with the view to extract Diesel or Ethanol and the leaves of the plants has been

* Yadav, Hridai R., "Cultivation of Jerusalem Artichoke in Waste Lands for Manufacturing of Ethanol and other by products: A Case Study of India." Research Paper presented by the World Environment Congress Tioz-2nd Conference and AOTs-3 Conference held at Cincinati, Ohio, USA, 26-29 October, 1996.

used for green manuring while the branches of the Jetropha plants have been used for fuel wood purpose by the villagers.

Ministry of Environment and Forests, Government of India and the Government of Uttar Pradesh including the Department of Forest and District level forest officials initiated large scale Jetropha plantation and Jetropha Plantation Mission has been launched which has been propagated and demonstrated for few years but after that the Jetropha Mission could not achieved the proper momentum. In reality, practically the Jetropha plantation has been found successful in reclaiming the wastelands, water logged lands and marshy lands and the leaves have been re-cycled for green mannuring and the branches of the plants have been used for the fuel wood and the Jetropha seeds has been used for extracting Ethanol or Diesel fuel for various purposes. Thus the Jetropha plants have been found suitable for improving eco-system and conserving the environmental conditions and generating the employment and increasing the income of the deprived rural poor multitudes living in the remote rural villages.

6. CONCLUSION

It is known fact that the forest is an important natural resource, which is a source of survival for cattle and human. Forest plays an important role in maintaining the Eco-System and improving the environmental conditions of the area. It has been observed that a centuries ago the district has been covered with extensive forests, flora, fauna and birds. The *jhils* and *tals* were famous for the birds sanctuary. Due to ever increasing cattle and human population, urbanization of the forest cover of the area had been destroyed and turned into cultivated lands and developed for the settlements. The patches of the forest cover has been observed along with in river banks of the district.

During the period of National campaign on "Grow More Food" following the Second World War the forests were recklessly cut down. During the early fifties extensive afforestation measures had been initiated by the Government, and Dhak, Khair, Babool, Shisham, Neem, Arjun, Kanji, Seras, Eucalyptus, Mango, Mahua, Jamun etc. trees were planted in the district. The major impact of deforestation and ecological degradation that has taken because half of our nation depends for its fuel wood, fodder on the common lands.

Due to the extensive jungles in the Amethi, village panchayat Ram Nagar was named as Jangal Ram Nagar covering one revenue village and 48 purwas, tolas, and hamlet villages. The village Jangal Ram Nagar is the largest village of Amethi in terms of geographical area and population. The area under forest has been 0.44 per cent of the total geographical area of the district. The per capita forest area of the district has been found 0.0021 hectares which is lower than the Rae Bareli district and higher than the neighbouring Sultanpur district. Musafirkhana range and Salon forest range has proportionately higher area under forest than other parts of the district.

The district was covered with the extensive forest which had been destroyed and turned into cultivated lands. During the early fifteen various afforestation programmes several varieties of trees, like Dhak, Khair, Babool, Shisham, Neem, Arjun, Kanji, Siras, Eucalyptus, Jamun, Mahua etc. plants were planted in the district with the view to meet the increasing demand of growing cattle and human population. The tiger, wild buffalows, used to be found in the jungles along the banks of rivers. The larger carnivore has disappeared and even wolf and hyena have become very rare in the area. The Indian Antelope has been protected species in the district. The main game birds found in the district are of several varieties of ducks, partridges, pigeons, geese, coots, jacana, moorhen, snipe and sand piper are also met in the area. During the cold weather season plenty of water fowl, snipe, quail are found in the district.

Afforestation has been an important plan for the integrated development of an area. The afforestation approach has been distinguished into 'Productive' and 'Social' Forestry. The social forestry has been considered as an asset for the rural economy, preventing soil erosion and sedimentation of reservoirs and increasing the soil moisture retentivity.

Social forestry programme has been launched along the road, rail road, small category forests, panchayat land, banjar land, and along with canal banks. The selection of need based appropriate tree species plays an important role for the success of afforestation schemes.

The suitable species for the social-forestry, like mahua, jamun, mango, kanji, eucalyptus, casurina, shisham, arjun, jetropha and various other tree species suitable to the different agro-climatic conditions have been promoted, propagated and planted by the various implementing agencies.

Thus, it can be concluded that afforestation or social forestry programmes are basic requirements of the cattle and human population of the district. The successful afforestation programmes are helpful in the reclamation of waste lands, conserving the environmental degradation, improving the eco-system of the area. The afforestation programmes is an asset to obtain fuel wood, fuel energy, fodder, fibre, fruits, wood and wood for the rural cottage industry and wood for the agricultural implements and household needs of the people. Afforestation programme also generates employment to the needy resourceless poor people and it has also been helpful in increasing the income of the deprived resourceless, rural poor multitudes surviving in the remote rural areas of the district.

Lastly, it can be said that the natural resource jungle has saved the land and it can improve the Agro-climatic conditions and helpful for the rainfall in the area. Therefore, the jungle (Forest) is more important than *Jameen* (Land) and *Jal* (Water) but to save the earth, cattle and human population along with *Jal*, *Jangal* and *Jameen* plays an important role in the development of rural economy and improving the socio-economic conditions of the rural poor. The forest is an asset for maintaining the Ecological balance and conserving the environmental degradation of the area and also improving the environmental crisis of the various villages of the Amethi.

6
Land Use

1. INTRODUCTION

The 'land' as a factor of production is very peculiar because it possesses some important features, which distinguishes it from other factors of production. Land (*Zameen*) is a free gift of nature like *Jal* and *Jungle* and we have to accept it as it is.

Due to the multi-dimensional use of land, the land has much significance and importance among the natural resources. The land provides space for economic and other activities whether they are related to Primary, Secondary or Tertiary sectors of economy. Everything we use can be innovated ultimately on the land.

"**Life without land is unconceivable**". However, more than 76 per cent of our population is dependent on primary sector of economy, i.e. surviving for their livelihood from agriculture which is inelastic and invaluable asset holds the key to rural and agrarian prosperity of the people living in the rural villages.

The nature has gifted the land with unique life sustaining potential which has made it invaluable and indispensable for various ecological needs. National Bureau of soil survey and land use planning, Ministry of Agriculture, Government of India has been engaged in developing land use plan and to control the changes in the land use pattern through State Governments and District Revenue Authorities.

The agricultural census and revenue records of the Tehsils and revenue officials consider mainly five land use categories which are—

1. Forest Land.
2. Land not available for cultivation.
3. Other uncultivated land excluding fallows.
4. Fallow land.
5. Net cultivated land.

From these five types of land use three categories, viz. land not available for cultivation, other uncultivated land excluding fallows and fallow lands has been subdivided.

A. Land not available for cultivation

i. Land put to non-agricultural use.
ii. Barren and uncultivable land.

B. Other cultivable land excluding fallows

i. Permanent pastures and grazing land.
ii. Culturable waste.

iii. Land under miscellaneous trees.

C. **Fallows**

i. Fallows other than current fallow.

ii. Current fallow.

The net cultivated land, i.e. net sown area contributes about 53.8 per cent area to the total geographical area of the district.

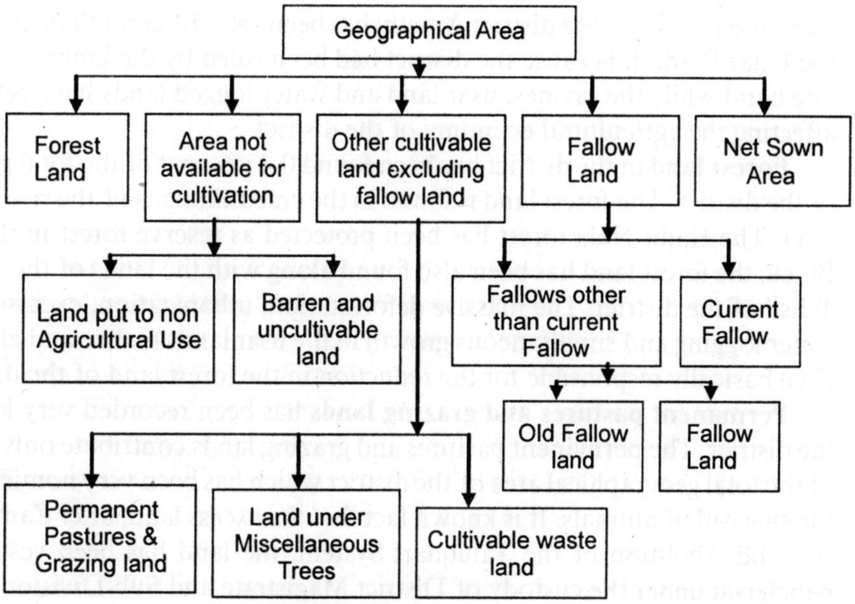

Fig. 6.12: Land Use

The land use classification is not the existing system of the government but it had been initiated during the period of Mughal Rule, Emperor "Akbar the Great" through Raja Todar Mal, categorized the land as per productivity and land quality with the view to collect the 'Malgujari' from the cultivators. The agricultural land was categorized as *Avval* (Good), *Doyam* (Medium) and *Soyam* (Low) quality land as per agricultural productivity. It has been assumed that *Avval* (good) quality land has been earmarked adjacent to the village settlement in which Rabi, Kharif and Jaid crops are sown. The *Doyam* (Medium) quality land, has been classified based on the fertility status of the soil, in which two crops of Kharif and Rabi cropping reasons are cultivated with medium productivity. The *Soyam* (low) quality land found far away from the village settlement having low productivity and low fertility status of the soil, in which only one crop had been sown with higher input cost.

The land quality in the village Khasra Map is still shown as *Avval, Doyam* and *Soyam* quality land with the objective of fixing land revenue charges from the farmers. The national soil survey and land use Board, Ministry of Agriculture, Government of India, classified the land as per the use of the land. The forest land, land not available for cultivation (Land put to non-agricultural use, Barren and uncultivable land), other uncultivated land excluding fallows (Permanent pastures and grazing land, culturable waste land under miscellaneous trees). Net cultivate land and fallow land (fallows other than current fallow and current fallow land) has been categorized and the Tehsil level revenue officials maintained and updated the village revenue khasara map and revenue khasara register as per the land use classification. The Malgujari of the land/revenue charges has been fixed as per the land quality. The changes in the land use in the Khasara Register or revenue record is being ensured as per existing status of the land by the Sub-divisional Magistrate of the concerned Tehsil depending on the reports and recommendation of the Lekhpal and Kanoongo of the concerned village of the Amethi block.

2. LAND USE PATTERN IN THE DISTRICT

Land use pattern of the district Amethi has been very different than the other districts of the Uttar Pradesh because the district had been ruled by the kings and emperor on the one hand while the ravines, usar land and water logged lands have been very seriously affecting the agricultural economy of the district.

Forest land in the district has been found 0.4 per cent of the total geographical area of the district. The forest land is found in the embankments of the river Gomati and Sai river. The Kadu Nala forest has been protected as reserve forest in the Musafirkhana Block, the forest land has been also found along with the lands of the Sai river in Salon Tehsil of the district. The massive deforestation, urbanization, excessive irrigation and water logging and simultaneous growth in the usar lands (saline and alkaline lands) has been basically responsible for the reduction in the forest land of the district.

Permanent pastures and grazing lands has been recorded very low percentage in the district. The permanent pastures and grazing lands contribute only 0.3 per cent area of the total geographical area of the district which has been very nominal proportion for the survival of animals. It is known fact that the excess land, after Zamindari Abolition Act and Abolition of the Taluqdari System, the land has been vested in the village panchayat under the custody of District Magistrate and Sub-Divisional Magistrates of the revenue tehsils. As per the policy of the Government such lands lying under the village panchayat has been allotted to the landless/resourceless poor people of the respective village and later on the patta allotted lands has been made regional and the land ownership rights has been given to the such patta land allottees during 2002 by the Government of Uttar Pradesh. The permanent pastures and grazing lands of the village has been permanently allotted to the village poor having land ownership rights, due to which the area under permanent pastures and grazing lands has been reduced.

Current fallow land in the district has been found 8.8 per cent of the total geographical area of the district. The current fallow in such type of land which has been under cultivation but the farmers has left the land out of cultivation for one or two cropping seasons, i.e. Rabi and Kharif cropping season for one year with the view to re-maintain the fertility status of the soil and to ensure the crop rotation system and to re-maintain agricultural productivity status of the soil.

Fallows other than the current fallow land include the old fallow land (2.5 years) and Fallow land (1-2 years), with the view to maintain the fertility status of the soil but due techno-scientific ignorance and the financial crisis the farmers could not apply the NPK fertilizers, insecticides and pesticides and organic and inorganic amendment to re-build the fertility status of the soil due to which the land has been left as current fallows in the subsequent years for one to two years has been categorized as fallow land and if the same land has been continued to leave the land as fallows from more than two years such land has been categorized as old fallow land. The fallow land is the district Amethi has been recorded 9.7 per cent of the total geographical area of the districts, which includes the proportionate distribution of the fallow land and old fallow land of the district.

Culturable waste lands in the district has been observed 3.6 per cent to the total area of the district. The cultivable waste lands are such lands which has been under for cultivation but due to usarization process or excess rain water or irrigation water the land has been affected adversely and it could not yield the input cost of the farmers due to which the land has been left out cultivation and the such lands under land use classification has been categorized as cultivable wastelands, i.e. the land is cultivable but it has been left out of cultivation and the land is lying waste due to some reason of the other.

Area under miscellaneous trees, crops and groves has been recorded as 1.9 per cent to the total geographical area of the district. The miscellaneous trees, groves etc. has been found on the very limited area because such lands has been utilized for the other purposes by the villagers. The land under miscellaneous trees are found near ponds, tanks, and around the village but no specific area has been found under the miscellaneous trees in the district. The growing population and cattle and infrastructural development has been mainly responsible for the reduction of the area under the miscellaneous trees and groves Ely in district Amethi.

The land not available for cultivation has been found 21 per cent to the total geographical area of the district. The area not available for cultivation includes barren and uncultivable land and the land not available for cultivation or agricultural use. The area under barren land and uncultivated land contributes 10.9 per cent of the total geographical area of the district Amethi.

The land not available for cultivation has been recorded as 10.1 per cent to total geographical area of the district. The area not available for cultivation includes settlement, rail, road, canal etc. Thus, there are 1/5 land of the district which is not available for cultivation. The general land use pattern of the district including wastelands, swamp, scrupt and grades etc. are depicted in Map 1.11.

The net sown area or net cultivated area has been found 53.8 per cent of total geographical area of the district.

Thus, it has been observed that there are proportionate distribution of land use pattern in the district Amethi excepting the area under forest which has been found very nominal in the district and the reduction of the area under the forest has very challenging task because in the rural economy of the down-trodden deprived rural poor multitudes are dependent on the forest products. The growing demand of the increasing population for the fuel wood, fibre, fruits etc. and the fodder for the cattle are fetched from the forest resources and it has created ecological imbalances of the area and developed Eco-socio-economic crisis among the deprived resourceless rural poor people surviving in the remote rural areas of the district Amethi.

3. LAND USE PATTERN IN AMETHI BLOCK

The land use pattern in the Amethi Block of the Amethi Tehsil of the district Amethi of Uttar Pradesh has been diversified in nature which varies from one field to another field and also varies from one village to another village within the Amethi Block. The recorded total geographical area of the Amethi has been 13252 hectares, covering 44 villages and 377 purwas or tolas or hamlet villages in the Amethi block.

It has been observed that the area under forest in Amethi Block is very negligible and the forest area in the revenue record, i.e. Khasara Register/Khasara maps are not shown except the area under the social forestry programmes being implemented on the community land in the various villages has been shown in the revenue records. The area under social forestry, horticulture (Bagh) etc. are categorically shown in the revenue records available Musafirkhana of the Tehsil Amethi.

The land not available for cultivation in the Amethi block has been recorded 1360 hectares which contributes 10.66 per cent land of the total geographical area of the Block, similarily the barren and uncultivable land has been found 442 hectares and it shares 3.47 per cent land of the total geographical area of the Block.

The net cultivated area or the net sown area of the Amethi Block has been recorded 6109 hectares which shares 47.90 per cent area under net sown area of the total geographical area of the Amethi Block. It has been observed that there is reduction in area under net cultivated land of the Amethi Block due to the changes in the cultivated lands use system for the development of settlements, chack roads, link roads and roads and development of the village infrastructural facilities in the villages. The net cultivated area has been reduced due to the excess does of chemical fertilizers, insecticides and pesticides and excess doses of canal irrigation water which had led to the development of salinity and alkalinity in the soil or usarization of the soil and water logging and development of calcareousness in the soil which causes hindrance in the growth of crops due to which such land has been left out of cultivation in the most of villages of the Amethi Block of the district Amethi.

The land which has been cultivable but it has been left out of cultivation due to some reason or the other has been classified as culturable waste lands which has been found 3172 hectares in the Amethi block and it contributes 24.87 per cent land of the total geographical area of the Block. The culturable wastelands includes the banjar land, usar land and water logged land and other types of wastelands which are presently lying unutilized and left out of cultivation due to various reasons including very poor fertility status of the land and the farmers are unable to restore the fertility status of the land due to techno-scientific ignorance and socio-economic backwardness. The culturable waste lands can be reclaimed and it can easily be brought under cultivation or agro-forestry or social forestry programmes to meet the increasing demand of fuel wood, fodder, fibre, fruits, food grains, fisheries (F^6), vegetables and wood for the growing cattle and human population of the Block. The proportionate distribution of the area under land use has been given in Table 6.1.

The area under permanent pastures and grazing lands have found 201 hectares and it shares 1.57 per cent land of the total geographical area of the Amethi Block. The percent- wise distribution of the land use pattern at village panchayat level has been depicted in Table 6.2.

There are 277 hectares land under the other uncultivated land excluding fallows which contributes 2.17 per cent land of the total geographical area of the Amethi.

The area under miscellaneous trees, groves etc. has been recorded 312 hectares which shares 2.45 per cent land of the total geographical area of the Amethi Block.

The fallows other than the current fallow lands have been recorded 578 hectares and it contributes 4.53 per cent land of the total geographical area of the Amethi Block. The fallow land other than the current fallow land includes old fallow land (2.5 years) and fallow land (1.2 years). The land which has been left out of cultivation to maintain the fertility status of the soil under the system of the changing cropping pattern the land which has been left out of the cultivation for one to two years has been classified as fallow land and if it is left out of cultivation for the subsequent years for more than two years up to five years has been categorized as old fallow land.

The current fallow lands are those lands which have been left out of cultivation for a year, i.e. Rabi, Kharif and Jaid cropping season, under the changing system of cropping

Table 6.1: Land Use (Amethi Block)

(Area in Hectares)

Sl. No	*Name of the Village Panchayat*	*Total Geographical Area*	*Land not available for cultivation*	*Barren and uncultivable land*	*Net Cultivated Area*	*Culturable Waste*	*Permanent Pastures and Grazing Land*	*Other Uncultivated Land excluding fallow*	*Area under Misc. Trees.*	*Fallows other than current fallow*	*Current Fallow*
1	*2*	*3*	*4*	*5*	*6*	*7*	*8*	*9*	*10*	*11*	*12*
1	Katra Maharani	134	12	7	72	23	1	5	2	6	3
2	Sarai Khema	557	67	20	240	138	15	10	10	30	25
3	Trilokpur	319	18	14	162	82	3	12	7	20	1
4	Ramdaipur	105	28	2	50	12	1	2	1	9	1
5	Jangal Ram Nagar	1198	69	24	603	316	13	10	3	134	10
6	Benipur	128	20	7	58	22	2	4	2	6	7
7	Parsawan	317	26	7	146	86	12	4	4	22	10
8	Mahmoodpur	320	28	8	150	87	2	6	10	20	9
9	Raipur Phulwari	285	34	5	130	86	2	3	5	18	2
10	Umapur Ganapatti	327	35	8	145	98	3	4	12	13	9
11	Bhaganpur	151	15	9	70	28	5	6	4	5	9
12	Tala	182	22	9	85	33	8	7	4	7	6
13	Kushi tali	341	32	6	170	99	2	3	13	13	3
14	Loharta	51	5	2	26	13	1	1	1	2	1
15	Darkha	300	22	6	158	92	2	4	1	11	4
16	Mochwa	166	12	5	80	55	1	3	1	6	3
17	Nuwawan	457	40	14	244	97	5	13	10	12	12
18	Loniapur	109	12	5	48	25	3	3	6	4	3
19	Kherauna	169	25	9	75	45	1	7	2	4	1
20	Katra Phulkunwar	175	22	8	78	41	3	6	5	7	5
21	Hathkila	216	25	8	125	38	1	6	8	4	1
22	Rebha	99	15	6	45	22	1	4	1	2	3
23	Raidaipur	260	23	10	115	86	1	5	5	7	8
24	Korari Girdhar Shah	383	35	17	170	107	6	9	12	15	12
25	Dedhpasar	314	40	5	140	92	7	3	10	11	6
26	Mahaso	86	18	2	40	21	1	1	1	3	1
27	Gangauli	380	42	20	175	97	5	16	10	8	7
28	Dhadhudhar	133	20	5	60	32	3	2	3	5	3
29	Gaderi	268	30	10	120	66	5	4	6	12	4
30	Bhusahari	235	29	4	125	49	6	2	10	4	6
31	Purabgaon	222	32	12	105	49	2	10	4	5	3
32	Chaturbhujpur	120	24	3	56	22	1	2	3	4	5
33	Mahrajpur	164	24	3	93	16	1	2	14	6	5
34	Aghar	217	22	12	112	29	2	3	14	5	8
35	Saidpur	180	30	6	80	40	5	3	8	5	3
36	Kakwa	368	32	22	160	90	5	20	14	20	5
37	Saraiya Duban	299	32	15	130	79	5	9	9	6	4
38	Kohra	573	52	22	275	144	12	12	21	22	12
39	Mahmadpur	445	40	20	210	112	8	15	12	13	7
40	Naraini	584	58	20	280	147	10	15	15	26	10
41	Nainaha	182	25	6	90	43	5	4	3	4	2
42	Ramgarh	658	72	29	295	164	12	11	13	25	7
43	Himmatgarh	340	64	4	192	94	10	2	10	12	2
44	Dehara	235	32	6	126	55	2	4	3	5	2
	Amethi Block	**13252**	**1360**	**442**	**6109**	**3172**	**201**	**277**	**312**	**578**	**242**

rotation pattern, to maintain the fertility status of the soil such lands has been classified as current fallow land in the revenue records including Khasara Register of the Tehsil. It has been observed that there are 242 hectares of current fallow land in the Amethi Block which contributes 1.96 per cent land of the total geographical area of the Amethi Block of the district Amethi.

It has been recorded that in the Amethi Block the reduction in the net cultivable area and growth in the culturable lands including usar lands, banjar lands, water logged lands and other types of waste lands are very alarming challenges to the cultivation and the policy makers of the rural development implementing institutions. The reclamation of wastelands with the involvement of the local people has to be initiated through techno-scientific methods for agriculture-agro-forestry-horticultural and social-forestry programmes to maintain the environment and ecological imbalance of the even and to improve the socio-economic crisis of the Amethi.

4. LAND USE PATTERN OF THE VILLAGE PANCHAYAT OF AMETHI

An attempt has been made in this study to collect, collate and analyze the land use pattern system of the Amethi Block at village panchayat level covering all 44 village panchayats of the Amethi because in the Panchayati Raj System after the District Level Panchayat the Block Panchayat (II-Tier) and Village Panchayat (IIIrd Tier) at local level has been strengthened and the Block Panchayat and at grossroots level Village Panchayat has been adopted as an unit for implementing of the various land use categories adopted by the land revenue records of the Tehsil Amethi. In fact there are 44 village panchayats in Amethi covering 82 revenue villages and 377 Purwas/tolas/hamlet villages in Amethi block of the district Amethi of Uttar Pradesh. In reality, it is very difficult to prepare village revenue khasara maps for all 82 revenue villages because most of the village khasra maps are very large and practically it is difficult to prepare map of an appropriate size, due to which only village panchayats of main villages have been taken to study and analyze the land use pattern based on the village khasara maps through geo-referencing of the khasara maps and available natural infrastructural village panchayat level information system especially the land use pattern. The digital data have been collected for the village panchayats along with the various revenue villages and purwas/tolas/hamlet villages of the respective village panchayat. The village panchayats of main villages maps have been very large in the size which has been very difficult to bring at the appropriate size level and revenue village including hamlet villages spatial mapping has not been done because at suitable level of base mapping, areawise information and location specific details could not be clarified.

Thus in this study village panchayat's main village khasara map has been taken as base map for spatial field level location specific and geo-referencing of land use pattern map has been prepared for all 44 village panchayats of the Amethi Block.

In view of the above, an effort has been made to analyze the land use pattern of each village panchayat separately with the view to study the natural infrastructural/resources information system at village panchayat level/grassroots level in Amethi Block of the district Amethi.

1. Village Panchayat Katara Maharani

The location specific and geo-referencing of the spatial information system fo the various categories of the land use system of the village panchayat Katara Maharni of the Amethi

Table 6.2: Land Use (Amethi Block)

(%age of total geographical area)

Sl. No	*Name of the Village Panchayat*	*Land not available for cultivation*	*Barren and uncultivable land*	*Net Cultivated Area*	*Culturable Waste*	*Permanent Pastures and Grazing Land*	*Other Uncultivated Land excluding fallow*	*Area under Misc. Trees*	*Fallows other than current fallow*	*Current Fallow*
1	*2*	*3*	*4*	*5*	*6*	*7*	*8*	*9*	*10*	*11*
1	Katra Maharani	8.9	5.2	53.7	18.7	1.4	3.0	2.6	4.3	2.23
2	Sarai Khema	12.0	3.59	43.1	24.78	2.7	1.8	1.8	5.9	4.5
3	Trilokpur	5.6	4.4	50.8	25.6	1.0	3.8	2.2	6.3	0.3
4	Ramdaipur	26.6	1.9	47.6	11.4	1.0	1.0	1.0	8.5	1.0
5	Jangal Ram Nagar	5.0	2.0	52.0	24.6	1.0	0.9	6.3	13.4	0.9
6	Benipur	15.6	5.0	45.0	17.9	1.0	3.0	2.0	3.8	5.0
7	Parsawan	17.0	2.0	46.0	20.2	3.0	1.0	1.0	6.8	3.0
8	Mahmoodpur	8.7	2.5	46.8	27.1	1.0	1.5	3.1	6.2	3.1
9	Raipur Fulwari	11.9	1.7	45.6	28.0	0.7	1.0	1.4	6.3	0.7
10	Umapur Ganapatti	10.7	2.4	44.3	28.1	0.9	1.2	3.6	4.9	3.3
11	Bhaganpur	9.9	5.9	46.3	18.8	3.4	3.9	2.6	3.2	6.0
12	Tala	12.0	4.9	46.7	17.2	4.3	3.8	4.1	3.8	3.2
13	Kushi tali	9.3	1.7	49.8	29.5	0.5	0.8	3.4	3.7	0.9
14	Loharta	9.4	3.0	51.0	25.2	2.0	2.0	2.0	4.8	2.0
15	Darkha	7.3	2.0	52.6	30.4	0.6	1.3	0.3	3.6	1.3
16	Mochwa	7.2	3.0	48.1	32.4	1.5	1.8	0.6	3.6	1.8
17	Nuwawan	8.7	3.0	53.3	22.97	1.0	2.8	2.3	3.5	2.5
18	Loniapur	11.0	4.5	44.0	21.4	4.1	3.2	5.5	3.6	2.7
19	Kherauna	14.7	5.3	44.3	26.0	1.8	4.1	1.1	2.2	0.5
20	Katra Phulkunwar	12.5	4.5	44.5	33.1	1.7	3.4	2.2	3.9	3.7
21	Hathkila	12.0	3.0	57.8	18.52	0.4	2.8	3.7	2.0	0.4
22	Rebha	15.1	6.0	45.0	21.0	1.0	4.0	1.9	3.0	3.0
23	Raidaipur	8.8	3.0	44.2	32.4	1.4	1.9	1.9	2.6	3.0
24	Korari Girdhar Shah	9.1	4.4	44.3	25.1	2.65	2.3	5.2	3.9	3.1
25	Dedhpasar	12.7	1.5	44.5	31.6	2.2	0.9	3.1	3.4	1.0
26	Mahaso	20.0	3.0	46.5	22.6	1.1	1.1	1.1	4.4	1.0
27	Gangauli	11.0	5.2	46.0	23.7	1.3	4.2	2.6	3.0	3.0
28	Dhadhudhar	15.0	3.7	45.1	22.3	3.1	2.5	2.2	3.7	2.4
29	Gaderi	11.1	3.7	44.7	24.7	2.8	2.4	3.2	4.3	3.1
30	Bhusahari	12.4	1.7	53.1	20.4	2.5	3.7	2.5	1.6	2.2
31	Purabgaon	14.4	5.4	47.2	21.7	0.9	4.5	1.8	2.2	1.3
32	Chaturbhujpur	20.0	2.5	46.6	18.6	0.8	1.6	2.5	2.4	5.9
33	Mahrajpur	14.6	1.8	56.7	12.9	0.6	1.2	6.5	3.6	2.0
34	Aghar	10.1	5.5	51.6	15.7	3.5	1.3	6.5	2.2	3.6
35	Saidpur	16.6	3.3	44.4	21.9	2.7	1.6	4.4	3.1	2.0
36	Kakwa	8.6	5.9	43.4	24.7	1.3	5.4	3.8	5.3	1.6
37	Saraiya Duban	10.7	5.0	43.4	25.1	1.6	5.0	5.0	1.9	2.0
38	Kohra	9.0	3.8	47.9	24.3	2.0	2.0	3.6	3.7	3.7
39	Mahmadpur	8.9	4.4	47.1	24.2	2.7	3.3	3.7	2.8	3.0
40	Naraini	9.9	3.4	47.8	23.6	3.1	2.5	2.5	4.4	2.7
41	Nainaha	13.7	3.2	49.4	23.5	2.7	2.1	1.6	2.0	1.8
42	Ramgarh	10.9	4.4	44.8	21.5	3.4	3.6	3.9	5.5	2.0
43	Himmatgarh	18.8	1.1	49.0	22.2	2.0	0.5	2.0	3.0	0.4
44	Dehara	13.6	2.5	53.6	23.0	0.8	1.7	1.2	2.1	1.5
	Amethi Block	**10.66**	**3.47**	**47.90**	**24.87**	**1.57**	**2.17**	**2.45**	**4.53**	**1.96**

Block has been depicted in Map 6.15. The total geographical area of the village panchayat Katara Maharani has been recorded 134 hectares out of which 72 hectares land has been found as net cultivated area of the Katara Maharani, Laukpur and Matiyar Baghaura revenue villages including Katara Maharani, Munshiganj, Chauraha and Usarapur hamlet villages of the Gram Sabha Katara Maharani.

It has been observed that there are 8 per cent land not available for cultivation which includes settlement, rail, road, canal etc. Proportionate distribution of Barren and uncultivable land including Banjar land/kankarili land/usar land and water logged land etc. has been found 5 per cent of the total geographical area of village panchayat Katara Maharani. The net cultivated area has been found 53 per cent of the total geographical area of the village. The cultivable waste land which has been previously under cultivation but left out of cultivation either due to water logging or usarization or due to the low fertility status of the soil has been recorded 18.7 per cent of the total geographical area of the village which is very challenging task to the agricultural planners. The permanent pastures and grazing lands has been synchronized having 1.4 per cent of the total geographical area of the village. The area under other uncultivable land excluding fallows has been recorded 3 per cent of the total geographical area of the village Katara Maharani. The area under miscellaneous trees has been found 1 per cent of the total area of the village. Fallows other than current fallow land has been found 4.3 per cent and the current fallow land has also been found only 2.23 per cent of the total geographical area of the village Katara Maharni of the Amethi Block.

The land use pattern of the village Katara Maharani has been proportionately distributed except the growth in the area under the culterable waste lands which is an alarming challenge to the agricultural planners and the environmental activities at the grassroots level. We have to control and maintain the fertility status of the Soil and increase the net cultivated area of the village.

2. Village Panchayat Sarai Khema

The total geographical area of the village panchayat Sarai Khema has been recorded 557 hectares including the Sarai Khema, Dandupur and Ram Nagar revenue villages and the Sarai Khema Khas, Munshiganj, Basuapur, Halapur, Beni Madhavpur, Kutia, Ahmad Nagar, Dandupur Khas, Pure Kutur, Jagapur, Pure Pandit and Mainapur hamlet villages of the village panchayat Sarai Khema of theAmethi Block. The land not available for the cultivation has been recorded 12.0 per cent of the total geographical area of the village which includes rail, road, chak road, schools, settlement area, kabristan, hadwari etc. the barren and uncultivable land area has been found 3.59 per cent and the area under the culturable wasteland has been recorded 43.1 per cent of the total geographical area of the village. The area under culturable waste lands has been found 24.78 per cent which includes banjar land, usar land, water logged land etc. which has been previously under cultivation but it has been left out of cultivation due to poor fertility status of the soil because of the one reason or the another reasons. The area specific land use pattern and land cultivation etc. at the field level has been depicted fieldwise/gata level in Map 6.16 of the village Sarai Khema panchayat. The area under the permanent pasture and grazing land has been found 2.7 per cent of the total geographical area of the village. The area under the other uncultivated land excluded fallows has been found 1.8 per cent and the area under miscellaneous tree has been recorded 1.8 per cent of the total geographical area of the village, while the area under the current fallow land has been observed 4.5 per cent of the total geographical area of the village panchayat Sarai Khema of the Amethi Block. and fallows other than current fallow land has been recorded 5.9 per cent of the total geographical area of the village.

3. Village Panchayat Trilokpur

The total geographical area of the village panchayat Trilokpur has been found 319 hectares out of which 82 hectares land has been recorded as net cultivated area of the village. The area specific field level or plot level land use system of the Trilokpur village panchayat has been depicted in Map 6.17. The land not available for cultivation has been recorded 5.6 per cent of the total geographical area of the village, while the barren and uncultivated land has been found 4 per cent of the total geographical area of the village. Net cultivated area of village Trilokpur has been observed 50.8 per cent of the total geographical area of the village. The area under the culturable waste lands which included banjar land, usar land, water logged land etc. has been observed 25.6 per cent of the total geographical area of the village. The permanent pastures and grazing lands has been observed 1 per cent and the other uncultivated land excluding fallows has been found 3.8 per cent while the area under miscellaneous trees etc. has been recorded 2.19 per cent of the total geographical area of the village. The area under fallows other than the current fallow land has been recorded 6.3 per cent and the area under the current fallow land has been found 0.3 per cent of the total geographical area of the village panchayat Trilokpur of the Amethi Block. The total geographical area of the village panchayat Trilokpur includes revenue village of the Trilokpur and Teri involving Pure Subedar, Pure Bhagwandin, Trilokpur, Pure Darshan, Dubepur and Teri and Kutur Ka Pura hamlet villages of the village panchayat Sarai Khema of the Amethi Block. The constant increase in the area under waste lands in the village panchayat Trilokpur has been very challenging task to the agricultural policy makers and planners. It has become an important task to reclaim the available waste lands of the village and to check and maintain the fertility status of the soil of the village.

4. Village Panchayat Ramdaipur

The total geographical area of the village panchayat Ramdaipur has been recorded 105 hectares which has been mentioned in Table 6.1. The proportionate distribution of the land use pattern has been stated in Table 6.2. The village panchayat Ramdaipur includes revenue villages of Ramdaipur, Bhupatipur, Nauganwa and Bahapur and it has Ramdaipur, Chhawani, Punnapur, Bhupatipur, Nauganwa, Bahapur, Sitarampur, Rajan Ka Purwa, and Gaudia hamlet villages.

The land not available for cultivation in the village panchayat Ramdaipur has been recorded 26.6 per cent of the total geographical area of the village which has been depicted in Map 6.18. The barren and uncultivable land has been found 1.9 per cent while permanent pastures and other uncultivated land excluding fallows and area under miscellaneous trees has been recorded 1.0 per cent respectively of the total geographical area of the village panchayat Ramdaipur. The area under current fallow has been recorded 1.0 per cent and the area under fallows other than the current fallow has been recorded 8.5 per cent of the total geographical area of the village. The culturable waste land area has been recorded 11.4 per cent of the total geographical area of the village which includes banjar land, usar land, water logged land etc. The net cultivable area of the village panchayat Ramdaipur has been found 47.6 per cent of the total geographical area of the village. The decrease in the net cultivable area of the village is a deep concern to the farmers and agricultural policy makers and planners.

5. Village Panchayat Jangal Ram Nagar

The distribution of land use pattern of the village panchayat Jangal Ram Nagar along with area specific locations and plot levels has been depicted in Map 6.19. The village panchayat Jangal Ram Nagar has been recorded as revenue village and 48 hamlet villages. The total geographical area of the village panchayat Ram Nagar has been recorded 1198 hectares. The village panchayat Jangal Ram Nagar has the largest size area in the Amethi Block.

The area under the land not available for cultivation has been found 5 per cent of the total geographical area of the village, while barren and uncultivated land has been recorded 2 per cent. The net cultivated area of the village panchayat Jangal Ram Nagar has been recorded 52 per cent of the total geographical area of the village. The culturable waste lands of the village has been seen 24.6 per cent and the area under permanent pasture and grazing lands has been found 1 per cent of the total geographical area of the village panchayat Jangal Ram Nagar. The other uncultivated land excluding fallows has been observed 0.9 per cent and the area under the miscellaneous trees, shrubs etc. has been found 0.3 per cent of the total geographical area of the village panchayat Jangal Ram Nagar.

The area under fallows other than the current fallow land has been recorded 13.4 per cent and the area under the current fallows has been found 0.9 per cent of the total geographical area of the village panchayat Jangal Ram Nagar.

The village panchayat Jangal Ram Nagar has been very famous due to its forest area but at present most of the forest area/jungles of the village have been vanished due to massive deforestation. Only the village Jangal Ram Nagar is existing in the record but the jungle areas of the village has been deforested which is a great danger to the environment and ecosystem of the area and socio-economics to the down-trodden, deprived sections of the society of the village panchayat Jangal Ram Nagar of the Amethi Block.

6. Village Panchayat Benipur

The total geographical area of the Village panchayat Benipur has been recorded 128 hectares which includes the revenue villages Benipur and Benipur New hamlet villages. The proportionate distribution of the land not available for cultivation has been recorded 15.6 per cent and barren and uncultivated land has been found 5 per cent of the total geographical area of the village. The land use system of the Village panchayat Benipur has been depicted in Map 6.20. It has been observed that there are 45 per cent net cultivable area in the Village panchayat Benipur, which has decreased due to poor fertility status of the soil and there is growth in the area of culturable waste lands. The area under the culturable wastelands has been recorded 7.2 per cent of the total geographical area of the Village panchayat Benipur. The permanent pasture and grazing lands have been 1.0 per cent of which other uncultivated land excluding fallows has been found 3 per cent of the total geographical area of the village. The area under miscellaneous tree has been found 2 per cent and the area under fallows other than current fallow land has been recorded 3.8 per cent and the current fallow land has been observed 5 per cent of the total geographical area of the Village panchayat Benipur. An increase in the area under culturable wasteland has been very challenging task to the farmers especially to check the decrease in the net sown area and also to reclaim the available waste lands of the Village panchayat Benipur of the Amethi Block.

7. Village Panchayat Parsawan

The total geographical area of the Village panchayat Parsawan has been recorded 317 hectares. The land use pattern of the village panchayat Parsawan has been depicted in Map 6.21. The land not available for cultivation which includes rail, road, canal, settlement etc. has been recorded 17 per cent of the total geographical area of the village. The barren and uncultivated land has been recorded 2 per cent and the permanent pastures and grazing land has been found 3 per cent of the total geographical area of the village. The net sown area has been recorded 46 per cent which has been recorded and the area under the culturable waste land has been observed 20.2 per cent of the total geographical area of the village panchayat Parsawan includes the Parsawan and Ram Nathpur Chhota revenue villages and besides this there are Pure Bhatani, Parsawan Khas, Parsawan Khalwa, Parssawan Ditwa and Ram Nathpur Chhota hamlet villages in the village panchayat Parsawan. The other uncultivated land excluding fallows has been recorded 1 per cent and area under miscellaneous trees has been found one per cent of the total geographical area of the village. The fallows other than the current fallows land has been found 6.8 per cent and the current fallow land has been recorded 3 per cent of the total geographical area of the village panchayat Parsawan of the Amethi Block.

8. Village Panchayat Mahmoodpur

The land use pattern of the village panchayat Mahmoodpur has been shown in Map 6.22. The total geographical area of the village panchayat Mahmoodpur has been recorded 320 hectares. There are 8.7 per cent area under the land not available for cultivation while there are 2 per cent area under the barren and uncultivated land. The net cultivated area of the village panchayat Mahmoodpur has been recorded 46.8 per cent and the culturable wastelands has been observed 27.1 per cent of the total geographical area of the village. The permanent pastures and grazing land has been found 1 per cent and other uncultivated land excluding fallows has been found 1.5 per cent of the total geographical arca of the village. The area under miscellaneous trees has been found 3.13 per cent and fallows other than the current fallow land has been found 6.2 per cent and the current fallow land has been recorded 3.07 per cent of the total geographical area of the village panchayat Mahmoodpur.

The village panchayat Mahmoodpur includes Mahmoodpur, Marfapur and Valipur revenue villages and there are Mahmoodpur, Ram Nathpur Bada, Munshiganj road, Lalcharia Khera Ke Pichhu, Awas Vikas Gopalpur, Marfapur, Mainapur, Valipur Khas and Valipur Aam purwas/tolas/hamlet villages in the village panchayat Mahmoodpur of the Amethi Block of the district Amethi.

9. Village Panchayat Raipur Phulwari

The total geographical area of the village panchayat Raipur Phulwari has been recorded 285 hectares which includes Raipur Phulwari and Katara Raj Himmat Singh revenue villages. The village panchayat Raipur Phulwari also includes Pure Ram Prasad Pandit,

Baniyani, Salamat Ali Ka Purwa, Ojha Ka Purwa, Datta Ka Purwa, Kakawa Mod and Katara Raja Himmat Singh hamlet villages/purwas/tolas in the village panchayat Raipur Phulwari. The land not available for cultivation has been found 11.9 per cent and barren and uncultivated land has been found 1.7 per cent of the total geographical area of the village. The net cultivated area of the village panchayat Raipur Phulwari has been recorded 45.6 per cent of the total geographical area of the village which has been depicted in Map 6.23. The culturable wastelands of the village panchayat Raipur Phulwari has been recorded 28.7 per cent of the total geographical area of the village which has been found very high and there is drastic growth in the area under culturable wastelands of the village. The reclamation of the waste lands and the check the decrease in the area under net sown area has been a great challenge to the farmers and agricultural policy makers and planners at the grassroots level. The permanent pastures and grazing land has been found 0.7 per cent and other uncultivated land excluding fallow and has been found 1 per cent of the total geographical area of the village. There are 1.4 per cent area under the miscellaneous trees and 6.3 per cent land has been seen under fallows other than current fallow land while the area under the current fallow land has been recorded 0.7 per cent of the total geographical area of the village panchayat Raipur Phulwari of the Amethi Block.

10. Village Panchayat Umapur Gana Patti

It has been observed that the total geographical area of the village panchayat Umapur Gana Patti has been recorded 327 hectares which includes Umapur Gana Patti revenue village and Bhagwanpur, Pure Guru Baksh, Pure Prem, Hati, Pure Ichha, Pure Nidhi and Sri Rampur hamlet villages/purwas/tolas of the village panchayat Umapur Gana Patti. The area under the land not available for cultivation has been found 10.7 per cent and the area under barren and uncultivated land has been seen 2.4 per cent of the total geographical area of the village has been recorded 44.3 per cent of the total geographical area of the village. The reduction in the sown area has been a great challenge to the farmers and it has become danger to the local environment of the village. There is substantial growth in the area under the culturable wastelands which has been recorded 28.1 per cent to the total geographical area of the village panchayat Umapur Gana Patti. The area under the permanent pastures and grazing lands has been found 0.9 per cent and the area under other uncultivated land excluding fallows has been found 1.2 per cent and the area under the miscellaneous trees and shrubs has been recorded 3.6 per cent to the total geographical area of the village panchayat Umapur Gana Patti. the area under the fallows other than the current fallow land has been observed 4.9 per cent and the area under the current fallow land has been recorded 3.3 per cent to the total geographical area of the village panchayat Umapur Gana Patti. The land use pattern of the village panchayat Umapur Gana Patti has been depicted in Map 6.24.

11. Village Panchayat Bhaganpur

The geographical area of the village panchayat Bhaganpur has been recorded 151 hectares. The land use system of the village panchayat Bhaganpur has been depicted in Map 6.25. The village panchayat Bhaganpur includes Bhaganpur and Jangal Tikari revenue village. The village panchayat Bhaganpur has 8 hamlet vilages, viz. Bhaganpur, Durgapur, Purelal, Pure Nand Mishra, Sringarpur, Pure Bhojai, Jangal Tikari and Dhamranwa purwas or tolas. The area under the land not available for cultivation has been recorded 9.9 per cent and barren and uncultivated land has been found 4.9 per cent of the total geographical area of the village panchayat Bhaganpur. The net cultivable area of village panchayat Bhaganpur has been seen 46.3 per cent and the culturable waste land has been found 18.8 per cent of the total geographical area of the village panchayat Bhaganpur. The area under the permanent pastures and grazing lands has been found 3.4 per and the other uncultivated land excluding fallows has been recorded 3.9 per cent of the total geographical area of the village. The area under miscellaneous has been recorded 2.6 per cent. The fallows other than the current fallow land has been recorded 3.2 per cent while the current fallow land has been recorded 6.0 per cent of the total geographical area of the village panchayat Bhaganpur. The decrease in the net sown area and the poor fertility status of the soil and the constant increase in the area under current fallow land and fallow other than current fallow land has been great concern to the agricultural farmers of the village panchayat Bhaganpur of the Amethi Block.

12. Village Panchayat Tala

The land use pattern of the village panchayat Tala has been depicted in Map 6.26. The total geographical area of the village panchayat Tala has been recorded 182 hectares which includes revenue village Tala along with Tala, Lodhani Patti, Bhaisupur, Pure Dina Malkata, Baripur, Mahadevan purwas/tolas/hamlet villages of the village panchayat Tala. The area under the land not available for cultivation has been observed 12.0 per cent and the cultivated land has been recorded 4.9 per cent to the total geographical area of the village panchayat Tala. The net cultivated area of the village panchayat Tala has been recorded 46.7 per cent which indicates reduction in the area under the net sown area due to poor fertility status of the soil. It has been observed that there is significant increase in the area under culturable waste lands. The proportionate area under the culturable waste lands has been observed 17.2 per cent to the total geographical area of the village panchayat Tala. The area under miscellaneous trees has been recorded 4.1 per cent while the area under fallows other than current fallow land has been observed 3.8 per cent while the area under the current fallow has been recorded 3.2 per cent of the total geographical area of the village panchayat Tala. The substantial increase in the area under fallows other than the current fallow land and the current fallow land has to be controlled and an effort should ensure to increase the fertility status of the soil of the village panchayat Tala of the Amethi Block.

13. Village Panchayat Kushi Tali

The specific area and location of the land use pattern at the plot level/field level/gata level based on the village khasara map has been presented in Map 6.27. The geographical area of the village panchayat Kushi Tali has been recorded 341 hectares, which includes Kushi Tali and Haripur revenue villages along with the Patakhapur, Pure Nardatt, Kushi Tali, Pure Ram Lal, Haripur, Loharata and Mali Ka Purwa tolas/purwas/hamlet villages of the village panchayat Kushi Tali of the Amethi Block. The proportionate distribution of the trees under the land not available for cultivation has been found 9.3 per cent and the barren and uncultivated land has been recorded 1.7 per cent of the total geographical area of the village panchayat Kushi Tali. The net cultivated area of the village panchayat Kushi Tali has been recorded 39.8 per cent and the area under the culturable wasteland has been observed 29.5 per cent of the total geographical area of the village panchayat

Kushi Tali. It has been observed that there is constant growth in the area under the culturable wastelands which has been a deep concern to the cultivators of the village. An attempt has to be initiated to reclaim the area under culturable waste lands and to control the reduction of the area under net cultivated land of the village panchayat Kushi Tali. The permanent pasture and the grazing lands has been found 0.5 per cent and other uncultivated land excluding fallows has been found 0.9 per cent and the area under miscellaneous trees, shrubs etc. has been found 3.8 per cent to the total geographical area of the village panchayat Kushi Tali. The area under the fallows other than the current fallow land has been observed 3.7 per cent and the area under the current fallow land has been recorded 0.9 per cent to the total geographical area of the village panchayat Kushi Tali which has been proportionately distributed.

14. Village Panchayat Loharata

The land use system of the village panchayat Loharata has been shown in Map 6.28. The total geographical area of the village panchayat Loharata has been recorded 51 hectares which includes along with the Loharata, Mali Ka Purwa, Pure Shiv Gulam, Lala Ka Purwa, Kalu Tewari Ka Purwa and Chak Belan Purwas/tolas/hamlet villages of the village panchayat Loharata.

The land not available for cultivation contributed 9 per cent of the total geographical area of the village panchayat Loharata. The land not available for cultivation includes the settlement area, rail, road, kabristan, etc. The barren and uncultivated land shares 3 per cent of the total geographical area of the village. The area under the culturable waste land has been found 25.2 per cent of the total geographical area of the village panchayat Loharata which has been found comparatively higher than the proportionate land use system. The permanent pasture and grazing lands and other uncultivated land excluding fallows and area under miscellaneous trees, shrubs etc. have been recorded 2 per cent for each category of the land use system of the village panchayat Loharata. The area under the fallows other than the current fallow land has been found 4.9 per cent and the current fallow land has been seen 2 per cent of the total geographical area of the village panchayat Loharata.

15. Village Panchayat Darakha

The area specific and plot/field level distribution of the land use system of the village panchayat Darakha has been shown in Map 6.29. The total geographical area of the village panchayat Darakha has been recorded 300 hectare land which includes revenue village Darakha along with the Pure Chandika, Pure Shiv Lal, Pue Bariyar, Pure Dutt, Darakha Diti, Pure Jagannath, Naya Purwa and Pure Shital purwas/tolas/hamlet villages. The area under the land not available for cultivation has been seen 7.3 per cent and the barren and uncultivated land has been recorded 2.0 per cent of the total geographical area of the village panchayat Darakha. The area under net cultivated land has been observed 52.6 per cent to the total geographical area of the village panchayat Darakha. The area under the culturable waste lands has been observed 30.4 per cent to the total geographical area of the village panchayat Darakha. The area under the culturable wastelands has been found significantly high which has to be controlled and an effort has to be initiated to reclaim the culturable wastelands with the view to increase the area under cultivation. The area under permanent pastures and grazing lands have been recorded 0.6 per cent and the area under miscellaneous trees etc. has been found

0.3 per cent and the other uncultivated land excluding fallow land has been recorded 1.3 per cent of the total geographical area of the village panchayat Darakha. The area under the fallow land other than the current fallow has been found 3.6 per cent and the current fallow land has been recorded 1.3 per cent of the total geographical area of the village panchayat Darakha of Amethi Block.

16. Village Panchayat Mochawa

The total geographical area of the village panchayat Mochawa has been recorded 166 hectares, which includes Mochawa and Sarai Hirmati, Jhaunei and Sultanpur revenue villages along with Mochawa, Maheshpur, Sarai Hirmati, Barsati, Jhaunei, Pure Daya Mishra, Sultanpur, Pure Daljit, Pure Dashrath, Pure Galhin, Pure Daya Mishra hamlet villages/purwas/tolas. Land not available for cultivation has been 7.2 per cent and the area under the barren and uncultivated land has been recorded 3.0 per cent of the total geographical area of the village panchayat Mochawa. The land use pattern of the village panchayat Mochawa has been depicted in Map 6.30. The net cultivated area of the village panchayat Mochawa has been recorded 48.1 per cent of the total geographical area of the village. The area under the culturable waste lands has been observed 32.4 per cent of the total geographical area of the village panchayat Mochawa. It has been observed that there is constant increase under the area in the culturable waste lands which has deep concern to the cultivators and agricultural policy makers and planners of the area. An effort must be initiated to reclaim the culturable wastelands of the village and to control the decrease in the cultivated land and also maintain the fertility status of the soil with the view to increase the agricultural productivity of the village panchayat Mochawa. The area under the permanent pastures and grazing land has been seen 1.5 per cent of the total geographical area of the village. The area under the other uncultivated land excluding fallows has been found 1 per cent of the total geographical area of the village. Very insignificant area under miscellaneous trees, shrubs etc. has been observed 0.6 per cent of the total area of the village panchayat Mochawa. The area under the fallow other than the current fallow land has been noted 3.6 per cent of the total geographical area of the village and fallow land has been recorded 1.8 per cent of the total geographical area of the village panchayat Mochawa. The area specific location-wise distribution of the land use pattern of the village panchayat Mochawa of the Amethi Block presents proportionate distribution of various land use categories significantly.

17. Village Panchayat Nuwawan

The detailed land use pattern of the village panchayat Nuwawan has been depicted in Map 6.31. The total geographical area of the village panchayat Nuwawan has been recorded 457 hectares which included Nuwawan revenue village and Nuwanwa Khas, Dubepur, Kodari, Saidpur, Teliyani, Pure Gopal and Pure Yamati tolas/purwas/hamlet villages of the main village panchayat Nuwawan. The area under the land not available for cultivation has been observed 8.7 per cent and the barren and uncultivated and has been found 3.0 per cent of the total geographical area of the village panchayat Nuwawan. The net cultivated land has been found 53.3 per cent of the total geographical area of the village panchayat Nuwawan. The significant area under the culturable wastelands has been recorded. The culturable waste land has been observed 22.97 per cent of the total geographical area of the village panchayat Nuwawan. The permanent pasture and grazing land has been observed 1.0 per cent and the area under the other uncultivated land excluding fallow land has been found 2.8 per cent of the total geographical area of the village panchayat Nuwawan. The area under the miscellaneous trees, shrubs etc. has been found 2.3 per cent and the area under the current fallow land has been seen 2.5 per cent of the total geographical area of the village panchayat Nuwawan. The area under the fallows other than the current fallow land has been recorded 3.5 per cent of the total geographical area of the village panchayat Nuwawan.

18. Village Panchayat Loniapur

The total geographical area of the village panchayat Loniapur has been recorded 109 hectares which includes Loniapur and Bharathipur revenue villages along with the Mattha Shukul Ka Purwa, Teliyani, Badhai Ka Purwa and Bharathipur tolas/purwa/hamlet villages. The area specific plotwise location and the distribution of the land use pattern of the village panchayat Loniapur has been depicted in Map 6.32. The area not available for cultivation contributes 11.0 per cent of the total geographical area of the village. The barren and uncultivated land has been found 4.5 per cent in the village panchayat Loniapur. The net cultivated area of the village panchayat Loniapur has been found 44.0 per cent of the total geographical area of the village which has been significantly reduced due to the poor fertility status of the soil. The area under the culturable waste land has been found 21.4 per cent of the total geographical area of the village panchayat Loniapur. Permanent pastures and grazing lands have been found 4.1 per cent and other uncultivated land excluding fallows has been observed 3.2 per cent of the total geographical area of the village panchayat Loniapur. The area under miscellaneous trees has been recorded 5.5 per cent and the area under the current fallow land has been observed 2.7 per cent of the total geographical area of the village. The area under the fallow other than current fallow land has been noted 3.6 per cent of the total geographical area of the village panchayat Loniapur of the Amethi Block.

19. Village Panchayat Kherauna

The area specific distribution of the land use system of the village panchayat Kherauna has been depicted in Map 6.33. The total geographical area of the village panchayat Kherauna has been recorded 169 hectares which includes the Kherauna, Sarai Raj Shah, Sarwanpur, revenue villages along with the Kherauna, Baramati, Kewalapur, Hardev Nagar Sarai Raj Shah, Sarwanpur, Keshaw Nagar, Gayatri Nagar, Gangaganj, Sringarpur and Pure Bhojai purwas/tolas/hamlet villages of the village panchayat Kherauna of the Amethi Block. The area under the land not available for cultivation has been recorded 14.7 per cent and the barren and uncultivated land has been noted 5.3 per cent of the total geographical area of the village panchayat Kherauna. Considerable decreased net cultivated area having 44.3 per cent of the total geographical area of the village panchayat Kherauna has been recorded in the revenue records. The area under the culturable waste land has been seen 26.0 per cent of the total geographical area of the village panchayat Kherauna which indicates significant increase in the area under the culturable wastelands. Permanent pastures and the grazing lands have been found 1.8 per cent and other uncultivated land excluding fallow land has been found 4.1 per cent to the total geographical area of the village panchayat Kherauna. The area under the miscellaneous trees, shrubs etc. has been found 1.1 per cent of the total geographical area of the village. The fallow land other than current fallow land has been observed 2.2 per cent and the current fallow land has been recorded 0.5 per cent of the total geographical area of the village panchayat Kherauna. Attempt must be made to check the decrease in the area under the net cultivated land and to maintain the fertility status of the soil with the view to increase the agricultural productivity of the village on the one hand while on the other effort should be initiated to reclaim the wastelands of the village through adopting latest techno-scientific method to improve the fertility status of the soil to grow agricultural crops.

20. Village Panchayat Katara Phulkunwar

The total geographical area of the village panchayat Katara Phulkunwar has been recorded 175 hectares which includes Katara Phulkunwar, Jagipatti, Baghwaria, Parsauli, revenue villages along with the Katara Phulkunwar, Bahorapur, Tewaripur, Kewalapur, Jagapati, Baghwaria and Parsauli purwa/tolas/hamlet village. The area specific land use pattern of the village panchayat Katara Phulkunwar has been depicted in Map 6.34. The area under the land not available for cultivation has been observed 12.5 per cent and the barren and uncultivated land has been seen 4.5 per cent of the total geographical area of the village panchayat Katara Phulkunwar. The net cultivated area of the village panchayat Katara Phulkunwar has been observed 44.5 per cent to the total geographical area of the village. The area under the culturable waste land has been seen 23.1 per cent of the total area of the village. The permanent pastures and grazing lands have been recorded 1.7 per cent and the area under other uncultivated land excluding fallows has been noted 3.4 per cent of the total geographical area of the village panchayat Katara Phulkunwar. The area under the miscellaneous trees, shrubs etc. has been found 2.8 per cent of the total geographical area of the village. The fallows other than the current fallow land has been recorded 3.9 per cent to the total geographical area of the village while the area under the current fallow land has been noted 3.7 per cent of the total geographical area of the village panchayat Katara Phulkunwar of the Amethi Block of the district Amethi.

21. Village Panchayat Katara Hathkila

The area specific locationwise land use pattern of the village panchayat Katara Hathkila has been presented in Map 6.35. The total geographical area of the village panchayat Katara Hathkila has been recorded 216 hectares which includes Hathkila revenue village along with the Hathkila, Dewaraka, Tola Dewatar, Baripur purwas/tolas/hamlet villages of the village panchayat Katara Hathkila. The land not available for cultivation has been found 12.0 per cent whch includes settlement area, rail, road, canal side Dhool land, barn, kabristan and Hadwar land etc. and such lands have been shown in the land use pattern map of the village panchayat Katara Hathkila. The area under the barren and uncultivated land has been observed 3.0 per cent of the total area of the village. The net cultivated land has been recorded 57.8 per cent of the total geographical area of the village panchayat Katara Hathkila, which has been comparatively higher than the other villages of the Amethi Block. The area under the culturable waste land has been found 18.52 per cent of the total geographical area of the village panchayat Katara Hathkila which has been noted comparatively lower than the other villages of the Amethi Block. The area under the permanent pastures and grazing land has been observed 0.4 per cent and other uncultivated land excluding fallows have been noted 2.8 per cent of the total geographical area of the village panchayat Katara Hathkila. The area under the miscellaneous trees has been observed 3.7 per cent of the total area of the village. The area under fallows other than the current fallow land has been found 2.0 per cent and the current fallow land has been recorded 0.4 per cent of the total geographical area of the village panchayat Katara Hathkila of the Amethi Block.

22. Village Panchayat Rebha

The total geographical area of the village Panchayat Rebha has been recorded 99 hectares which includes Rebha and Mandharpatti revenue villages along with the Rebha, Pure Pothi and Mandharpatti purwas/tolas/hamlet village. The area specific land use system of the village Panchayat Rebha has been depicted in Map 6.36. The land not available for cultivation has been recorded 15.1 per cent and the area under the barren and uncultivated land has been founded 6.0 per cent of the total geographical area of the village Panchayat Rebha. The net cultivated land of the village Panchayat Rebha has been found 45 per cent of the total area of the village. The culturable waste lands have been observed 21 per cent of the total area of the village Panchayat Rebha. The permanent pastures and grazing lands have been noted 1.0 per cent and the area other than the uncultivated land excluding fallow land has been recorded 4.0 per cent of the total geographical area of the village Panchayat Rebha. The area under the miscellaneous trees has been recorded 1.9 per cent of the total geographical area of the village. The area under the fallows other than the current fallow land has been found 3.0 per cent and the current fallow land has been noted 3.0 per cent of the total geographical area of the village Panchayat Rebha of the Amethi Block. It has been observed that the area under the net cultivated land has been comparatively lower than the other villages of the Amethi. Therefore, re-maintaining the fertility status of the soil to increase the fertility status of the soil has become very important task to the farmers of the village and reclamation of the wastelands for agricultural activities has to be initiated in the integrated system in the village Panchayat Rebha of the Amethi Block.

23. Village Panchayat Raidaipur

The land use pattern of the village panchayat Raidaipur has been depicted in Map 6.37. The total geographical area of the village panchayat Raidaipur has been recorded 260 hectares which includes Raidpur, Chachakapur, Birahimpur revenue villages along with the Chachakapur, Birahimpur Kundwa, purwas/tolas/hamlet villages of the village panchayat Raidaipur. The land not available for cultivation has been found 8.8 per cent and the barren and uncultivated land has been recorded 3.8 per cent of the total geographical area of the village panchayat Raidaipur. The net cultivated area of the village has been recorded 44.2 per cent of the total geographical area of the village which has been found comparatively lower due to the low fertility status of the soil. The area under the culturable wastelands has been found 32.4 per cent of the total geographical area of the village which has been noted very high due to poor fertility status of the soil and an increase in the area under usar and water logging problem in the village Raidaipur. The permanent pastures and grazing land have been observed 1.4 per cent and the area under other uncultivated land excluding fallows has been found 1.9 per cent and the area under permanent trees, shrubs etc. has been recorded 2.6 per cent of the total geographical area of the village panchayat Raidaipur. The area under the fallows other than the current fallow land has been noted 2.6 per cent and the area under the current fallow land has been recorded 3.0 per cent of the total geographical area of the village panchayat Raidaipur. Attempt has to be made to reclaim the available waste lands and also to control the reduction in the net cultivated area of the village through restoring the fertility status of the soil to increase the agricultural producing of the village panchayat Raidaipur of the Amethi Block.

24. Village Panchayat Korari Girdhar Shah

The total geographical area of the village panchayat Korari Girdhar Shah as per the available revenue records of the Tehsil Amethi has been recorded 383 hectares which includes Korari Girdhar Shah, Mangalpur and Dariyapur revenue villages along with the Pure Matadin, Ghatampur, Pure Bhagwan, Mangalpur and Tewari Ka Purwa, Gohanwa, Dariyapur, Bikham Bherpur, Mahadewa, Pure Roshani, Baba Ka Purwa, Pure Talin, Pure Soschiman Manharia purwas/tolas/hamlet villages in the village panchayat Korari Girdhar Shah. The land not available for cultivation has been found 9.1 per cent and the barren and uncultivated land has been observed 4.4 per cent of the total geographical area of the village panchayat Korari Girdhar Shah. The area specific and plotwise location of the land use pattern of the village panchayat Korari Girdhar Shah has been shown in Map 6.38. The net cultivated area of the village panchayat Korari Girdhar Shah has been recorded 44.3 per cent of the total geographical area of the village. The culturable waste land has been found 2.65 per cent of the total geographical area of the village. The permanent pasture and the grazing lands have been found 2.3 per cent of total reporting area of the village. The area under the other uncultivated land excluding fallows has been noted 5.2 per cent of the total area of village. The fallows other than current fallow land has been seen 3.9 per cent and the area under the current fallow land has been recorded 3.1 per cent of the total geographical area of the village panchayat Korari Girdhar Shah of the Amethi Block.

25. Village Panchayat Dedhpasar

The area specific distribution pattern of the land use system of the village Panchayat Dedhpasar has been depicted in Map 6.39. The total geographical area of the village panchayat Dedhpasar has been noted 314 hectares land which includes Dedhpasar and Jamal Patti revenue villages along with the Tulapur, Parashuram, Dedhpasar, Sarai Kodha, Paware Dakshin, Pware Uttar, Jamal Patti purwas/tolas/hamlet villages of the village panchayat Dedhpasar. The land not available for cultivation has been recorded 12.7 per cent and the barren and uncultivated land has been noted 1.5 per cent of the total geographical area of the village panchayat Dedhpasar. New cultivated area of the village has been recorded 44.5 per cent of the total reporting area of the village. The culturable waste lands has been observed 31.6 per cent of the total geographical area of the village panchayat Dedhpasar which has been found significantly very high due to substantial increase in the poor quality and deteriorating fertility status of the soil which has not allowed proper growth of agricultural crops. The permanent pastures and grazing lands have been found 2.2 per cent and the area under uncultivated land excluding fallows has been noted 0.9 per cent of the total geographical area of the village. The area under miscellaneous trees has been found 3.1 per cent and the area under fallows other than the current fallow land has been seen 3.4 per cent of the total geographical area of the village panchayat Dedhpasar. The area under the current fallow land has been recorded 1.0 per cent to the total geographical area of the village panchayat Dedhpasar of the Amethi Block.

26. Village Panchayat Mahaso

The total geographical area of the village panchayat Mahaso has been recorded 86 per cent land which includes Mahaso and Gyanchandpur revenue villages along with Mahaso, Basuput, Tolas Dewatar and Paripur purwas/tolas/hamlet village of the village panchayat Mahaso. The distribution pattern of the land use system of the village panchayat Mahaso has been depicted in Map 6.40. The land not available for cultivation has been found 20.0 per cent of the total area of the village. The barren and uncultivated land has been recorded 3.0 per cent of the total area of the village. The net cultivated area of the village Panchayat Mahaso has been recorded 46.5 per cent of the total geographical area of the village. The area under the culturable waste lands has been found 22.6 per cent of the total area of the village panchayat Mahaso. The area under the permanent pastures and grazing lands, other uncultivated land excluding fallows and the area under miscellaneous trees etc. have been recorded 1.0 per cent each respectively of the total geographical area of the village panchayat Mahaso. The area under fallows other than the current fallow land has been noted 4.4 per cent and the area under the current fallow land has been recorded 1.0 per cent to the total geographical area of the village panchayat Mahaso of the Amethi Block.

27. Village Panchayat Gangauli

The detailed land use pattern of the village panchayat Gangauli has been depicted in Map 6.41. The total geographical area of the village has been recorded 380 hectares which includes revenue village Gangauli along with Pandit Ka Purwa, Subedar Ka Purwa, Gangauli Matera, Pure Diggaj Ka Purwa, Bari Ka Pura, Tulsi Pandey Ka Purwa, Agaha, Pure Kurai, Dina Ka Purwa and Soiya purwas/tolas and hamlet villages of the village panchayat Gangauli. The land not available for cultivation has been recorded 11.0 per cent of the total area of the village. The barren and uncultivated land has been found 5.2 per cent of the total geographical area of the village panchayat Gangauli. The net cultivated area of the village panchayat Gangauli has been recorded 46.0 per cent of the total area of the village. The area under the culturable wastelands has been noted 23.7 per cent of the total area of the village. The permanent pastures and grazing lands have been found 1.3 per cent and the other uncultivated land excluding fallows has been observed 4.2 per cent of the total geographical area of the village panchayat Gangauli. The area under miscellaneous trees, shrubs etc. has been recorded 2.6 per cent of the total area of the village. The areas under the fallows other than the current fallow land has been found 3.0 per cent and the current fallow land has been recorded 3.0 per cent of the total geographical area of the village panchayat Gangauli of the Amethi Block.

28. Village Panchayat Dhandhudhar

The area specific distribution of the land use system of the village panchayat Dhandhudhar has been shown in Map 6.42. The total geographical area of the village panchayat Dhandhudhar has been recorded 133 hectares including Dhandhudhar and Pithipur revenue villages along with Dhandhudhar, Pure Upadhyaya, Pure Hanuman Tewari, Pure Ojhas, Pure Mishra, Bangarwa purwas/tolas/hamelet villages of the village panchayat Dhandhudhar. The area under the land not available for cultivation has been seen 15.0 per cent and the barren and uncultivated land have been found 3.7 per cent of the total geographical area of the village panchayat Dhandhudhar. The net cultivated area of the village panchayat Dhandhudhar has been noted 45.1 per cent of the total area of the village. The culturable waste lands of the village panchayat Dhandhudhar has been noted 22.3 per cent of the total area of the village. The area under the permanent pastures and grazing lands have been recorded 3.1 per cent and the area under the other uncultivated land excluding fallows has been seen 2.5 per cent and the area under the miscellaneous trees, shrubs etc. has been found 2.2 per cent of the total geographical area of the village panchayat Dhandhudhar. The area under the fallows other than the current fallow land has been noted 3.7 per cent and the area under the current fallow land has been recorded 2.4 per cent of the total geographical area of the village panchayat Dhandhudhar of the Amethi Block of the district Amethi.

29. Village Panchayat Gaderi

The total geographical area of the village panchayat Gaderi has been recorded 268 hectares which includes Gaderi revenue village along with Gaderi, Sadwa, Bhanpurwa, Pure Hardatt, Pure Dutan, Pure Kalika, Pure Tula Ram, Rampur Pure Ganga, Pure Bal, Pure Bithasia, Samari, Pure Bhagwandin, Pure Hawaldar, Pure Chandra Mani purwas/tolas/hamlet villages. The land use pattern of the village panchayat Gaderi had been depicted in Map 6.43. The land not available for cultivation has been noted 11.1 per cent of the total geographical area of the village. The area under the barren and uncultivated land has been found 3.7 per cent of the total area of the village Gaderi. The area under the culturable wasteland has been recorded 24.7 per cent of the total geographical area of the village panchayat Gaderi. The area under the permanent pastures and grazing lands has been found 2.8 per cent and the other uncultivated land excluding fallow land has been noted 2.4 per cent of the total geographical area of the village panchayat Gaderi. The area under the miscellaneous trees, shrubs etc. has been observed 3.2 per cent of the total area of the village. The area under the fallows other than current fallow land has been noted 4.3 per cent and the area under the current fallow land has been recorded 3.1 per cent of the total geographical area of the village panchayat Gaderi of the Amethi Block.

30. Village Panchayat Bhusahari

The area specific distribution of the land use pattern has been depicted in Map 6.44. The total area of the village panchayat Bhusahari has been recorded 235 hectares which include revenue village Bhusahari along with the Bhusahari Khas, Barain Tola, Pure Gaderian, Karnaipur, Makdumpur, Patipur, Jhan Singh Ka Purwa tolas/purwas/hamlet villages of the Village Panchayat Bhusahari. The area under the permanent pasture and grazing lands has been found 12.4 per cent to the total area of the village. The area under barren and uncultivated land has been found 1.7 per cent of the total reporting area of the village panchayat Bhusahari. The net cultivated area of the village panchayat Bhusahari has been recorded 53.1 per cent of the total area of the village. The culturable waste lands have been noted 20.4 per cent of the total geographical area of the village panchayat Bhusahari. The area under permanent pastures and grazing lands has been seen 2.5 per cent and area under the other uncultivated land excluding fallows has been found 3.7 per cent of the total geographical area of the village panchayat Bhusahari. The area under the miscellaneous trees, shrubs etc. has been observed 2.5 per cent of the total area of the village. The area under the fallows other than the current fallow land has been noted 1.6 per cent and the area under the current fallow land has been recorded 2.2 per cent of the total geographical area of the village panchayat Bhusahari of the Amethi Block.

31. Village Panchayat Purabgaon

The total geographical area of the village Panchayat Purabgaon has been noted 222 hectares which includes Purabgaon revenue and hamlet villages. The land use pattern of the village Panchayat Purabgaon has been shown in Map 6.45. The land not available for cultivation has been noted 14.4 per cent and the barren and uncultivated land has been recorded 5.4 per cent of the total geographical area of the village panchayat Purabgaon. The area under the net cultivated land has been noted 47.2 per cent of the total geographical area of the village panchayat Purabgaon. The area under the culturable waste lands has been found 27.7 per cent of the total reporting area of the village. Permanent pastures and grazing lands have been found 0.9 per cent and other uncultivated land excluding fallows has been recorded 4.5 per cent of the total geographical area of the village panchayat Purabgaon. The area under the miscellaneous trees etc. has been seen 1.8 per cent of the total area of the village. The area under the current fallow land has been recorded 1.3 per cent of the total geographical area of the village panchayat Purabgaon of the Amethi Block of the distrit Amethi.

32. Village Panchayat Chaturbhujpur

The total geographical area of the village panchayat Chaturbhujpur has been recorded 120 hectares as per the revenue records which include Chaturbhujpur revenue village along with the Pure Shuklan, Hanumanganj, Jagat Gaon, Chaturbhujpur, Pure Harsahai and Pailepur purwas/tolas/hamlet villages of the village panchayat Chaturbhujpur. The area specific plot level locationwise land use pattern of the village panchayat Chaturbhujpur has been shown in Map 6.46. It has been observed that the land not available for cultivation has been found 20.0 per cent to the total geographical area of the village panchayat Chaturbhujpur. The barren and uncultivated land has been observed 2.5 per cent to the total reporting area of the the village. The net cultivated area of the village panchayat Chaturbhujpur has been recorded 46.6 per cent of the total geographical area of the village. The area under the culturable wastelands has been found 18.6 per cent to the total geographical area of the village panchayat Chaturbhujpur. The area under the permanent pastures and grazing lands have been seen 0.8 per cent and area under other uncultivated land excluding fallows has been noted 1.6 per cent of the total geographical area of the village panchayat Chaturbhujpur. The area under the miscellaneous trees, shrubs etc. has been observed 2.5 per cent to the total reported area of the village. The area under the fallows other than the current fallow land has been noted 2.4 per cent and the area under the current fallow land has been recorded 5.9 per cent of the total geographical area of the village panchayat Chaturbhujpur of the Amethi Block. The area under the current fallow land has been significantly increased due to poor fertility status of the soil for growing the agricultural crops due to which the cultivators has been forced to leave the land as current fallow. In view of decrease in the cultivated land attempt has to be ensured to reclaim the waste lands and to maintain the fertility status of the soil with the view to increase the agricultural productivity of the village panchayat Chaturbhujpur of the Block Amethi.

33. Village Panchayat Maharajpur

The total geographical area of the village panchayat Maharajpur has been noted 164 hectares which includes Maharajpur revenue village along with the Maharajpur Khas, Kumataran Ka Purwa and Khusadi purwas/tolas/hamlet villages of the village panchayat Maharajpur. The detailed land use pattern of the village panchayat Maharajpur has been depicted in Map 6.47. The land not available for cultivation has been found 14.6 per cent and the barren and uncultivated land has been observed 1.8 per cent to the total geographical area of the village panchayat Maharajpur. The net cultivated area of the village panchayat Maharajpur has been noted 56.7 per cent of the total area of the village which has been found proportionately significant than the other villages of the Amethi Block. The area under the culturable waste lands has been observed 12.9 per cent to the total area of the village which has been found lower than the other villages of the Amethi Block. The area under the permanent pastures and grazing lands has been seen 0.6 per cent and the area under the other uncultivated land excluding fallows has been noted 1.2 per cent of the total geographical area of the village panchayat Maharajpur. The area under the miscellaneous trees, shrubs etc. has been found 6.5 per cent of the total reporting area of the village panchayat Maharajpur. The area under the fallows other than the current fallow land has been noted 3.6 per cent and the area under the current fallow land has been recorded 2.0 per cent of the total reporting area of the village panchayat Maharajpur of the Amethi Block.

34. Village Panchayat Agahar

The area specific land use pattern of the village panchayat Agahar has been depicted in Map 6.48. The total geographical area of the village panchayat Agahar has been recorded 217 hectares which includes Agahar revenue village along with the Agahar Dih, Pure Ram Sewak, Pure Bhawani, Pure Gosnai and Pure Patoli purwas and hamlet villages of the village panchayat Agahar. The area under the land not available for cultivation has been found 10.0 per cent and the barren and the uncultivated land has been noted 5.5 per cent of the total geographical area of the village panchayat Agahar. The net cultivated area of the village panchayat Agahar has been recorded 5.16 per cent of the total reporting area of the village. The area under the culturable wastelands has been recorded 15.7 per cent of the total area of the village. The permanent pastures and the grazing land has been seen 3.5 per cent and the area under the other uncultivated land excluding fallows has been noted 1.3 per cent of the total reporting area of the village panchayat Agahar. The area under miscellaneous trees, shrubs etc. has been found 6.5 per cent of the total reporting area of the village. The area under fallows other than the current fallow land has been noted 2.2 per cent and the area under the current fallow land has been recorded 3.6 per cent of the total geographical area of the village panchayat Agahar of the Amethi Block.

35. Village Panchayat Saidpur

The total geographical area of the village panchayat Saidpur has been recorded 180 hectares which includes Saidpur, Dalshahpur revenue villages along with the Pure Shivalal, Pure Dala, Pure Vir Singh, Saidpur, Dalshahpur, Ram Nagar and Sadakpur purwas/tolas/hamlet villages of the village panchayat Saidpur. The area specific distribution of the village panchayat Saidpur has been depicted in Map 6.49. The land not available for cultivation has been recorded 16.6 per cent and the barren and uncultivated land have been found 3.3 per cent of the total geographical area of the village panchayat Saidpur. The net cultivated land has been noted 44.4 per cent of the total reporting area of the village while the area under the culturable wastelands has been recorded 21.9 per cent of the total geographical area of the village panchayat Saidpur. The permanent pastures and the grazing land have been recorded 2.7 per cent and the area under the other cultivated land excluding fallows has been found 1.5 per cent and the area under the miscellaneous trees etc. has been observed 4.4 per cent of the total geographical area of the village panchayat Saidpur. The area under the fallow other than the current fallow lands has been noted 3.1 per cent and the area under the current fallow land has been recorded 2.0 per cent of the total geographical area of the village panchayat Saidpur of the Amethi Block.

36. Village Panchayat Kakawa

The area specific plotwise location and distribution of land use pattern of the village panchayat Kakawa has been depicted in Map 6.50. The total geographical area of the village panchayat Kakawa has been recorded 368 hectares which includes Kakawa and Vidhsiya revenue villages along with the Kunwarganj, Kakawa, Pure Ganesh Baksh Singh, Pure Ram Sahai, Pure Girdhari, Shahadara, Kakaralia, Voda Nagar, Pure Valdi Tewari, Pure Gaderian, Vidhasiya, Pure Dhanduria, Pure Valdu, Pure Tada Tewari purwas/tolas or hamlet villages of the village panchayat Kakawa. The land available for cultivation has been found 8.6 per cent and the barren and uncultivated land have been seen 5.9 per cent of the total geographical area of the village panchayat Kakawa. The Net cultivated area of the village panchayat Kakawa has been recorded 43.4 per cent of the total geographical area of the village which has been significantly reduced to the detrioraing fertility status of the soil and very low agricultural productivity capability of the soil. The area under the culturable waste lands has been found 24.7 per cent of the total area of the village panchayat Kakawa. The culturable wastelands of the village Kakawa has been significantly distributed which has a deep concern to the farmers and agricultural policy plannes. It is necessary to reclaim the wastelands of agricultural activity in the village Kakawa. Permanent pastures and the grazing lands have been found 1.3 per cent and the other uncultivated land excluding fallows has been seen 5.4 per cent of the total geographical area of the village panchayat Kakawa. The area under miscellaneous trees has been observed 3.8 per cent of the total reporting area of the village Kakawa. The area under the fallows other than the current fallow land has been noted 5.3 per cent and the area under the current fallow land has been recorded 1.6 per cent of the total geographical area of the village panchayat Kakawa of the Amethi Block.

37. Village Panchayat Saraiya Duban

The total geographical area of the village panchayat Saraiya Duban has been recorded 229 hectares which include revenue village Saraiya Duban along with the Saraiya Duban, Jaggubari, Pure Lal Ram Prasad, Pure Mankunwar, Pure Brahmahari and Murarpur purwas/tolas/hamlet villages of the village panchayat Saraiya Duban. The area specific land use pattern and its distribution at plot level has been depicted in Map 6.51 for the village panchayat Saraiya Duban of the Amethi Block. The area under the land not available for cultivation has been observed 10.7 per cent and the barren and uncultivated land have been seen 5.0 per cent of the total geographical area of the village panchayat

Saraiya Duban. 43.4 per cent of the total reporting area of the village has been significantly reduced due to very poor fertility status of the soil. The area under the culturable waste lands has been recorded 25.1 per cent of the total area of the village. The area under the permanent pastures and the grazing land have been observed 1.6 per cent and the area under the other uncultivated land excluding fallow land has been found 5.0 per cent of the total geographical area of the village panchayat Saraiya Duban. The area under the miscellaneous trees, shrubs etc. has been found 5.0 per cent of the total reporting area of the village. The area under the fallows other than the current fallow land has been noted 3.7 per cent and the current fallow land has also been recorded 3.7 per cent of the total geographical area of the village panchayat Saraiya Duban of the Amethi Block of the district Amethi.

38. Village Panchayat Kohara

The area specific plot level location of the village panchayat Kohara has been shown in Map 6.52. The total geographical area of the village panchayat Kohara has been recorded 573 hectares land which includes revenue village Kohara along with Harakhpur, Dodhai Ka Purwa, Dihawan, Vishyapur, Devigarh and Kohara Khas, Ahiran Ki Saraiya, Aditya Nagar (Chamarauti), Pure Jogdhar, Ulapur, Pure Shivdas, Pure Hanuman Mishra, Pure Jagannath, Baniyan Ka Purwa, Ganga Mishra Ka Purwa, Pure Shiv Bhikh, Malik Ka Purwa, Ojha Ka Purwa, Pure Badosian, Mukunpur, Pure Ghisiya purwas/tolas/hamlet villages of the village panchayat Kohara. The area under the land not available for cultivation has been seen 9.0 per cent and the barren and uncultivated land have been found 3.8 per cent of the total geographical area of the village panchayat Kohara. The net cultivated area of the village panchayat Kohara has been recorded 47.9 per cent of the total reporting area of the village panchayat Kohara. The area under the culturable wastelands has been observed 24.3 per cent of the total area of the village panchayat Kohara. The area under permanent pastures and the grazing land has been observed 2.0 per cent and the area under the other uncultivated land excluding fallows has been noted 2.0 per cent of the total geographical area of the village panchayat Kohara. The area under the miscellaneous trees etc. has been found 3.6 per cent of the total area of the village panchayat Kohara. The area under the fallow other than the current fallow land has been noted 3.7 per cent and area under the current fallow land has been recorded 3.7 per cent of the total geographical area of the village panchayat Kohara of the Block Amethi.

39. Village Panchayat Mahmadpur

The total geographical area of the village panchayat Mahmadpur has been recorded 445 hectares land including revenue village Mahmadpur along with the Mahmadpur, Shankarpur, Dudhai Ka Purwa, Kalu Ka Purwa, Sitaram Nagar, Chaubepur, Kurmi Ka Purwa, Pure Baksh Singh Hulasi Ka Purwa, Bhawani Baksh Ka Purwa, Pure Pandit ka Purwa and Pure Hari Baksh purwas/tolas/hamlet villages of the village panchayat Mahmadpur. The area specific distribution of the land use pattern has been depicted in Map 6.53. The land not available for cultivation has been recorded 8.9 per cent and the barren and uncultivated land have been found 4.4 per cent of the total geographical area of the village panchayat Mahmadpur. The net cultivated area of the village panchayat Mahmadpur has been recorded 47.1 per cent of the total reported area of the village. The culturable wasteland has been found 24.2 per cent of the total geographical area of the village panchayat Mahmadpur. The area under permanent pastures and the grazing land have been seen 2.7 per cent and the area under uncultivated land excluding fallows other than current fallow land has been observed 3.3 per cent of the total geographical area of the village panchayat Mahmadpur. The area under miscellaneous trees, shrubs etc. has been found 3.7 per cent of the total area of the village. The area under the fallows other than the current fallow land has been noted 2.8 per cent and the current fallow land has been recorded 3.0 per cent of the total geographical area of the village panchayat Mahmadpur of the Amethi Block.

40. Village Panchayat Naraini

The area specific distribution of the land use pattern of the village panchayat Naraini has been depicted in Map 6.54. The total geographical area of the village panchayat Naraini has been recorded 584 hectares land including Naraini revenue village along with the Narainidih, Pure Pachhim, Pure Matlu, Pure Walta, Dai Ka Purwa, Pure Gangeran, Pure Ayodhya and Pure Khojhawa tolas/purwas/hamlet villages of the village panchayat Naraini. The land not available for cultivation has been seen 9.9 per cent and the barren land and uncultivated land have been found 3.4 per cent of the total geographical area of the village panchayat Naraini. The net cultivated area of the village panchayat Naraini has been recorded 47.8 per cent of the total area of the village. The area under the culturable waste lands has been noted 23.6 per cent of the total area of the village panchayat Naraini. The area under the permanent pastures and the grazing land has been seen 3.1 per cent and the area under other uncultivated land excluding fallows have been observed 2.5 per cent of the total geographical area of the village panchayat Naraini. The area under the miscellaneous trees, shrubs etc. has been found 2.5 per cent of the total area of the village. The area under the fallows other than the current fallow land has been noted 4.4 per cent and the area under the current fallow land has been recorded 2.7 per cent of the total geographical area of the village panchayat Naraini of the Amethi Block.

41. Village Panchayat Nainaha

The total geographical area of the village panchayat Nainaha has been recorded 182 hectares land which includes Nainaha, Bastali and Rajapur Kohara revenue villages and including Nainaha Khas, Pure Tewari, Pure Bhisen, Danapur, Pure Jangali, Damgarha, Rajapur, Ajayab Sinda and Sadarpur purwas/tolas/hamlet villages of the village panchayat Nainaha. The land use pattern of the village panchayat Nainaha has been depicted in Map 6.55. The area under the land not available for cultivation has been found 13.7 per cent and the barren and uncultivated land has been observed 3.2 per cent of the total geographical area of the village panchayat Nainaha. The net cultivated area of the village panchayat Nainaha has been recorded 49.4 per cent of the total reporting area of the village panchayat Nainaha. The area under the culturable wastelands has been found 23.5 per cent of the total geographical area of the village panchayat Nainaha. The area under the permanent pastures and the grazing lands has been observed 2.7 per cent and the area under the other uncultivated land excluding fallows has been observed 2.1 per cent of the total geographical area of the village panchayat Nainaha. The area under the miscellaneous trees, shrubs etc. has been observed 1.6 per cent of the total reported area of the village panchayat Nainaha. The area under the fallows other than the current fallow land has been observed 2.0 per cent and the area under the current fallow land has been recorded 1.8 per cent of the total geographical area of the village panchayat Nainaha. It has been observed in general that there is decrease in the net cultivated area and an increase in the area under the culturable waste lands has been significantly noted not only in the village panchayat Nainaha but in most of the villages of the Amethi.

42. Village Panchayat Ramgarh

The area specific distribution of the land use pattern of the village panchayat Ramgarh has been depicted in Map 6.56. The total geographical area of the village panchayat Ramgarh has been recorded 658 hectares land which includes Ramgarh revenue village along with the Badi Soiya, Chhoti Soiya, Pure Bhoop, Rampur, Bahni, Maloka, Sarriya Purwa, Pure Kotwa and Ramgarh Khas hamlet villages/purwas/tolas of the village panchayat Ramgarh. The land not available for cultivation has been found 10.9 per cent and the barren and uncultivated land has been observed 4.4 per cent to the total geographical area of the village panchayat Ramgarh. The net cultivated area of the Village Panchayat Ramgarh has been recorded 44.8 per cent of the total geographical area of the village. The area under the culturable wastelands has been found 21.5 per cent of the total reporting area of the village panchayat Ramgarh. The area under the permanent pastures and the grazing land has been found 3.4 per cent and the area under the other uncultivated land excluding fallow land has been recorded 3.6 per cent of the total reported area of the village panchayat Ramgarh. The area under the miscellaneous trees, shrubs etc. has been noted 3.9 per cent of the total area of the village panchayat Ramgarh. The area under the fallows other than current fallow land has been observed 5.5 per cent and the current fallow land has been recorded 2.0 per cent of the total geographical area of the village panchayat Ramgarh of the Amethi Block of the district Amethi.

43. Village Panchayat Himmatgarh

The location and area specific distribution of the village panchayat Himmatgarh has been depicted in Map 6.57. The total geographical area of the village panchayat Himmatgarh has been recorded 340 hectares land which includes the revenue village Himmatgarh along with the Pure Bhoop, Madaripur, Pure Ajit, Jhali, Visheshganj, Pure Upadhyay purwas/tolas/hamlet villages of the village panchayat Himmatgarh. The land not available for cultivation has been seen 18.8 per cent and the barren and uncultivated land has been found 1.1 per cent of the total geographical area of the village panchayat Himmatgarh. The net cultivated area of the village panchayat Himmatgarh has been recorded 49.0 per cent of the total geographical area of the village. The area under the culturable waste lands has been observed 22.2 per cent of the total reporting area of the village panchayat Himmatgarh. The area under the permanent pastures and the grazing land has been noted 2.0 per cent and other uncultivated land excluding fallows has been recorded 0.5 per cent of the total geographical area of the village panchayat Himmatgarh. The area under the miscellaneous trees etc. has been observed 2.0 per cent of the total reporting area of the village panchayat Himmatgarh. The area under the fallow other than the current and the area under the current fallow land has been recorded 0.4 per cent of the total geographical area of the village panchayat Himmatgarh.

44. Village Panchayat Dehara

The total geographical area of the village panchayat Dehara has been recorded 235 hectares land which includes revenue villages Dehara and Kudwa Mahmadpur along with the Dehara Khas, Pure Darjiyan, Pure Shiv Pal, Pure Pandey, Shanichara, Gulabgarh, Shukla Ka Purwa, Daya Garha, Pure Jhingur, Kudwa, Mahmadpur, purwa/ tolas/hamlet villages of the village panchayat Dehara. The area specific and plotwise location of the village panchayat Dehara has been depicted in Map 6.58. The land not available for cultivation has been observed 13.6 per cent and the barren and uncultivated land have been observed 2.5 per cent to the total geographical area of the village panchayat Dehara. The proportionate distribution of the net cultivated area has been found 53.6 per cent of the total geographical area of the village panchayat Dehara. The area under the culturable waste lands has been noted 23.0 per cent of the total reported area of the village panchayat Dehara. The area under permanent pastures and the grazing lands has been observed 0.8 per cent and the area under the other uncultivated land excluding fallows has been found 1.7 per cent of the total geographical area of the village panchayat Dehara. The area under the miscellaneous trees, shrubs etc. has been observed 1.2 per cent of the total geographical area of the village panchayat Dehara. The area under the fallows other than the current fallow land has been noted 2.1 per cent and the area under the current fallow land has been recorded 1.5 per cent of the total geographical area of the village panchayat Dehara.

It has been recorded that there is subsequent decrease in the area under the cultivation on the one hand while there is subsequent increase in the wasteland area due to very poor fertility status of the soil leading towards very poor agricultural productivity due to which the farmers has been forced to leave the cultivated land out of cultivation which has led to the environmental crisis as well as socio-economic crisis for the people of this area. Thus it has become an important task to plan scientifically for the reclamation of available waste lands for Agro-afforestation to meet the increasing demand of the cattle and human population for the fuel wood, fodder, fibre, fruits, fisheries, food grains (F^6), wood and vegetables. Attempts should be initiated to control the growth in the wastelands and to maintain the fertility status of the soil to increase the agricultural productivity. Adoption of the techno-scientific process of the waste lands reclamation measures for water logged lands, banjar lands, usar lands, kankarili lands, old fallow lands, fallow lands and other types of waste lands must be initiated and implemented at war-footing level through ensuring people's participation to make the reclamation process of waste lands more successful. Appropriate irrigation system should be developed and excess water must be drained out to control the water logging problem of the proper soil testing must be carried out to know the NPK fertilizer requirements of the soil to produce the agricultural crops. Requisite varieties of high yielding varieties crops during the Kharif, Rabi and Jaid cropping seasons should be sown which can tolerate the salt and grow proper agricultural crops. Proper application of the insecticides and pesticides should be ensured as per the requirement of the crops. Agri-Horti-Social-Forestry programmes must be propagated to fulil the increasing demand of ever increasing population. No doubt the various villages of the Amethi are endowed with the natural infrastructural resources but techno-scientifc appropriate management of these natural resources will be an asset for an integrated development of Amethi which will ensure environmental conservation and maintenance of the eco-system of the area and also improving the socio-economic crisis of the people of the Amethi.

5. CONCLUSION

The land is a free gift of the nature like *Jal*, *Jungle* and *Jameen* (water, forest and land). The land has much importance and significance among the natural resources. These provides space for economic and other activities whether they are related to primary, secondary or tentiary sector of economy. Everything we use, can be ultimately innovated on the land.

"Life without land is unconceivable"

The nature has gifted the land with unique life sustaining potential which has made it invaluable and indispensable for the various ecological needs.

Agricultural census and National Bureau of Soil Survey and Land Use Planning, Government of India considered forest land, land not available for cultivation, other uncultivated land excluding fallows, fallow land and net cultivated land which has been re-classified as per the needs and suitability of revenue system.

Village level revenue record system including village Khasara Register, Khasara Map, land Khetauni, and other land revenue records has been the basic source of the land use system and natural infrastructure of the various village panchayats of Amethi.

The land use pattern of the district Amethi indicates that the forest area has been found 0.4 per cent, and permanent pastures and the grazing land has been noted 0.3 per cent of the total geographical area of the district. The current fallow land has been recorded 8.8 per cent and the fallow land has been noted 9.7 per cent of the total area of the district. The culturable wastelands has been observed 3.6 per cent of the total area of the district. The area under the miscellaneous trees etc. has been noted 1.9 per cent of the total area of the district. The land not available for cultivation has been observed 21.0 per cent to the total area of the district. The barren and uncultivated land shares 10.9 per cent of the total area of the district. The net cultivated area of the district has been recorded 53.8 per cent of the total geographical area of the district Amethi.

There are 44 village panchayats, 82 revenue villages and 377 purwas/tolas/hamlet villages in the Amethi Block. The land not available for cultivation has been noted 10.66 per cent and the barren and uncultivated land has been observed 3.47 per cent of the total geographical area of the Amethi Block. The net cultivated area of the Amethi Block has been recorded 6109 hectares which contributes 47.90 per cent land to the total geographical area of the Amethi Block. The area under the culturable waste lands contributes 24.87 per cent of the total geographical area of the Amethi Block. The area under permanent pasture contributes 1.57 per cent and other uncultivated land excluding fallows shares 2.17 per cent of the total reporting area of the Amethi Block. The area under miscellaneous trees shares 2.45 per cent and the area under the fallows other than the current fallow land has been noted 4.53 per cent and the area under the current fallow to the total geographical area of the village panchayat Ramdaipur, Mahso and Chaturbhujpur has been noted while very low percentage less than 8 per cent are under the land not available for cultivation has been recorded in the Trilokpur, Jangal Ram Nagar, Village panchayats of the Amethi Block.

Very low percentage less than 2 per cent area under the barren and uncultivated land has been observed in the Ramdaipur, Raipur Phulwari, Kushi Tali, Dedhpasar, Bhusahari, Maharajpur and Himmatgarh village panchayats while very high percentage more than 6.0 per cent has been recorded in the village panchayat Rebha of the Amethi Block.

The proportionate distribution of the net cultivated area having high percentage more than 50.0 per cent has been observed in the Trilokpur, Jangal Ram Nagar, Loharata,

Darakha, Nuwanwa, Bhusahari and Dehra village panchayats while very high more than 55.0 per cent net cultivated area has been noted in the Hathkila and Maharajpur village panchayats. The very low percentage less than 45.0 per cent area under net cultivated land has been observed in the Sarai Khema, Umapur Gana Patti, Loniapur, Kherauna, Katara Phulkunwar, Raidaipur, Korari Girdhar Shah, Dedhpasar, Gaderi, Saidpur, Kakawa, Saraiya Duban and Ramgarh village panchayats of the Amethi Block has been recorded 1.96 per cent of the total geographical area of the Amethi Block.

It has already been analyzed that there are 44 village panchayats and 82 revenue villages and 377 purwas/tolas/hamlet villages in the Amethi Block. It has been found very difficult to analyze the natural resources including land use pattern categorically for all 82 revenue villages and 377 purwas/tolas/hamlet villages adopting the revenue khasara maps as a base map for area specific description of the land use pattern. It is known fact that Village Panchayats, Block Panchayats and District Panchayats are the three-Tier system of strengthening the Panchayati Raj Institutions. At the grassroots level village panchayats has been taken as lowest/grassroots level rural development units. In view of this village panchayats of the Amethi have been taken as a base to analyse the area specific distribution of natural resources including land use pattern and other socio-economic-institutional structures and development planning in Amehti.

It has been observed that proportionately very high percentage more than 20 per cent of the Amethi Block.

Very high percentage of the culturable wastelands more than 30.0 per cent has been observed in the Darakha, Katara Phulkunwar, Mochawa, Dedhpasar and Raidaipur village panchayats while very low percentage less than 15.0 per cent of the area under the culturable wastelands has been recorded in the Ramdaipur village panchayat of the Amethi Block.

It has been observed that the very high percentage more than 4.0 per cent area under the permanent pastures and the grazing land has been noted in the village panchayat Tala and Loniapur while very low percentage less than 1.0 per cent has been recorded in the village panchayat Raipur Phulwari, Umapur Gana Patti, Kushi Tali, Darakha, Hathkila, Purabgaon, Chaturbhujpur, Maharajpur and Dehara of the Amethi Block.

Other uncultivated land excluding fallows has been found very high more than 4.0 per cent of the total geographical area in the Kherauna, Rebha and Saraiya Duban village panchayats while very low percentage area less than 1.0 per cent has been observed in the Jangal Ram Nagar, Kushi Tali, Dedhpasar and Himmatgarh village panchayats of the Amethi.

It has been noted that the area under the miscellaneous trees etc. has been found very high more than 6.0 per cent in the Jangal Ram Nagar and Agahar Village Panchayat while very low more than 1.0 per cent area under the miscellaneous trees etc. has been recorded in the Darakha, Mochawa village panchayats of the Amethi Block.

It has been observed that the very high per centage 10.0 area under the fallows other than the current fallow land has been recorded in the Jangal Ram Nagar village panchayat while very low percentage less than 2.0 per cent area under the fallows other than current fallow land has been noted in the village panchayat Bhusahari of the Amethi Block.

Proportionate distribution of the area under the current fallow land more than 6.0 per cent area has been noted in the village panchayat Bhaganpur while very low percentage less than 1.0 per cent area under the current fallow land has been recorded in the Trilokpur, Jangal Ram Nagar, Raipur Phulwari, Kushi Tali, Kherauna, Hathkila and Himmatgarh village panchayats of the Block Amethi of the district Amethi.

Thus, it has been observed that there is significant decrease in the area under the net cultivated area or the net sown area in most of the village panchayats of the Amethi Block. Subsequently, there is increase in the area under the culturable wastelands and the area under the uncultivated in the most of the village panchayats of the Amethi Block of the District.

Misconception of over irrigation, excess doses of fertilizer and insecticides and pesticides, careless over cultivation etc. has led to deterioration in the fertility status of the soil of the area. Appropriate scientific testing of the soil and as per the soil requirement the application of organic and inorganic amendments and use of NPK fertilizer and insecticides and pesticides etc. may be helpful in controlling the deterioration of the soil fertility status of the soil.

Thus, it has become an important task to the cultivators and agricultural policy makers and planners including the various agricultural implementing agencies to control the deterioration of the fertility status of the soil and to apply organic and inorganic amendments as the requirements of the soil. To ensure proper drainage system of the water logged areas especially to drain out the excess accumulated water. To yield high salt tolerant varieties of the agricultural crops which can easily grow in the salt affected land. To ensure high yielding varieties of the crops.

Proper techno-scientific demonstration of the reclamation of wastelands must be ensured with the continued participation of the farmers. Appropriate scientific techniques should be demonstrated to the farmers for agricultural activities including the systematic cropping system and crop rotation system. The requisite agro-forestry, horticulture and social-forestry programmes should be practically demonstrated to the farmers as per the fertility status of the soil to meet the increasing demand of the ever growing cattle and human population for the fuel wood, fodder, fibre, fruits, fisheries, food grains (F^6), wood and vegetables. The excessive and over exploitation of the available land, Agri-Horti-Social Forestry has to increase to check land degradation and control deforestation and to maintain the eco-system and to improve environmental conditions of the various villages of the Amethi.

7
Wastelands

1. INTRODUCTION

There is symbiotic relationship between man and nature from immemorial time because man has lived in the harmony with nature. The nature gave its bounty to the people and was, in return recompensed adequately by the self re-generating process of eco-development, built into the cultural, socio-economic traditions of the human life. During the recent past, regrettably, the relationship between the man and nature has been shattered causing incalculable damage to the nature and considerable hardship to the man.

The problem of the wastelands has been appropriately documented. The National Commission on Agriculture stated in its report of 1972 that as much as 175 million hectares out of the 266 million hectares which are available for agricultural activities, are wastelands. The explosive increase of the population and increasing demand of the people to exploit marginal and sub-marginal lands for a medium of returns and unmindful further degradation by the people which causes pressure from industry for raw material requirements, defective land use system and cropping patterns and in-efficient water management has emerged to as the main factors. The result is, multiple and escalating damage to natural resources, due to both by over-exploitation and mismanaged agricultural practices.

It has been observed that there has been alarming challenge for the loss of our forest cover. Out of the 75 million hectare forest area 40 million hectares forest land has been degraded. Our nation is loosing 1.5 million hectares of forest land and 12000 million tonnes of top soil is eroded every year due to deforestation and soil erosion and run off. The rehabilitation process is indeed very hard and laborious. It takes anything from 500 to 1000 years to restore one inch of top soil and up to 100 years to re-establish a good natural forest.

Therefore, it is imperative that the rehabilitation process be taken up on a war-footing through mobilizing every conceivable natural resources and adopting imaginative strategies of the reclamation of the wastelands. The traditional rights of the landless resourceless deprived rural poor multitudes depend much more on the forest produce, to fodder grasses, and fallen dry wood for fuel have kept the rural communities going through the centuries. It seems that the relationship between man and the nature now stands threatened. The legal and otherwise restrictions, increase, its results in an innovated invitation to the rural poor to cause further degradation, simply because they have no alternative. It has been realized that successful afforestation cannot be achieved only by the government implementing agencies alone. The main targets of afforestation have to be perceived as the individual's goals. Therefore, such an endeavour is rendered more difficult because the needs of the community and individual's are at variance with each other. It cannot be reconciled unless poverty amelioration programmes raise the level of living of those surviving below the poverty line. There has been a slackening of afforestation and absence of the support of the local communities for the plantation, its protection and augmentation of the local forests.

It has been noted that various, scientific, research and development agencies has also been to direct their energies and skills towards result-oriented research on improving the ecology, ecosystem and tackling the problems of afforestation on wastelands. Tissue culture techniques for quick multiplication of seedlings work on location specific species has to be adopted at larger extent.

In view of the above observations, everyone has to strive to create a true consciousness amongst the entire people of the area. The urgency of the problems of the wastelands and its reclamation process must be taken as a challenge to meet the increasing demand of growing cattle and human population for **fuel wood, fodder, fisheries, food grain, fibre, fruits** (F^6), wood and vegetables etc. and to restore ecological imbalances and environmental degradation of the area and to improve environmental crisis and also to improve socio-economic crisis and to ensure eco-friendly sustainable development.

2. WASTELANDS

The wastelands are not a new phenomenon. The term wastelands has been used for common lands usually for the less fertile soil which failed to yield crops and to give any return to the medieval farmers. The common man may define the wastelands as the land lying uninhabited and uncultivated and the such land is no longer serving any purpose. Now, the abjective waste has been disappeared because in many cases these Common lands are much valued as open spaces. The lands which has been left out of cultivation for one to two years to maintain the fertility status of the soil which has been termed as fallow land after two to five years if not re-cultivated, as old fallow land and after a successive five years if the same land is not being brought under cultivation such uncultivated land has been turned as Banjar land.

The wastelands has been appropriately defined as the land which has been previously used for cultivation or other agri-horti-social-forestry activities but which has been abandoned and no further use has been found for it.

A broad definition of the wastelands would be "**Land which is uncultivable for it is lying uncultivated but due to different constraints but it has been used previously, which is giving very low actual return, i.e. has low economic potential, is ecological unstable, or whose top soil has completely lost its fertility status, which has developed toxicity and is, therefore, unfit for the growth of crops and trees due to environmental or anthropogenic problems, has been abandoned and no further use has been found for it.**"

The land which could not produce the bio-mass as per the soil and the water availability, such land may be defined as wastelands.

The land which is lying unuseful due to improper care to maintain the fertility status of the soil and the land is lying vacant or due to excessive grazing, soil erosion, land degradation and due to very poor fertility status of the soil and in such lands the agricultural cropping expenditure has been higher than the profit accrued out of it, such lands has been termed as wastelands.

The land which is unuseful due to improper agricultural activities and mismanagement or due to over-grazing, land degradation and due to salinity/alkalinity the land is non-profitable and the cost of the agricultural cropping is higher than the productivity, such land has been termed as wastelands.

The land which is ecologically stable, which is eroded and its agricultural productivity capability provided very low productivity has been termed as wastelands.

The land, whose ecological system is not in accordance with the production capability, may be called as wastelands.

The land which has lost its agricultural production capability due to various reasons including due to improper agricultural practices, mismanagement and in which the cost of production is higher than the profit accrued out of it, such types of land may be termed as wastelands.

3. TYPES OF THE WASTELANDS

The wastelands classification are being done on the point of view of its utilization, which are useful in the different forms of application and the cultivation of the such lands.

The classification of the wastelands has been derived on the basis of the land use classification proposed by the Directorate of Economics and Statistics, Ministry of Agriculture, Government of India. The culturable wastelands of Amethi have been classified into the following broad categories from the land use categories.

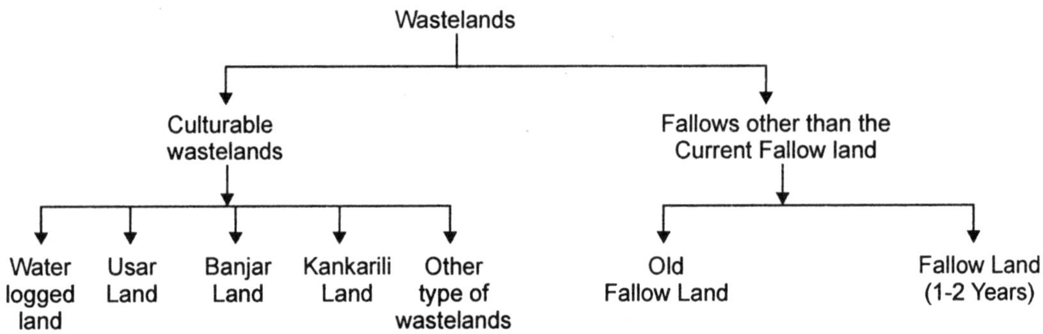

Fig. 7.13: Wastelands

The culturable wastelands has been divided into water logged land, usar land, banjar land, kankarili land and other type of wastelands while the fallows other than current fallow land has been further subdivided into the old fallow land (2-5 years) and fallow land (1-2 years).

1. **Kankarili Land:** The land which has excess calcariousness in the soil and hard kankar pan is observed below the top soil upto a depth of one metre, which does not allow water retentivity in the soil and provide hindrance in the growth of agricultural crops such lands have been termed as Kankarili land.
2. **Usar Land:** The fluffy soils which have a whitish or grayish appearance deposited on the surface and are not easily cultivable. Such lands are commonly known as Reh or Rehar or Usar lands.
   The land which have excessive proportions of soluble salts due to which crop production has been seriously hampered are known as salt affected soils and locally such lands are known as usar land.

3. **Water Logged Land**: Waterlogged land is that land in which water table is at or near the earth's surface and water stands for most of the year and soil pores in the root zones of the crop and get saturated with water which inhibits the growth and germination activity of plants or crops, as a result such lands are left out of cultivations and termed as water logged land.
4. **Banjar Land**: The land which has been under cultivation but are presently left out of cultivation for a period of not less than five years. Such lands are left uncultivated for a long period are termed as Banjar Land.
5. **Old Fallow Land**: The land which has been under cultivation but presently left out of cultivation for a period of not less than two years and not more than five years may be termed as Old Fallow land.
6. **Fallow Land**: The land which cannot be taken up for cultivation but temporarily left out of cultivation for not less than one year and not more than two years due to poor fertility status of the soil, such lands are termed as Fallow land.
7. **Other types of wastelands**: The other types of wastelands include such lands which cannot as definition be put under any other above types of wastelands. Such lands are mentioned in village land book. The other types of wastelands are those lands which has been under cultivation but has reason or the other reasons or due to deteriorated fertility status of the soil. Such lands are termed as other types of wastelands.

4. WASTELANDS OF THE DISTRICT

The wastelands of the district Amethi has been proportionately distributed but the problems of water logging and usarization has been significantly recorded which has hampered the agricultural cropping pattern of the district.

Fig. 7.14: Percentage of Wastelands to geographical area

Sl. No.	*Wastelands*	*%age of total geographical area.*
1	Kankarili Land	1.98
2.	Banjar Land	4.87
3.	Usar Land	8.32
4.	Water logged land	8.86
5.	Old Fallow Land	2.67
6.	Fallow Land	2.19
7.	Other types of wastelands	6.31

It has been observed that the Kankarili land contributes 1.98 per cent area of the total geographical area of the district.

The area under the Banjar land has been recorded 4.87 per cent of the total geographical area of the district.

It has been noted that the usar land significantly contributes 8.32 per cent land of the total geographical area of the district.

The water logged land has significantly contributed 8.86 per cent of the total geographical area of the district.

The old fallow land has been noted 2.67 per cent of the total geographical area of the district.

The fallow land contributes 2.19 per cent land of the total geographical area of the district Amethi.

The other types of wastelands has been recorded 6.31 per cent of the total geographical area of the district Amethi.

The water logging and usarization has been recorded as the acute problem of the district which has caused poor agricultural productivity due to the poor fertility status of the soil.

5. WASTELANDS OF AMETHI

The wastelands of the Amethi has been very serious problems of the Amethi. The salinity/alkalinity and the water logging has been noted as acute problems for the cultivators of the Amethi Block. The water logged land and usar lands are significantly distributed in the various village panchayats of the Amethi Block. The proportionate distribution of the various categories of wastelands to the total geographical area of the respective village panchayats and wastelands percentage of the total area under wastelands has been collected through revenue records, collated and computed and analyzed at the village panchayat level and also for the Amethi Block of district Amethi.

1. Kanarili Lands

The kankarili land has been recorded very insignificantly in the various village panchayats of the Amethi Block. The area under the kankarili wastelands has been noted 5.65 per cent of the total wastelands area of the Amethi Block, while the kankarili land has been recorded 1.87 per cent land of the total geographical area of the Amethi Block, district Amethi.

2. Banjar Lands

The banjar lands are such lands which are left out of cultivation for more than five years due to very low fertility status of the soil. It has been observed that there are 14.88 per cent banjar land of the total wastelands of the Amethi Block, while the banjar land has been recorded 4.67 per cent of the total geographical area of the village panchayat of the Amethi Block.

3. Usar Lands

The reality, the usar land has been noted as the acute problem of the agricultural farmers of the Amethi Block. It has been very famous and well know historical line of the Amethi that "if there would not have been usar land in Amethi, then the peoples of the Amethi would have been next to the God." It means that the usar land had been historically noted and the problem of usar land is not a current problem of the district. Several attempts have been made to reclaim the usar lands of the Amethi which could not get success to achieve the requisite goals and its objectives due to some reason or the other. The problem of usar land has been found still challenging task not only to the agricultural farmers but also to the agricultural policy makers and planners of the area. It has been observed that the usar land has been found significantly distributed in most of the village panchayats of the Amethi Block. The usar lands of the Amethi Block has been recorded 22.51 per cent of the total wastelands of the Amethi Block, while the usar land of the

Amethi Block has been noted 7.45 per cent of the total geographical area of the Amethi Block.

4. Waterlogged Land

The water logged land has been found very serious problem of the Amethi. Due to excess rainy water and excess canal irrigation water which get accumulated in the low lying areas has created hindrance in the growth of agricultural crops. The problems of the water logging has been noted in most of the village panchayats of the Amethi Block. It has been recorded that there are 22.16 per cent water logged land of the total wastelands of the Amethi Block, while the proportionate distribution of the wastelands in the Amethi Block has been recorded 7.53 per cent of the total geographical area of the Amethi Block of the district Amethi.

5. Old Fallow Land

The land which has been left out of cultivation for more than two years and less than five years due to very poor fertility status of the soil has been termed as old fallow land. The area under the old fallow land has been noted 6.63 per cent of the total wastelands area of the Amethi Block, while the area under the old fallow land has been recorded 2.19 per cent of the total geographical area of the Amethi Block.

6. Fallow Land

The fallow land is such land which has been left out of cultivation due to very low and deteriorated fertility status of the soil for more than a year and less than two years. The area under the fallow land has been found 6.89 per cent area of the total wastelands of the Amethi Block. The proportionate distribution of the fallow land has been recorded 2.82 per cent land of the total geographical area of the Block Amethi.

7. Other Types of Wastelands

The area under the other types of wastelands has been significantly distributed in most of the village panchayats of the Amethi Block. Other types of lands are noted as such agricultural lands which are lying along with the rail, road, canal and other sides due to some reason or the other but agricultural activities along with the Horticultural and Social-Forestry Programmes could be initiated to conserve the available wastelands and also to improve the degraded eco-system and environmental conditions of the area. The area under the other types of wastelands has been significantly distributed in most of the village panchayats of the Amethi Block. It has been noted that there are 22.6 per cent other types of wastelands of the total area under the wastelands of the Amethi Block. The other types of wastelands has been recorded 7.51 per cent land of the total geographical area of the Amethi Block. It has been observed that the other types of the wastelands has also been an acute problem of the farmers because it contributes significant share than the various other categories of the wastelands of the Amethi Block. Therefore, an attempt must be initiated to reclaim the other types of wastelands at war footing in an integrated manner through ensuring the people's participation and ensure the agro-afforestation on other types of wastelands as people's participatory programmes and people's movement to ensure successful implementation of the plantation programmes.

8. Wastelands

The wastelands of the Amethi Block has been a great problem to the people of the Amethi. It has been noted that there are 4223 hectares wasteland in the Amethi Block covering all 44 village panchayats and 82 revenue villages and 377 purwas/tolas/hamlet villages of the Amethi Block. The area under the total wastelands has been critically realized by the local people and also the agricultural implementing agencies agricultural policy makers and planners because the total wastelands of the Amethi Block alone contributes one-third area of the total geographical area of the Amethi Block. It has been observed that there are 33.11 per cent wastelands in the Amethi Block to the total geographical area of the Block Amethi. The one-third significant proportionate distribution of the wastelands has been great challenge to the people of the Amethi. An attempt must be initiated and ensured to reclaim the one-third area under the wastelands especially to maintain the ecological imbalances of the area and also to check the environmental degradation. It has become an important task to reclaim the wastelands to improve the environmental conditions of the Amethi in an eco-friendly system through people's involvement and participation.

6. DISTRIBUTION OF WASTELANDS AT VILLAGE PANCHAYAT LEVEL

The substantial increase in the area under the wastelands at village level in Amethi has been a great challenge to the people of Amethi from very long time. The usar lands of the Amethi has been very well known critical problem of the farmers from the historical time. The salinity/alkalinity problem of soil and existence of the calcareousness in the soil creates hindrance in the growth of the plants and agricultural crops.

The high dose of irrigation and excess irrigation water including rain water gets accumulated in the low-lying areas and it has no proper drainage system to drain out the excess accumulated water also creates problems in the agricultural activities. The water logging has also been an acute problem of the farmers of the Amethi.

The existence of the calcareousness and kankar pan upto a depth of one metre of the soil and emergence of the kankarili soil does not permit the water retentivity and underground water discharge due to which the plants and the agricultural crops do not grow.

The problem of the banjar land has been also a critical problem of the Amethi. The farmers oftenly leave the land fallow due to poor fertility status of the soil and rotation in the cropping pattern. Due to the techno-scientific ignorance of economic hardship the farmers are unable to re-maintain the fertility status of the soil for more than five years the such left out land lying uncultivated are termed as banjar land.

The old fallow lands are left out of cultivation for more than two years but less than five years with the view to maintain the fertility status of the soil and ensuring the crop rotation, such left out lands uncultivated has been termed as old fallow land and the fallows other than the current fallow land category of the land use pattern of the revenue records.

The land which has been left out uncultivated with the view to maintain the fertility status of the soil for more than one year but less than two years by the farmers, such

Table 7.3: Waste Lands (Amethi Block)

(Area in Hectares)

Sl. No.	*Name of the Village Panchayat*	*Kankarili Land*	*Banjar Land*	*Usar Land*	*Water Logged Land*	*Old Fallow Land*	*Fallow Land*	*Other Types of Waste land*	*Total Waste Lands*
1	*2*	*3*	*4*	*5*	*6*	*7*	*8*	*9*	*10*
1	Katra Maharani	1	4	7	10	2	4	8	36
2	Sarai Khema	10	23	36	42	16	14	47	188
3	Trilokpur	2	16	18	24	10	10	36	116
4	Ramdaipur	1	2	4	6	1	1	8	23
5	Jangal Ram Nagar	20	43	88	89	72	62	100	474
6	Benipur	2	5	10	8	5	1	4	35
7	Parsawan	5	7	20	25	12	10	36	115
8	Mahmoodpur	6	8	18	19	11	9	36	107
9	Raipur Fulwari	2	11	18	22	10	8	38	109
10	Umapur Ganapatti	4	10	24	20	5	8	40	113
11	Bhaganpur	1	8	11	7	1	4	10	42
12	Tala	1	7	17	11	2	5	6	49
13	Kushi tali	4	10	23	30	5	8	38	118
14	Loharta	1	4	5	3	1	1	2	17
15	Darkha	4	14	27	29	5	6	24	109
16	Mochwa	2	9	14	18	2	4	15	66
17	Nuwawan	3	14	23	50	8	4	21	123
18	Loniapur	1	6	8	10	3	1	5	34
19	Kherauna	2	9	13	12	2	2	18	58
20	Katra Phulkunwar	2	10	15	12	3	4	10	56
21	Hathkila	2	8	15	15	2	2	6	50
22	Rebha	1	5	8	10	1	1	4	30
23	Raidaipur	4	16	22	32	4	3	22	103
24	Korari Girdhar Shah	42	22	27	33	5	10	30	139
25	Dedhpasar	2	13	16	25	6	5	41	108
26	Mahaso	1	4	8	7	1	2	3	26
27	Gangauli	10	22	32	33	5	3	20	125
28	Dhadhudhar	2	8	15	8	1	4	4	42
29	Gaderi	6	12	21	15	4	8	22	88
30	Bhusahari	2	10	14	12	1	3	15	57
31	Purabgaon	2	12	22	13	1	4	12	66
32	Chaturbhujpur	1	4	10	8	2	2	2	29
33	Mahrajpur	1	3	5	8	1	5	2	25
34	Aghar	2	11	18	7	3	2	3	46
35	Saidpur	2	10	18	8	2	3	8	51
36	Kakwa	5	22	38	18	12	8	27	130
37	Saraiya Duban	4	20	28	20	1	5	22	100
38	Kohra	19	30	40	35	10	12	42	188
39	Mahmadpur	11	25	33	33	5	8	30	145
40	Naraini	18	31	35	37	16	10	40	187
41	Nainaha	2	7	12	11	2	2	18	53
42	Ramgarh	20	63	78	40	10	15	45	271
43	Himmatgarh	4	8	20	46	8	4	20	110
44	Dehara	2	9	17	15	1	4	18	66
	Amethi Block	**239**	**595**	**951**	**936**	**280**	**291**	**958**	**4223**

uncultivated lands has been termed as fallow land or *Parati Bhoomi* in the land revenue records and also for the purpose of revenue collection, i.e. *Malgujari* from the farmers.

The other types of wastelands are those lands which have been under cultivation but it have been left out of cultivation due to encroachment of rail, road, canals, settlements etc. due to which the farmers are forced to leave such lands uncultivated and these lands are categorized as other types of wastelands.

It has been recorded that the overall distribution of wastelands in the Amethi has been very critical because the problem of the wastelands contributes significant proportion of the total geographical area of the Amethi. It has also been noted that the problem of the wastelands shares more than one-third area of the respective village panchayats to the total reporting area of the village.

The acute problem of the wastelands of the Amethi has led to environmental degradation, and disturbance in the eco-system of the area. The wastelands of the Amethi has caused ecological imbalances and led to the environmental crisis of the area. In view of the constant increase in the area under the wastelands of the Amethi as natural infrastructural resources has to be properly innovated and planned to conserve the available wasteland resources of the Amethi for sustainable development.

An attempt has been made in this study to analyse the various wasteland categories of the Amethi at village panchayat level for all 44 village panchayats of Amethi including 82 revenue villages along with 377 purwas/tolas/hamlet villages of the Amethi Block.

Village panchayatwise information for all the categories of wastelands has been collected through the revenue records from the Tehsil Amethi and the collected informations related to the wastelands have been compared through field observations/ field survey for all the village panchayats of the Amethi to ensure the authenticity of the available wastelands innovated information on natural infrastructures of the area.

Village panchayatwise distribution of the wastelands and the percentage of the wastelands to the total wastelands of the respective village panchayats has been analyzed. An effort has also been made to analyse the proportionate distribution of the wastelands of village panchayat level to the total geographical area of the respective village panchayat in Amethi.

(1) Kankarili Land

The proportionate distribution of the kankarili land has been noted very low and insignificant but the problem of the calcareousness and hard kankar pan below one metre depth of the soil does not allow the water percolation which creates hindrance in the growth of plants and the affected area through calcareousness also develops the alkalinity in soil due to which the lands are lying waste and left out of cultivation. It has been observed that only 239 hectares of land has been affected by the Kankar pan or calcareousness or Kankarili land. The area under Kankarili land has been recorded 42 hectares in the Korari Girdhar Shah village panchayat, while in most of the villages very nominal kankarili land has been recorded in the village panchayat of Amethi Block.

The distribution of the wastelands and the kankarili land at the village panchayat level has been mentioned in Table 7.3.

The proportionate distribution of the kankarili land to the total wastelands of the village panchayats has been mentioned in Table 7.5. The distribution of the kankarili land and its percentage of the total area under the wastelands at village panchayat level has been depicted in Map 7.59.

Very high percentage more than 9.0 per cent distribution of the kankarili land to the area under the total wastelands has been recorded in the Korari Girdhar Shah, Gangauli, Kohara, Naraini village panchayats of Amethi, while medium percentage between 6.0 to 8.0 per cent distribution of Kankarili land has been noted in the village Ramgarh, Mahmoodpur, Gaderi village panchayats of the Amethi Block. Low percentage distribution of the kankarili land ranging from 4.0 to 6.0 per cent of the total wasteland has been observed in the Sarai Khema, Benipur, Parsawan, Mahmoodpur, Jangal Ram Nagar, Ramdaipur, Hathkila, Loharata, Saraiya Duban, Agahar, Dhandhudhar, village panchayats of the Amethi Block. Very low percentage varying less than 4.0 per cent to the total wastelands of the kankarili land has been observed in the Katara Maharani, Trilokpur, Bhaganpur, Tala, Kushi Tali, Darkha, Mochawa Nuwawan, Umapur Ganapatti, Raipur Phulwari, Katara Phulkunwar, Rebha, Ramdaipur, Loniapur, Dedhpasar, Mahaso, Bhusahari, Himmatgarh, Saidpur, Nainaha and Kakawa village panchayats of the Amethi Block.

The proportionate distribution of the kankarili land percentage to the total geographicál area of the village panchayats of the Amethi Block has been mentioned in Table 7.4. The very high percentage distribution of the kankarili land more than 3.0 per cent to the total geographical area of the village panchayat of Amethi Block has been observed in the Naraini, Ramgarh, and Korari Girdhar Shah village panchayats. The proportionate distribution of the kankarili land to the total geographical area of the respective village panchayats has been depicted in Map 7.60. Medium percentage distribution of the kankarili land to the total geographical area of the ranging from 2.0 to 3.0 per cent has been noted in the Mahmadpur, Gaderi and Gangauli village panchayats of the Amethi Block. Low percentage of the kankarili land varying from 1.0 to 2.0 per cent of the total geographical area has been recorded in the Sarai Khema, Jangal Ram Nagar, Kushi Tali, Loharata, Darakha, Mochawa, Umapur Gana Patti, Raidaipur, Mahaso, Katara Phulkunwar, Kherauna, Parsawan, Mahmoodpur, Himmatgarh, Saidput, Saraiya Duban and Kakawa village panchayats of the Amethi Block. Very low percentage distribution of the kankarili land less than 1.0 per cent to the total geographical area of the village panchayats has been observed in the Raipur Phulwari, Rebha, Dedhpasar, Hathkila, Agahar, Chaturbhujpur, Maharajpur, Purabgaon, Nainaha, Dehara, Bhusahari, Dedhpasar, Hathkila, Katara Maharani, Trilokpur, Ramdaipur, Bhaganpur, Tala, Nuwawan village panchayats of the Amethi Block of the district Amethi.

The reclamation of the kankarili land has to be initiated with the involvement of the local people and their continued, sustained participation. The scientific reclamation of the kankarili land has been ensured through breadking the hard kankarpan of the soil and application of the organic and inorganic amendments as per requirement of the soil social forestry and other types of plantation has been initiated for survival of the plants. Leaching and flushing of the calcareous soil and appropriate drainage system to drain out the affected water must be ensured so that the other lands are not affected. The reclamation of the kankarili land requires techno-scientific techniques and it requires long term planning process for its appropriate reclamation measures.

(2) Banjar Land

The land which has been left out of cultivation due to very poor and deteriorated fertility status of the soil from more than five years has been classified as Banjar Land. The distribution of banjar land area has been mentioned in Table 7.3. The percentagewise

Table 7.4: Waste Lands (Amethi Block: Village-wise)

(%age to Total Geographical Area)

Sl. No.	*Villages*	*Kankarili Land*	*Banjar Land*	*Usar Land*	*Water Logged Land*	*Old Fallow Land*	*Fallow Land*	*Other Types of Waste Land*	*Total Waste Lands*
1	*2*	*3*	*4*	*5*	*6*	*7*	*8*	*9*	*10*
1	Katra Maharani	0.7	2.9	5.2	7.4	1.4	2.9	5.9	26.8
2	Sarai Khema	1.7	4.1	6.4	7.5	2.8	2.5	8.4	33.75
3	Trilokpur	0.6	5.0	5.6	7.5	3.1	3.1	11.2	36.3
4	Ramdaipur	0.9	1.9	3.8	5.7	0.9	0.9	7.6	21.0
5	Jangal Ram Nagar	1.6	3.5	7.3	7.4	6.0	5.1	8.3	39.5
6	Benipur	1.5	3.9	7.8	6.2	3.9	0.7	3.1	27.3
7	Parsawan	1.5	2.2	6.3	7.8	3.7	3.1	11.3	36.0
8	Mahmoodpur	1.8	2.5	5.6	5.9	3.4	2.8	11.2	33.4
9	Raipur Fulwari	0.7	3.8	6.3	7.7	3.5	2.8	13.3	38.2
10	Umapur Gana Patti	1.2	7.3	6.1	6.1	2.5	2.4	12.2	34.5
11	Bhaganpur	0.6	5.2	7.2	4.6	0.6	2.6	6.6	27.8
12	Tala	0.5	3.8	9.3	6.0	1.1	2.7	3.3	26.9
13	Kushi tali	1.1	2.9	6.7	8.7	1.4	2.3	11.1	34.6
14	Loharta	1.9	7.8	9.8	5.8	1.9	2.9	3.9	33.3
15	Darkha	1.3	4.6	9.0	9.6	1.6	2.0	8.0	36.8
16	Mochwa	1.2	5.4	8.4	10.8	1.2	2.4	9.0	39.7
17	Nuwawan	0.6	3.0	5.0	10.9	2.7	0.8	4.5	26.9
18	Loniapur	0.9	5.5	7.3	9.1	2.7	0.9	4.5	31.1
19	Kherauna	1.1	5.3	7.6	7.1	1.1	1.1	10.6	34.3
20	Katra Phulkunwar	1.1	5.7	8.5	6.8	1.7	2.2	5.7	32.0
21	Hathkila	0.9	3.7	6.9	6.9	0.4	0.9	2.7	23.1
22	Rebha	1.0	5.0	8.0	10.1	1.0	2.0	4.0	30.3
23	Raidaipur	1.5	6.1	8.4	12.3	1.5	1.1	8.4	39.6
24	Korari Girdhar Shah	3.1	5.7	7.0	8.6	1.3	2.6	0.7	36.2
25	Dedhpasar	0.6	4.1	5.0	7.9	1.9	1.5	13.0	34.3
26	Mahaso	1.1	4.6	9.3	8.1	2.16	2.3	3.4	30.2
27	Gangauli	2.6	5.7	8.4	8.6	2.3	0.7	5.2	32.8
28	Dhadhudhar	1.5	6.0	11.2	6.0	0.7	3.0	3.0	31.5
29	Gaderi	2.2	4.4	7.8	5.5	1.4	2.9	8.2	32.8
30	Bhusahari	0.8	4.2	5.9	5.1	0.4	1.2	6.3	24.2
31	Purabgaon	0.9	5.4	9.9	5.8	0.4	1.8	5.4	29.7
32	Chaturbhujpur	0.8	3.3	8.3	6.6	1.6	0.8	1.6	24.1
33	Mahrajpur	0.6	1.8	3.0	4.8	0.6	3.0	1.2	15.2
34	Aghar	0.9	5.0	8.2	3.2	1.3	0.9	2.7	21.1
35	Saidpur	1.1	5.5	10.0	19.0	0.5	2.6	4.4	28.3
36	Kakwa	1.3	5.9	10.3	4.8	3.2	2.1	7.3	35.3
37	Saraiya Duban	1.3	6.6	9.3	6.6	0.3	1.6	7.0	33.4
38	Kohra	3.3	4.3	6.9	6.1	1.7	2.0	7.3	32.8
39	Mahmadpur	2.4	5.6	7.4	7.4	1.1	1.7	6.7	32.5
40	Naraini	3.0	5.3	5.9	6.3	2.7	1.7	6.8	32.0
41	Nainaha	0.5	3.8	6.5	6.4	1.0	1.0	9.8	29.1
42	Ramgarh	3.0	9.5	11.8	5.8	3.4	2.1	6.5	41.1
43	Himmatgarh	1.0	2.0	5.8	11.7	2.0	1.0	5.1	32.3
44	Dehara	0.8	3.8	7.2	6.3	0.4	1.7	7.6	28.0
	Amethi Block	**1.87**	**4.67**	**7.45**	**7.53**	**2.19**	**2.82**	**7.51**	**33.11**

distribution of banjar land to the total wastelands at village panchayat level in Amethi Block has been depicted in Table 7.5. The proportionate distribution of the banjar land to the total wastelands has been shown in Map 7.61.

It has been observed that very high percentage of the banjar land to the total wastelands more than 20 per cent has been seen in the Agahar, Saraiya Duban and Loharata village panchayats of the Amethi Block. The medium percentage of the banjar land of the total wastelands ranging from 15.0 to 20.0 per cent has been recorded in the Saidpur, Kakawa, Kohara, Mahmadpur, Naraini, Purabgaon, Dhandhudhar, Korari Girdhar Shah, Gangauli, Bhusahari, Bhaganpur, Loniapur, Rebha, Katara Phulkunwar, Raidaipur, Kherauna and Hathkila village panchayats. Low percentage distribution of banjar land of the total wastelands ranging from 10.0 to 15.0 per cent has been recorded in the Katara Maharani, Sarai Khema, Trilokpur, Benipur, Raipur, Phulwari, Tala, Darakha, Mochawa Nuwawan, Gaderi, Chaturbhujpur, Maharajpur, Nainaha and Dehra village panchayats. Very low percentage of the banjar land of the total wastelands less than 10.0 per cent has been noted in the Himmatgarh, Dedhpasar, Parsawan, Mahmoodpur, Kushi Tali, Umapur Gannapatti, Jangal Ram Nagar and Ramdaipur village panchayats of the Amethi Block.

The proportionate distribution of the banjar land to the total geographical area of the village panchayats has been depicted in Map 7.62. It has been noted that very high percentage of the banjar land more than 6.0 per cent of the total geographical area has been found in the Loharata, Saraiya Duban, Ramgarh, Dhandhudhar, Raidaipur and Loharata village panchayats of the Amethi Block. Medium percentage of the banjar land ranging from 5.0 to 6.0 per cent of the total geographical area has been noted in the Trilokpur, Bhaganpur, Kherauna, Katara Phulkunwar, Loniapur, Korari Girdhar Shah, Gangauli, Mochawa, Bhaganpur, Purabgaon, Naraini, Agahar, Saidpur, Kakawa and Mahmadpur village panchayats of the Amethi Block. Low percentage distribution of the banjar land to the total geographical area varying from 4.0 to 5.0 per cent has been observed in the Darakha, Sarai Khema, Gaderi, Kohara, Bhusahari, Mahaso, Rebha, Dedhpasar village panchayats of the Amethi Block. Very low percentage distribution of the banjar land of the total geographical area less than 4.0 per cent has been found in the Katara Maharani, Ramdaipur, Jangal Ram Nagar, Tala, Kushi Tali, Nuwawan, Umapur Gana Patti, Mahmoodpur, Benipur, Parsawan, Hathkila, Raipur Phulwari, Himmatgarh, Dehara, Nainaha, Chaturbhujpur and Maharajpur village panchayats of the Amethi Block.

Reclamation measures for banjar land may be initiated through levelling, bunding, leaching, flushing and top dressing of the fertilizer deep ploughing green manuring, application of the organic and inorganic amendments, ensuring uniform distribution of the irrigation, sowing, high yield variety Kharif and Rabi crops as per suitability of the soil. The construction of the surface drainage to drain out the excess water mechanical processing, application of gypsum and pyrite and application of the NPK fertilizer and also bio-fertilizer and micro-organic based bio-fertilizer and bio-pesticides can be applied for the successful reclamation of the banjar land for agro-afforestation programmes.

(3) Usar Lands

The land which has fluffy soils having whitish or grayish appearance deposited on the surface and/ or not easily culturable. Such lands are commonly know as *reh* or *rehar* or usarland. The land which has excessive proportions of soluble salts due to which crop production has been seriously hampered are known as salt affected soils and locally such lands are known as usar land in Amethi. The proportionate distribution of usar land to the total wastelands at village panchayat level in Amethi Block has been depicted in Map 7.63.

It has been observed that more than 30.0 per cent area under usar land to the total wastelands at village panchayat level has been found in the Saidpur, Agahar, Chaturbhujpur, Dhandhudhar, Mahaso and Tala village panchayats of the Amethi Block. The medium percentage of usar land varying from 25.00 to 30.00 per cent to the total wastelands has been seen in the Bhaganpur, Loharata, Benipur, Katara Phulkunwar, Rebha, Hathkila, Loharata, Kakawa and Gangauli village panchayats of the usar land to the total wastelands varying from 20.0 to 25.0 per cent has been noted in the Umapur Gana Patti, Kherauna, Loniapur, Raidaipur, Darakha, Mochawa, Bhusahari, Gaderi, Purabgaon, Nainaha, Dehara, Saraiya Duban, Maharajpur, Kohara and Himmatgarh village panchayats of the Amethi Block. Very low percentage distribution of the usar land of the total wastelands less than 20.0 per cent has been recorded in the Katara Maharani, Trilokpur, Sarai Khema, Ramdaipur, Jangal Ram Nagar, Mahmoodpur, Parsawan, Raipur Phulwari, Kushi Tali, Dedhpasara, Korari Girdhar Shah, Nuwawan, Himmatgarh, Ramgarh and Naraini village panchayats of Amethi Block.

It has been noted that proportionate distribution of the usar land to the total geographical area of the village panchayats has been recorded high more than 9.0 per cent in the Mahaso, Dhandhudhar, Purabgaon, Ramgarh, Darakha, Loharata, Tala, Saidpur, Kakawa and Saraiya Duban village panchayats of the Amethi Block. Medium percentage distribution of usar land varying from 8.0 to 9.0 per cent of the total geographical area of the village panchayats has been observed in the Katara Phulkunwar, Rebha, Raidaipur, Gangauli, Mochawa, Agahar and the Chaturbhujpur village panchayats of the Amethi Block. Low percentage distribution of the usar land of the total geographical area ranging from 7.0 to 8.0 per cent has been recorded in the Mahmadpur, Dehara, Gaderi, Umapur Gannapatti, Bhaganpur, Jangal Ram Nagar, Kherauna, Loniapur, Korari Girdhar Shah village panchayats of the Amethi Block while very low percentage distribution of usar land of the total geographical area of the village panchayats less than 7.0 per cent has been noted in the Katara Maharani, Sarai Khema, Ramdaipur, Mahmoodpur, Parsawan, Raipur Phulwari, Trilokpur, Nuwawan, Dedhpasar, Hathkila, Kushi Tali, Bhusahari, Himmatgarh, Nainaha, Naraini and Kohara village panchayats of the Amethi Block. The proportionate distribution of usar land to the total geographical area of the village panchayats has been depicted in Map 7.64.

The reclamation of the usar land adopting all necessary techno-scientific agricultural, horticultural, afforestation practices and provision of the good quality water for irrigation is an essential prerequisite for the reclamation of the usar land. Proper levelling, bending, top dressing and deep ploughing, leaching and flushing of the salt affected soils are necessary for the reclamation of the salt affected soils are necessary for the reclamation of the usar land, mechanical processing, green manuring and application of the organic and inorganic amendments, use of gypsum and pyrite and application of the bio-fertilizer and agro-waste composting is required for the reclamation of the usar land, the salt tolerant plant species for agro-forestry, social-forestry and agricultural crops can be grown after reclamation of the usar land through adopting agricultural scientific system to ensure proper growth of the plants and sustenance of the agricultural crops and plants. Bio-remediation of the agro-afforestation management and recycling of the afforested plants. leaves etc. for the bio-fertilizer and requisite appropriate good quality water irrigation and proper application of organic and inorganic amendments and use of insecticides and pesticides should be ensured as per requirement of the plants and the reclaimed usar soil.

Table 7.5: Waste Lands (Amethi Block)

(%age of Total Wastelands)

Sl. No.	*Villages*	*Kankarili Land*	*Banjar Land*	*Usar Land*	*Water Logged Land*	*Old Fallow Land*	*Fallow Land*	*Other Types of Waste Land*
1	*2*	*3*	*4*	*5*	*6*	*7*	*8*	*9*
1	Katra Maharani	2.7	11.1	19.4	27.7	5.5	11.2	22.2
2	Sarai Khema	5.3	12.2	19.2	22.3	8.5	7.4	25.0
3	Trilokpur	2.7	13.7	15.5	20.6	8.6	8.6	32.0
4	Ramdaipur	4.9	8.6	17.3	26.0	4.3	4.9	34.7
5	Jangal Ram Nagar	4.2	9.0	10.5	18.7	15.1	13.0	22.0
6	Benipur	5.7	14.2	28.5	22.8	14.2	2.8	11.4
7	Parsawan	5.0	6.0	17.3	22.7	10.4	8.6	32.3
8	Mahmoodpur	5.6	8.0	16.0	17.7	10.2	8.4	33.6
9	Raipur Fulwari	1.8	10.0	16.5	20.1	9.2	7.3	24.8
10	Umapur Gana Patti	3.5	8.0	22.2	19.4	4.4	7.0	35.3
11	Bhaganpur	2.3	19.0	26.2	16.6	2.3	9.5	23.8
12	Tala	2.0	14.2	34.6	22.4	4.0	10.2	12.2
13	Kushi tali	3.3	8.4	19.4	25.4	4.2	6.7	32.2
14	Loharta	5.8	23.5	29.4	17.6	5.8	5.8	11.7
15	Darkha	3.6	12.8	24.7	26.6	4.5	5.5	22.0
16	Mochwa	3.0	13.6	21.2	27.2	3.0	6.0	22.7
17	Nuwawan	2.4	11.3	18.6	40.6	6.5	3.2	17.0
18	Loniapur	2.9	17.6	23.5	29.4	8.8	2.9	14.7
19	Kherauna	3.4	15.5	22.4	20.6	3.4	3.4	32.0
20	Katra Phulkunwar	3.5	17.8	26.7	22.4	5.3	7.14	17.8
21	Hathkila	4.0	16.0	30.0	30.0	4.0	4.0	10.7
22	Rebha	3.3	16.6	26.6	33.3	3.3	3.3	13.3
23	Raidaipur	3.8	15.5	22.3	32.0	3.8	2.9	21.3
24	Korari Girdhar Shah	8.6	15.8	19.4	23.4	3.5	7.1	21.5
25	Dedhpasar	1.8	12.0	14.8	23.2	5.5	4.6	32.9
26	Mahaso	3.8	15.3	3.7	26.9	3.8	7.6	11.5
27	Gangauli	8.0	17.6	25.6	26.4	4.0	2.4	16.0
28	Dhadhudhar	4.7	19.0	35.7	19.0	2.3	9.5	9.5
29	Gaderi	6.8	13.6	23.8	17.0	4.5	9.0	25.0
30	Bhusahari	3.5	17.5	24.5	23.0	2.7	5.2	26.3
31	Purabgaon	3.0	15.2	21.2	19.6	1.5	6.0	18.1
32	Chaturbhujpur	3.4	13.7	34.4	27.5	6.8	6.8	6.8
33	Mahrajpur	4.0	12.0	20.0	32.0	4.0	20.0	8.0
34	Aghar	4.3	23.9	39.1	15.2	6.5	4.3	6.5
35	Saidpur	3.9	19.6	35.2	15.6	3.9	5.8	15.6
36	Kakwa	3.8	16.9	29.2	13.8	9.2	6.2	20.7
37	Saraiya Duban	4.0	20.0	25.0	20.0	1.0	5.0	22.0
38	Kohra	10.2	15.9	22.2	18.6	5.3	6.3	22.3
39	Mahmadpur	7.5	17.2	22.7	22.7	3.4	5.5	20.6
40	Naraini	9.6	17.1	18.7	19.7	8.5	5.3	21.3
41	Nainaha	3.7	13.2	22.6	20.7	3.7	3.7	33.9
42	Ramgarh	7.3	23.2	28.7	14.7	3.6	5.5	16.6
43	Himmatgarh	3.6	7.2	18.2	41.8	7.2	3.6	18.2
44	Dehara	3.0	13.6	25.7	22.7	5	6.0	27.2
	Amethi Block	**5.65**	**14.08**	**22.51**	**22.16**	**6.63**	**6.89**	**22.6**

(4) Water Logged Land

The water logged land is that land where water table is at or near the earth's surface and water stands for the most of the year and soil pores in the root zones of the crop get saturated with the water which inhibits the growth and survival activity of the plants. As a result such lands are left out of cultivation and locally termed as water logged land.

The proportionate distribution of the water logged land to the total waste lands of the village panchayats of the Amethi Block has been depicted in Map 7.65. It has been observed that high percentage of water logged land to the total wastelands is more than 30.0 per cent has been observed in the Nuwawan, Raidaipur, Rebha, Maharajpur and Himmatgarh village panchayats of the Amethi. Medium percentage of the water logged land of the total wastelands varying from 25.0 to 30.0 per cent has been recorded in the Kushi Tali, Ramdaipur, Katara Maharani, Loniapur, Darakha, Mochawa, Chaturbhujpur, Mahaso and Hathkila village panchayats of the Amethi Block. Low percentage distribution of the water logged land of the total wastelands ranging from 30.0 to 25.0 per cent has been noted in the Trilokpur, Sarai Khema, Katara Phulkunwar, Dedhpasar, Korari Girdhar Shah, Tala, Bhusahari, Ramgarh, Dehara, Nainaha, Saraiya Duban and Mahmadpur village panchayats of the Amethi Block. Very low percentage distribution of the water logged land to the wastelands less than 20.0 per cent has been recorded in the Mahmoodpur, Jangal Ram Nagar, Umapur Gana Patti, Loharata, Bhaganpur, Dhandhudhar, Purabgaon, Gaderi, Naraini, Kohara, Saidpur and Kakawa village panchayats of the Amethi Block.

The proportionate distribution of the water logged land to the total geographical area of the village panchayats of the Amethi Block has been shown in Map 7.66. Very high percentage of the water logged land of the total geographical area of the village panchayats more than 12.0 per cent has been found in the Raidaipur and Saidpur village panchayats of the Amethi Block. Medium percentage of the water logged land of the total geographical area varying from 10.0 to 12.0 per cent in the Mochawa, Nuwawan, Rebha and Himmatgarh village panchayats of the Amethi Block. Low percentage distribution of the water logged land of the total geographical area of the village panchayats ranging between 8.0 and 10.0 per cent has been recorded in the Kushi Tali, Darakha, Gangauli, Mahaso and Loniapur village panchayats of the Amethi Block. Very low percentage less than 8.0 per cent of the total geographical area of the village panchayats has been noted in the Katara Maharani, Trilokpur, Ramdaipur, Sarai Khema, Benipur, Mahmoodpur, Parsawan, Kherauna, Raipur Phulwari, Dedhpasar, Hathkila, Jangal Ram Nagar, Umapur Gana Patti, Bhaganpur, Tala, Loharata, Agahar, Maharajpur, Chaturbhujpur, Naraini, Purabgaon, Dhandhudhar, Gaderi, Bhusahari, Ramgarh, Dehara, Naraini, Nainaha, Kakawa, Saraiya Duban, Kohara and Himmatgarh village panchayats of the Amethi Block.

The reclamation of the water logged land can be ensured through proper drainage system to drain out the accumulated and water logged land in the low-lying area. The construction of the pond and the development of the water logged land for fisheries including soil testing and quality of water including the underground water must be ensured. The drainage system for the inlet and outlet of the water and construction of the bunds of the ponds and appropriate development of the pond as per its feasibility must be ensured. The development of the pond for the fisheries culture, and mini-hatchery including development of the high yielding variety fisheries seed must be ensured. The horticulture and social forestry plantation for fodder, fuel wood and horticultural plants should be planned along the bunds of the pond. Demonstration of fish cropping and recycling of fisheries culture and proper seeding of the fishes must be appropriately ensured. Maintenance and management cf fish ponds along with the horticultural and social-forestry plants should be ensure scientifically for successful and sustainable development of the ponds and fisheries culture.

(5) Old Fallow Land

The old fallow land is that uncultivated land which has been previously under cultivation but presently, it has been left out of cultivation due to some reason or when for a period of not less than two years but not more than five years with the view to maintain the fertility status of the soil, such land has been termed as old fallow land.

The proportion of the old fallow land to the total wastelands at the village panchayat level in Amethi Block has been shown in Map 7.67 very high percentage of the old fallow land of the total wastelands more than 10.0 per cent has been recorded in the Jangal Ram Nagar, Mahmoodpur, Benipur, Parsawan village panchayats of the Amethi Block. Medium per cent of the old fallow land of the total wastelands ranging between 7.0 and 10.0 per cent has been noted in the Sarai Khema, Trilokpur, Raipur Phulwari, Loniapur, Kakawa, Naraini and Himmatgarh village panchayats of the Amethi Block. Low percentage of the old fallow land of the total wastelands ranging between 4.0 and 7.0 per cent has been noted in the Katara Maharani, Ramdaipur, Kushi Tali, Loharata, Darakha, Umapur Gannapatti, Nuwawan, Katara Phulkunwar, Dedhpasar, Agahar, Chaturbhujpur, Gaderi, Ramgarh, Agahar and Kohara village panchayats of the Amethi. Very low percentage less than 4.0 per cent area under the old fallow land of the total wastelands has been recorded in the Bhaganpur, Tala, Kherauna, Hathkila, Rebha, Mahaso, Korari Girdhar Shah, Mochawa Dhandhudhar, Gangauli, Bhusahari, Dehara, Maharajpur, Purabgaon, Nainaha, Saidpur, Saraiya Duba and Mahmadpur village panchayats of Amethi Block.

The proportionate distribution of the old fallow land to the total geographical area of the village panchayats of the Amethi Block has been depicted in Map 7.65. It has been noted very high percentage of the old fallow land of the total geographical area more than 3.0 per cent has been recorded in the Trilokpur, Jangal Ram Nagar, Benipur, Parsawan, Mahmoodpur, Raipur Phulwari, Ramgarh and Kakawa village panchayats of the Amethi Block. Medium percentage of the old fallow land of the total geographical area varying from 2.0 to 3.0 per cent has been observed in the Sarai Khema, Umapur Gana Patti, Loniapur, Nuwawan, Gangauli, Mahaso and Naraini village panchayats of the Amethi. Low percentage of the old fallow land ranging between 1.0 and 2.0 per cent of thc total geographical area of the village panchayats has been recorded in the Katara Maharani, Tala, Kushi Tali, Loharata, Darakha, Mochawa, Korari Girdhar Shah, Ramdaipur, Katara Phulkunwar, Dedhpasar, Himmatgarh, Gaderi, Agahar, Kohara and Mahmadpur village panchayats of the Amethi. Very low percentage of the old fallow land of the total geographical area less than 1.0 per cent has been observed in the Bhaganpur, Kherauna, Rebha, Hathkila, Bhusahari, Rebha, Dehara, Nainaha, Purabgaon, Maharajpur, Bhaganpur, Dhandhudhar and Saraiya Duban village panchayats of the Amethi Block of district Amethi of Uttar Pradesh.

The reclamation of the old fallow land can be ensured through levelling, bunding, deep ploughing, cultivation, top dressing of the fertilizer, leaching and flushing of the soil, irrigation management and proper drainage system for the inlet and outlet of the water green manuring, application of the organic and inorganic amendments, use of appropriate insecticides and pesticides and application of the bio-fertilizer including

NPK fertilizer and micro-organism based bio-fertilizers and bio-pesticides can be used as per the requirement of the soil. High yielding variety of the agricultural crops and horticultural, social-forestry or agro-forestry plants can be planted as per tolerance of the soil and requirements of the local participating people ensuring people's participation with the view to achieve the need based reclamation measures of the old fallow land.

6. Fallow Land

The land which has been left out of cultivation due to very poor or low fertility status of the soil, or such lands has been left out of cultivation to re-maintain the fertility status of the soil and ensuring cropping rotation system. Initially the land has been left uncultivated for a year but if such land has been left uncultivated for more than one year but less than two years such uncultivated land has been locally termed as fallow land.

The appropriate distribution of the fallow land of the total wastelands at village panchayat level has been depicted in Map 7.69. Very high percentage of the fallow land of the total wastelands more than 8.0 per cent has been recorded in the Katara Maharani, Trilokpur, Jangal Ram Nagar, Tala, Bhaganpur, Mahmoodpur, Parsawan, Dhandhudhar, Gaderi, Ramgarh and Maharajpur village panchayats of the Amethi Block. Medium percentage of the fallow land of the total wastelands varying from 6.0 to 8.0 per cent has been recorded in the Sarai Khema, Kushi Tali, Raipur Phulwari, Umapur Gana Patti, Katara Phulkunwar, Mahaso, Korari Girdhar Shah, Kohara, Kakawa, Chaturbhujpur village panchayats of the Amethi. Low percentage between 4.0 and 6.0 per cent fallow land of the total wastelands has been noted in the Ramdaipur, Hathkila, Dedhpasar, Bhusahari, Dehara, Purabgaon, Naraini, Agahar, Mochawa, Darakha, Loharta, Saraiya Duban, Saidpur and Mahmadpur village panchayats of the Amethi Block. Very low percentage less than 4.0 per cent area under fallow land of the total wastelands has been recorded in Benipur, Kherauna, Loniapur, Raidaipur, Rebha, Gangauli, Himmatgarh, Nainaha and Nuwawan village panchayats of the Amethi Block.

The distribution of the fallow land of the total geographical area of the village panchayats has been depicted in Map 7.70.

Proportionately very high percentage more than 3.0 per cent fallow land to the total geographical area of the village panchayat has been noted in the Trilokpur, Jangal Ram Nahar, Parsawan of the Amethi. Medium percentage of the fallow land varying from 2.0 to 3.0 per cent of the total geographical area has been found in the Katara Maharani, Sarai Khema, Mahmoodpur, Raipur Pulwari, Katara Phulkunwar, Mahaso, Korari Girdhar Shah, Umapur Gana Patti, Bhaganpur, Tala, Kushi Tali, Loharata, Mochawa, Gaderi, Ramgarh, Dehara, Saidpur and Kakawa village panchayats. Low percentage distribution of the fallow land of the total geographical area ranging between 1.0 to 2.0 per cent has been found in the Darakha, Kherauna, Rebha, Bhusahari, Raidaipur, Purabgaon, Naraini, Saraiya Duban, Mahmadpur and Kohara village panchayats of the Amethi Block. Very low percentage fallow land less than 1.0 per cent to the total reporting area of the village panchayats has been recorded in the Hathkila, Loniapur, Ramdaipur, Benipur, Nuwawan, Gangauli, Himmatgarh, Nainaha, Chaturbhujpur and Agahar village panchayats of Amethi Block.

The reclamation of the fallow land can be initiated through proper levelling, bunding, leaching, flushing and construction of proper draining system for the inlet and outlet of the water, green manuring, application of the organic and inorganic amendments including bio-fertilizer and application of the insecticides and pesticides as per requirements of the agricultural crops. The use of high yielding variety of the crops and tissue-culture raised agro-forestry, horticultural and social forestry plants can be planted in the reclaimed fallow land. Proper management including irrigation and NPK fertilizer applications should be ensured as per the need of the plants and agricultural crops to ensure successful production and also to ensure eco-system, ecological and environmental management of the area.

7. Other Types of Wastelands

The agricultural land which has been left out of cultivation due to some reason or the other but other parts of the cultivable land is under agricultural use. Such lands are lying uncultivated along with the rail, road, canal sides and also along the settlement areas or along with the forest land or horticultural land due to which proper field has not been carried out due to poor fertility status of the soil, such part of land which has been left out of cultivation has been locally termed as other types of wastelands.

The proportionate distribution of the other type of wastelands of the total wastelands at Village panchayat level has been depicted in Map 7.71. Very high percentage distribution of the other types of wastelands more than 30.0 per cent of the total wastelands has been found in the Trilokpur, Ramdaipur, Kushi Tali, Umapur Gana Patti, Mahmoodpur, Parsawan, Kherauna, Raipur Phulwari, Dedhpasar, and Nainaha village panchayats of the Amethi Block. Medium percentage of the other type of wastelands to the total wastelands ranging from 25.0 to 30.0 per cent has been noted in the Bhusahari and Dehara village panchayats of the Amethi Block. Proportionately low percentage distribution of the other type of wastelands to the total wastelands varying between 20.0 to 25.0 per cent has been observed in the Katara Maharani, Sarai Khema, Bhaganpur, Jangal Ram Nagar, Korari Girdhar Shah, Raidaipur, Darakha, Mochawa, Gaderi, Naraini, Kakawa, Saraiya Duban, Kohara and Mahmadpur village panchayats. Very low percentage distribution of the other type of wastelands to the total wastelands less than 20.0 per cent has been observed in the Tala, Loharata, Benipur, Katara Phulkunwar, Hathkila, Loniapur, Rebha, Nuwawan, Gangauli, Mahaso, Himmatgarh, Ramgarh, Agahar, Purabgaon, Chaturbhujpur, Maharajpur and Saidpur village panchayats of the Amethi Block.

The proportionate distribution of the other type of wastelands to the total geographical area of the village panchayats of the Amethi Block has been depicted in Map 7.72. Very high percentage of the other type of wastelands of the total geographical area more than 10.0 per cent has been noted in the Trilokpur, Kushi Tali, Mahmoodpur, Parsawan, Raipur Phulwari, Kherauna, Dedhpasar, Umapur Ganapatti village panchayats of the Amethi. Medium percentage of the other type of wastelands ranging between 8.0 and 10.0 per cent of the total geographical area has been recorded in the Sarai Khema, Jangal Ram Nagar, Raidaipur, Mochawa and Nainaha village panchayats of the Amethi Block. Low percentage distribution of the other type of wastelands of the total geographical area varying between 6.0 and 8.0 per cent has been recorded in the Ramdaipur, Bhaganpur, Darakha, Bhusahari, Ramgarh, Dehara, Kakawa, Naraini, Saraiya Duba, Kohara, Mahamadpur village panchayats of Amethi Block. Very low percentage less than 6.0 per cent distribution of the other type of wastelands of the total geographical area has been recorded in the Katara Maharani, Benipur, Katara Phulkunwar, Hathkila, Rebha, Mahaso, Gangauli, Korari Girdhar Shah, Dhandhudhar, Nuwawan, Loharata, Tala, Purabgaon, Himmatgarh, Chaturbhujpur, Maharajpur, Agahar and Saidpur village panchayats of Amethi Block.

The reclamation of the other type of wastelands can be successfully implemented through proper levelling, bunding, deep ploughing, leaching, flushing, construction of the drainage system, top dressing of the fertilizer application of the organic and inorganic amendments and use of bio-fertilizer and appropriate water management system has been properly arranged and implemented. It is necessary to adopt agro-forestry, horticultural and social-forestry plantation programmes which can easily survive in such other types of wastelands left out cultivation along with the sides, rail, road, canal and settlements. The agro-afforestation programmes can be successfully implemented with the view to obtain fuel wood, fodder, fibre, fruits, wood and other needs of the local people to ensure environmental conservation and maintain the ecological imbalances of the area.

8. Total Wastelands

The proportionate distribution of the total wastelands to the total geographical area of the village panchayats has been depicted in Map 7.73. It has been observed that the distribution of the total wastelands of the total geographical area of the village panchayats having more than 35.0 per cent has been observed in the Trilokpur, Jangal Ram Nagar, Raipur Phulwari, Parsawan, Korari Girdhar Shah, Raidaipur, Darakha, Mochawa, Ramgarh and Kakawa village panchayats of the Amethi Block. Medium percentage distribution of the total wastelands of the total geographical area of the village panchayats ranging between 30.0 and 35.0 per cent has been found in the Sarai Khema, Mahmoodpur, Kherauna, Katara Phulkunwari, Dedhpasar, Loniapur, Umapur Ganapatti, Kushi Tali, Loharata, Dhandhudhar, Gaderi, Gangauli, Himmatgarh, Naraini, Saraiya Duban, Kohara and Mahmadpur village panchayats of the Amethi Block. Low percentage distribution of the total wastelands ranging from 25.0 to 30.0 per cent of the total geographical area of the village panchayats has been noted in the Katara Maharani, Benipur, Rebha, Mahaso, Nuwawan, Tala, Bhaganpur, Purabgaon, Dehara, Nainaha and Saidpur village panchayats of the Amethi Block. Very low per cent distribution of the total wastelands less than 25.0 per cent to the total geographical area of the village panchayats has been recorded in the Hathkila, Bhusahari, Agahar, Chaturbhujpur, Maharajpur and Ramdaipur village panchayats of the Amethi Block of the district Amethi.

The wastelands are an important natural resources and infrastructure of the village. An effort must be made to reclaim the available natural wastelands infrastructure to meet the increasing demand of growing cattle and human population because more than two-third population living in the villages are surviving on the land resources especially the fuel wood, fodder, wood and other day-to-day requirements of the local people. Thus, it has become an important task of the local people to initiate the reclamation of wastelands at the war-footing level through ensuring the people's participation and developing a people's movement because the one-third land of the village which was supported to be under greenery through forest cover is lying waste which brought the people face to face for the environmental and ecological crisis. The forest land has been degraded especially there is no forest land in most of the villages or very insignificant forest land has been noted in the rare villages of the Amethi. The reclamation of the Kankarili, banjar, usar, water logged, old fallow, fallow land and other types of wastelands should be initiated through adopting techno-scientific techniques and systems. As per suitability, feasibility of the land based on the fertility status of the soil, the agricultural cropping agro-forestry, horticultural and social-forestry plantation programmes must be started. The application of the organic, and inorganic amendments, bio-fertilizers and use of the NPK fertilizer as per the soil requirements and needs of the crops and plants should be applied to ensure successful reclamation of the wastelands. Need base agricultural cropping and agro-afforestation plantation programmes, ensuring peoples movement and involvement should be properly ensured. With the view to conserve the degraded environmental conditions and to maintain the ecological imbalances of the area and to protect and conserve the natural wastelands infrastructure resources of the various village panchayats of the Amethi.

7. CONCLUSION

The wastelands of the Amethi contributes approximately one-third area of the total geographical area. The unutilized leftout uncultivated land of the area has also been an important natural wasteland resources of the Amethi. The appropriate utilization and conservation of the available wasteland resources has become an important task to the policy makers and planners at the grassroots level as well as the district level.

Therefore, it is imperative that the rehabilitation process be taken up on a war-footing through mobilizing every conceivable natural wasteland resources and adopting imaginative strategies of the reclamation of wastelands natural infrastructural resources. Thus everyone has to strive to create a true consciousness amongst the entire people of the Amethi. The urgency of the problems of the wastelands and its reclamation process must be taken as a challenge to meet the increasing demand of the growing cattle and human population for fuel wood, fodder, fibre, fisheries, food grains (F^6), wood and vegetables etc. and to restore the ecological imbalance and environmental degradation of the area and to improve environment crisis of Amethi.

The wastelands has been used for the common land usually for the less fertile soil which failed to yield crops and to give any return to the medieval farmers. The common man may define the wastelands as the land lying uninhibited and uncultivated and the such land is no longer serving any purpose. The land which has been left out of cultivation from one to two years to maintain the fertility status of the soil, has been termed as fallow land and if such land is left out of cultivation from two to five years has been termed as old fallow land and of the uncultivated land is left out of cultivation lands has been termed as banjar land.

The land which has lost its agricultural production capability due to various reasons and due to improper agricultural practices, mismanagement and in which the cost of production is higher than the profit occurred out of it, such types of the land may be termed as wastelands.

The wastelands of the Amethi has been classified as Kankarili land, Banjar land, usar land, old fallow land, fallow land, waterlogged land and other types of wastelands.

It has been noted that 33.11 per cent area of the Amethi has been lying as wasteland and uncultivated out of which the Kankarili land shares 1.87 per cent, Banjar land 4.67 per cent noted in the Saraiya Duban village panchayat of the Amethi Block.

Proportionately very high percentage of fallow land, 3.0 per cent area to the total geographical area has been noted in the village Dhandhudhar, Maharajpur and 3.1 per cent in the Parasawan village panchayats recorded in the Gangauli and Benipur Village Panchayats of the Amethi Block.

It has been observed that very high, 13.3 per cent, other types of wastelands have been noted in the village panchayats Raipur Phulwari and very low, 1.2 per cent, land of the total geographical area has been found in the village panchayats Maharajpur of the Amethi Block.

The total wastelands of the various village panchayats of the Amethi Block has been found very high in the Ramgarh, Raidaipur, Sarai Khema, Trilokpur, Parsawan, Raipur Phulwari, Umapur Gangapatti, Kushi Tali, Loharata, Darakha, Mochawa and Kherauna village panchayats of Amethi Block.

Thus, it has been observed that the various types of wastelands contribute significant proportion at village panchayat level of the total geographical area of the village panchayats of the Amethi. The significant availability of wastelands in the Amethi Block and usar land 7.45 per cent, water logged land 7.5 per cent land. The old fallow land contributes 2.9 per cent, fallow land 2.82 per cent and other types of the wastelands share 7.51 per cent to the total geographical area of the Amethi Block.

The proportionate distribution of the Kankarili land at village panchayat level varies from 4.0 to 6.0 per cent of the total geographical area in Tala and 3.3 per cent in the Kohara village panchayats of the Amethi Block.

Highest percentage of the Banjar land has been recorded 9.5 per cent in the Ramgarh and the lowest percentage of the Banjar land 1.8 per cent of the total geographical area has been noted in the Maharajpur village panchayats of the Amethi Block.

Proportionately very high percentage of usar land has been found 11.2 per cent in the Dhandhudhar and very low 3.0 per cent to the total geographical area of the village panchayats Maharajpur of the Amethi Block.

Very high percentage of the water logged 19.0 per cent in the Saidpur and very low 3.2 per cent of the total geographical area in the Agahar village panchayat of the Amethi Block.

It has been recorded that very high 6.0 per cent of old fallow land has been found in the village Jangal Ram Nagar and very low 0.3 per cent old fallow land to the total geographical area has been various village panchayats of Amethi Block has been great concern to the village people and the agricultural policy makers and planners at the grassroots level.

Thus it has become an important task to initiate the reclamation of wastelands at war-footing level. The reclamation of the wastelands should be ensured as per the suitability and feasibility of the wastelands, to meet the increasing demand of the growing cattle and human population for the fuel wood, food grain, fodder, fibre, fruits, fisheries (F^6), wood and vegetables. The techno-scientific reclamation measures of the wastelands may be helpful in the conservation of available natural wastelands resources. The reclamation of the wastelands may be an asset to conserve the environmental degradation and maintain the ecological imbalances of the Amethi. The reclamation of the wastelands will be an asset in maintaining the eco-system of the area and regenerating the environmental conditions of the area and also to maintain the environmental crisis as well as socio-economic crisis of the people of the Aemthi.

8

Land Capability

1. INTRODUCTION

The land capability is a broad term. The land capability can be defined as the capability of the land to produce the crop or other types of agricultural activities or agro-forestry or social-forestry use or for the multi-cropping pattern. The term capability of the land has been perceived by the users in their own terms, i.e. the agricultural scientists or the cultivators may define the land capability in terms of the agricultural cropping capability of the land, while the animal husbandry workers may define the capability of the land for pastures and grazing use of the land. The plantation workers may define the land capability in terms of the social-forestry and agro-forestry plantation purpose of the land. The commercial agricultural cultivators may define the land capability for commercial cropping capability of the land. The fisherman may define the land capability in terms of the fisheries culture. Thus the land capabilities classification has also been viewed in terms of the use of the land for different purposes by the land users.

The capability of the land or the land quality has been classified in terms of the cultivable land and the revenue collection. As per the quality of the land the revenue charges are fixed by the revenue officials. The revenue collection and the revenue charges are determined by the revenue authorities at the Tehsil level under the control of the district level authorities.

The capability of the land and the quality of the land had been initiated from the historical era. The land quality classification for the cultivated land as per the land ownership has been fixed during the Mughal period. During the regime of the Akbar Badshah the land quality had been fixed under the supervision of Raja Todar Mal for the collection of the land revenue charges from different cultivated land owners of their kingdom.

During the Mughal period the land capability classification was done on the basis of Aana value. The cultivated land was valued in 16 Aana value. The very good quality cultivable land having multiple cropping including Kharif, Rabi and Jaid cropping season having good agricultural productivity. The less agricultural productive land was valued with very low Aana value of the land. The land productivity value in the Aana was done to fix the land revenue charges. As per the agricultural productivity the king used to fix the land revenue charges in the form of the Aana value because the one rupee has been valued in 16 Aana value and accordingly the land revenue charges had been fixed.

The revenue collection of the agricultural land continued from the Mughal period to Zamindari system and Taluqdari system of the land revenue collection in the form of cash and kind. The Aana value collection of the land revenue charges continued in the British period. The revenue collection from the agricultural land on the basis of Aana value continued after independence. The Aana value system of the land revenue charges fixed as per the agricultural productivity of the land including multiple cropping and gross cropping including Rabi, Kharif and Jaid agricultural cropping season. The agricultural land adjacent to the village settlement with good quality soil and good fertility status of the soil having proper irrigation system has value very high having 12 Aana land value upto 16 Aana land revenue value. Such high Aana value land has been termed by the revenue officials as the Avval land quality. The agricultural land valuing from 8 Aana value to 12 Aana land revenue value has been termed as Doyam land quality land, having medium agricultural productivity land and medium soil fertility status of the land. The agricultural land value ranging from 4 Aana land value to 8 Aana land revenue value has been termed as Soyam land quality having low agricultural productivity and low fertility status of the soil to produce agricultural crops. Presently the land revenue charges are fixed on the basis of the Aana value of the land based on the Avval, Doyam, Soyam quality of land on the basis of agricultural productivity of the land and fertility status of the soil to produce the good quality and quantity of the agricultural crops. Very low land Aana value revenue collection charges less than 4 Aana value land has been fixed as very low or poor quality land and such land may be termed as other types of the very poor land quality.

Thus, the revenue department of the State Government has fixed the land revenue charges on the basis of the following quality or capability of the land to produce the agricultural crops:

- Avval land quality.
- Doyam land quality.
- Soyam land quality.
- Other types of very poor land quality.

Other types of very poor land quality value in the form of Aana value less than 4 Aana value land quality has been classified based on the very poor agricultural productivity and also very poor fertility status of the soil including poor irrigation facilities and other agricultural cropping support system.

Thus, in this study agricultural land capability has been classified as Avaal, Doyam, Soyam and other types which has been categorized by the revenue officials of the Government based on the agricultural productivity and fertility status of the soil to produce the agricultural crops. The Avval, Doyam and Soyam land quality classifications are mentioned in the revenue records to fix the land revenue charges based on the agricultural productivity. The land capability classifications are noted in the revenue Khasara Register and revenue Khasara Maps and also in other revenue records for the revenue collection or fixing the Malgujari of the land. The Malgujari of the land, i.e. revenue charges of the land is being updated during the each agricultural cropping season to fix the revenue charges as per the agricultural productivity including the natural hazards and calamities.

2. LAND CAPABILITY CLASSIFICATION

The land capability classification based on the various land capability classes using the criteria of soil characteristics and limitations to produce the agricultural crops the land capability classification has been classified by the Soil Survey Manual, All India Soil and Land Use Survey Organization, Indian Agricultural Research Institute (IARI), New Delhi.

Fig. 8.15: Land Capability Classification

Land Capability Class	*Sub-class*	*Land Characteristics*	*Agricultural productivity limitations*
Class-I	No Sub-Class	Nearly land level with deep, well, drained, satisfactory soil fixture, and soil structure, not affected by land harmful factors like Kankar, granules soil erosion, soil wetness and salinity and alkalinity of soil.	No agricultural productivity limitations, suitable for intensive and multi agricultural cropping, of the Agro-climatic conditions adapted for agricultural crops.
Class-II	Sub-class designed IIe; erosion hazard II; Soil wetness IIS; root zone limitations.	Gently to moderately sloping land with deep soils of fairly satisfactory texture and structure, affected by moderate erosion wetness and alkalinity and salinity.	Productivity limitations due to moderate erosion, drainage impedance, and wetness, droughtness, salinity and alkalinity, cultivation of crops adopted to soil limitation.
Class-III	IIe; erosion hazard III3; Soil wetness IIIw; root zone limitations.	Gently to strongly sloping land with moderately deep soils having very slowly permeable substratum, affected by marked erosion wetness presence of shallow water table and salinity and alkalinity.	Productivity limitations due to moderate erosion, drainage impedance and wetness droughtness and salinity and aikalinity cultivation of with precautions against permanent land damage and poor water holding capacity and careful selection of crops adapted to soil limitations.
Class-IV	IVe; erosion hazard IVw; Soil wetness IVs; root zone limitations.	Strongly sloping to sheep land, deep to shallow sandy or concretionary soil with low moisture holding capability, affected by severe erosion, pronounced marshiness/droughtness and very high salinity and alkalinity.	Productivity limitations due to very severe erosion, pronounce marshiness/ droughtness and very severe root zone limitations, occasional cultivations of crops in rotation with grasses.

It has been observed that Class-I land capability class occupies 25 to 35 per cent area in Amethi Block having no agricultural productivity limitation and suitable for intensive agricultural cropping of agro-climatically adapted agricultural crops.

The land capability Class-II occurs in 25 to 35 per cent in the various village panchayats of the Amethi Block. The land capability Class II land areas have limitations to agricultural crop productivity due to moderate drainage impedance and soil sodicity singly and in combination. The land capability class II areas with coarse textured soils are available in the various village panchayats of the Amethi Block.

The land capability Class III occupies approximately 30 per cent area in the Amethi Block. The land capability Class III land areas have limitations to crop productivity due to marked drainage impedance, high sodicity and marked depth limitations to crop rooting in sub-soil kankar layers and inundation with water logging.

The land capability Class IV occupies about 5 to 10 per cent area in the various village panchayats of the Amethi Block. The land capability Class IV land has limitations to agricultural crop productivity. In these areas water logging and pronounce marshiness crises due to lying and poor drainage system to drain out the accumulated water. Very high sodicity and severe root zone restrictions due to soil concretion area observed in the class capability Class IV land in the various villages of the Amethi Block of the district Amethi.

In reality the land capability classes based on the agricultural productivity, limitations to agricultural productivity, soil characteristics, agro-climatic suitability to the agricultural productivity and natural calamities/hazards including other factors responsible for the classifications of the land capability classes classified as above has not been accepted and adopted by the revenue officials. The revenue officials adopted the land capability classes based on the agricultural productivity and soil fertility status of the land to produce agricultural crops. The land revenue charges are fixed based on the Aana value of the cultivable land. As per the Avval, Doyam and Soyam land quality the Malguzari of the land is charged for the collection of the revenue charges. The Aana value of the land as per agricultural productivity and soil fertility status of the land capabilities are classified by the revenue authorities.

3. LAND CAPABILITY OF AMETHI

The land capability classified by the revenue authorities of the Tehsil Amethi and adopted in the revenue records including Khasara Map, Khasara Register and fixation of the Malguzari, i.e. land revenue charges as per agricultural productivity and fertility status of the soil.

It has been recorded from the revenue records that agricultural land capability has been classified as:

1. Avval Land (Good Quality Land)
2. Doyam Land (Medium Quality Land)
3. Soyam Land (Low Quality Land)

The Avval, Doyam and Soyam land quality classes has been depicted in the Khasara Maps of the respective revenue villages of the Amethi Block of Amethi Tehsil.

It has been recorded that there are 1538 hectares Avval/Good Quality cultivable land in the Amethi Block covering all 44 village panchayats and 82 revenue villages along with 377 purwas/tolas/hamle villages.

There are 2707 hectares Doyam quality land in the Amethi Block and 1067 hectares land are recorded as Soyam quality land which is very low and poor quality soil in the Amethi Block of the district Amethi.

It has been noted that 25.17 per cent area has been covered by the Avval/Good quality agricultural land to the total net cultivated area of the Amethi Block. The Doyam agricultural land quality shares 44.31 per cent land to the total net cultivated area of the Amethi Block. It has been observed that Soyam/Low agricultural land quality contributes 17.46 per cent land to the total net cultivated area of the Amethi Block. It has been observed that other types of very poor quality agricultural land shares 13.33 per cent land of the total net cultivated area of the Amethi Block.

The cultivable land capability of the Block Amethi indicates that there are 11.61 per cent Avval/Good quality land of the total geographical area of the Amethi Block. It has been recorded that there are 20.43 per cent Doyam/medium cultivable land quality of

Table 8.6: Land Capability (Block Amethi)

Sl. No.	*Name of the Villages Panchayat*	*Area in Hectare*				*%age to Total Geographical Area*				*%age to Net Cultivated Area*			
		Avval Land	*Doyam Land*	*Soyam Land*	*Others Types*	*Avval Land*	*Doyam Land*	*Soyam Land*	*Others Types*	*Avval Land*	*Doyam Land*	*Soyam Land*	*Others Types*
1	*2*	*3*	*4*	*5*	*6*	*7*	*8*	*9*	*10*	*11*	*12*	*13*	*14*
1	Katra Maharani	22	35	10	5	16.4	26.1	7.4	3.1	30.5	48.6	13.8	6.9
2	Sarai Khema	55	90	50	40	9.8	16.1	8.9	7.1	22.1	37.5	20.8	16.6
3	Trilokpur	45	72	30	15	14.1	22.5	9.4	4.7	27.7	44.7	18.5	9.2
4	Ramdaipur	15	25	7	3	14.2	23.8	6.6	2.8	30.0	50.0	14.0	6.0
5	Jangal Ram Nagar	160	222	120	101	13.3	18.5	10.0	8.4	26.5	36.8	19.9	16.7
6	Benipur	13	35	7	3	10.1	27.3	5.4	2.3	22.4	60.3	12.0	5.17
7	Parsawan	35	70	26	15	11.0	22.0	8.2	4.7	23.9	47.9	17.8	10.2
8	Mahmoodpur	40	70	25	15	12.5	21.8	7.8	4.6	26.6	46.6	16.6	10.0
9	Raipur Fulwari	30	60	25	15	10.5	21.0	8.7	5.2	23.0	46.1	19.2	11.5
10	Umapur Gana Patti	35	75	20	15	10.7	22.9	6.1	4.5	24.1	51.1	13.7	10.3
11	Bhaganpur	20	35	10	5	13.2	23.1	6.6	3.9	28.5	50.0	14.2	7.1
12	Tala	20	40	15	10	11.0	22.0	8.2	5.5	23.5	47.0	17.6	11.7
13	Kushi tali	45	75	30	20	13.1	22.9	8.7	5.8	26.4	44.1	17.1	11.7
14	Loharta	6	15	3	2	11.7	29.0	5.0	3.0	23.0	57.0	11.0	7.0
15	Darkha	45	70	25	18	15.0	23.3	8.3	6.0	28.4	44.3	15.8	11.3
16	Mochwa	20	40	15	5	12.0	24.0	9.0	3.0	25.0	50.0	18.7	6.2
17	Nuwawan	70	124	30	20	15.3	27.1	6.8	4.3	28.6	50.8	12.2	4.3
18	Loniapur	12	25	8	3	11.0	22.9	7.3	2.7	25.0	52.0	16.6	7.0
19	Kherauna	20	35	15	5	11.8	20.7	8.8	2.9	26.6	46.6	20.0	6.6
20	Katra Phulkunwar	22	35	16	5	11.4	20.0	9.14	2.8	28.2	44.8	20.5	6.4
21	Hathkila	35	50	25	15	16.2	23.1	11.5	6.9	28.0	40.0	20.0	12.0
22	Rebha	10	25	7	3	10.0	25.0	7.0	3.0	22.0	55.0	15.5	6.0
23	Raidaipur	30	50	20	15	11.0	19.2	7.6	5.7	26.0	43.4	17.3	13.0
24	Korari Girdhar Shah	45	70	35	20	11.7	18.2	9.1	5.2	26.4	41.1	20.5	11.7
25	Dedhpasar	40	70	20	10	12.7	22.2	6.3	3.1	28.5	50.0	14.2	7.1
26	Mahaso	10	20	5	5	11.6	23.2	5.8	5.8	25.0	50.0	12.5	12.5
27	Gangauli	40	75	35	20	10.5	19.7	9.2	5.2	22.8	42.8	20.0	11.4
28	Dhadhudhar	15	30	10	5	11.2	22.5	7.5	3.7	25.0	50.0	16.6	8.3
29	Gaderi	30	50	25	15	11.1	18.6	9.3	5.5	25.0	41.6	20.8	12.5
30	Bhusahari	30	55	25	15	12.7	23.4	10.6	6.3	24.0	44.0	20.0	12.0
31	Purabgaon	25	55	15	10	11.2	24.7	6.7	4.5	23.8	44.0	14.2	9.5
32	Chaturbhujpur	15	25	10	6	12.5	20.8	8.3	5.0	26.7	44.6	17.8	10.7
33	Mahrajpur	25	40	18	10	15.2	24.3	10.9	6.0	26.8	43.0	19.3	10.7
34	Aghar	31	40	20	15	14.2	18.4	9.2	6.9	27.6	35.7	17.8	13.3
35	Saidpur	20	40	15	5	11.1	22.2	8.3	2.7	25.0	50.0	18.75	6.2
36	Kakwa	40	70	30	20	10.8	19.0	8.1	5.4	25.0	43.7	18.7	12.5
37	Saraiya Duban	40	60	20	10	13.3	20.6	6.6	3.3	30.7	46.1	15.3	7.6
38	Kohra	70	130	45	30	12.2	22.6	7.8	5.2	25.0	47.0	16.3	10.9
39	Mahmadpur	40	120	30	20	8.9	26.9	6.7	4.4	19.0	57.1	14.2	9.5
40	Naraini	80	130	40	30	13.6	22.2	6.8	5.1	28.5	46.4	14.2	10.7
41	Nainaha	25	40	15	10	13.7	21.9	8.2	5.4	27.7	44.4	16.6	11.1
42	Ramgarh	80	140	50	25	11.6	20.4	7.2	3.6	27.1	47.4	16.9	8.4
43	Himmatgarh	50	80	40	22	12.8	20.5	10.2	5.6	26.0	41.0	20.8	11.4
44	Dehara	30	54	25	15	12.7	22.9	10.6	6.3	23.8	42.8	19.8	11.9
	District Amethi	1538	2707	1067	817	11.61	20.43	8.01	6.17	25.17	44.31	17.46	13.33

the total geographical area of the Amethi Block. There are 8.01 per cent Soyam land quality to the total geographical area of the Amethi Block. The other types/very poor quality cultivable land quality has been recorded 6.17 per cent of the total geographical area of the Amethi Block.

It has been observed that the Doyam quality land contributes very significant area under the net cultivated land. The Avval/good quality land also contributes more than one-tenth part of the total geographical area of the Amethi Block.

Thus it has been observed that the net cultivable area in the Amethi Block has been found comparatively low out of which Doyam land quality has been observed significantly distributed while Avval land quality is very insignificantly distributed in the Amethi Block. It indicates that proportionately the net cultivated land has been found comparatively low out of which Avval land quality contributes very insignificant proportion than Doyam and Soyam cultivable land quality. Very low occurrence of the Avval land quality has adversely affected the agricultural productivity of the Amethi Block which has led to the very poor agricultural productivity and socio-economic crisis for the people of Amethi.

4. LAND CAPABILITY OF THE VILLAGE PANCHAYATS

The land capability classification has been done by the various institutions as per the requirements. The agricultural institutions and soil survey organizations and other institutions, has classified the land capability classes based on the different criteria especially the soil characteristics, fertility status of the soil to produce crops and plants, salinity/alkalinity status of the soil, agro-climatic conditions, the limitations to the agricultural productivity, natural hazards/calamities etc. But the actual and realistic classification of the cultivable land etc. has been done by revenue authorities based on the land use system. The land capability classes or the quality of the land has been valued in the form of the Aana value. The land value has been considered upto one Rupee, i.e. 16 Aana value. The very good fertile land has been valued upto 16 Aana value based on the multiple agricultural cropping of the land and very good fertility status of the soil to produce the crop. The multi-agricultural cropping land, cropped during all three Kharif, Rabi, Jaid agricultural cropping of the land has been classified under the Avval/very good quality land while very poor fertility status of the soil, low agricultural productivity land and single agricultural cropped land quality has been classified as Soyam land in the revenue system.

The land capability Class-I, Avval land quality has been valued in rupees from 12 Aana value to 16 Aana value land in which the multi-cropping of the land including Kharif, Rabi and Jaid cropping season including multi-agricultural cropping and commercial cropping is being cropped in the Avval/good quality land. Avval land quality has good soil characteristics and very good fertility status of the soil. The Avval land quality has no agricultural productivity limitations and the Avval land is suitable for intensive, multi-cropping and commercial cropping of the cultivable land.

The IInd Class capability land has been classified as Doyam land quality which has been valued in terms of the Rupee Aana value varying between 8 Aana land value to 12 Aana land value quality. The Doyam quality land is double cropped land and medium agricultural productivity on the Doyam land. The doyam land quality has medium fertility status of the soil. The doyam land quality has medium agricultural productivity limitations due to agro-climatic conditions and medium soil fertility limitation to produce the agricultural crops.

Fig. 8.16: Land Capability Classifications (Amethi)

Sl. No.	*Class*	*Land Capability*	*Aana Value of the land*	*Agricultural Productivity*	*Soil Characteristics*	*Productivity Limitations*
1.	Class-I	Avval Land	12 Aana to 16 Aana Value land	Multi-cropping good agricultural productivity	Very good fertility status of the soil	No productivity limitations suitable for intensive and multi-agricultural cropping.
2.	Class-II	Doyam Land	8 Aana to 12 Aana Value land	Double cropped medium productivity	Medium fertility status of the soil	Productivity limitations due to agro-climatic conditions and soil limitations.
3.	Class-III	Soyam Land	4 Aana to 8 Aana Value land	Single cropped medium productivity	Low fertility status of the soil	Productivity limitations due to moderate agro-climatic conditions selection of crops adapted to soil limitations.
4.	Class- IV	Other types/poor	Less than 4 Aana value land.	Single cropping very low productivity	Very low fertility status of the soil.	Productivity limitations due to very severe agro-climatics conditions occasional cultivation of crops.

The IIIrd Class land capability class has been termed as Soyam land quality which has been valued from 4 Aana value to 8 Aana value land quality. The Soyam quality land has been single/double agricultural cropped land. The agricultural productivity of the land has been found low productive and agricultural fertility status of the soil has been found low. Limitations of the agricultural productivity of the Soyam land has been observed due to various agro-climatic conditions and the selection of the agricultural crops has been adopted due to the soil limitations.

The land capability Class IV has been classified as other types or poor quality land and the land has been valued in Rupee from less than 4 Aana value of the land. The land has been noted single agricultural cropping land having very low agricultural productivity of the land. Such land has very low fertility status of the soil. The agricultural productivity limitations has been recorded due to the very severe agro-climatic conditions and occasional cultivation of the agricultural crops.

The various land capabilities classes as classified by the revenue authorities based on the agricultural productivity status of the land and fertility status of the soil in the Aana value basis has been classified as Avval, Doyam, Soyam and other types of the land quality. The classified land capability classes and the area under the different land quality and its percentage of the total net cultivated area and the land quality area percentage of the total geographical area of the respective village panchayat of the Amethi Block has been analyzed because the land capability classification based on the agricultural productivity and the fertility status of the soil etc. has been explained at village panchayat level in the Amethi Block of the district Amethi.

I. Avval Land Capability

The proportionate distribution of the Avval land to the total net cultivated land has been depicted in Map 8.74. The Avval land capability class has been significantly noted in the various village panchayats of the Amethi Block. It has been observed that more than 30 per cent area under Avval quality lands of the total net cultivated land has been found in the Katara Maharani and Saraiya Duban village panchayats of the Amethi Block. Medium percentage between 25 per cent to 30 per cent area under Avval quality land of the total net cultivated land has been noted in the Trilokpur, Ramdaipur, Bhaganpur, Jangal Ram Nagar, Kushi Tali, Mahmoodpur, Kherauna, Katara Phulkunwar, Hathkila, Dedhpasar, Lonaipur, Raidaipur, Korari Girdhar Shah, Nuwawa, Darakha, Agahar, Chaturbhujpur, Maharajpur, Naraini, Nainaha, Ramgarh and Himmatgarh village panchayats of Amethi Block. Low percentage of the Avval land quality ranging between 20.0 per cent to the 25.0 per cent avval land of the total net cultivated land has been observed in the Benipur, Sarai Khema, Parsawan, Raipur Phulwari, Umapur Gana Patti, Loharata, Tala, Mochawa, Dhandhudhar, Purabgaon, Kohara, Saidpur and Kakawa village panchayats of the Amethi Block. The very low percentage of the Avval land quality less than 20.0 per cent of the net cultivated land has been recorded in the Mahmadpur village panchayat of the Amethi Block.

The Avval quality land of the total geographical area of the village panchayats has been depicted in Map 8.75. Very high percentage of the Avval land of the total geographical area of the village panchayats having more than 15.0 per cent has been recorded in Katara Maharani, Hathkila, Nuwawan and Maharajpur village panchayats of the Amethi Block. The medium percentage ranging 13.0 to 15.0 per cent avval land of the total geographical area has been noted in the Trilokpur, Ramdaipur, Jangal Ram Nagar, Kushi Tali, Agahar, Naraini, Nainaha and Saraiya Duban village panchayats of the Amethi Block. It has been observed that low percentage of Avval land, varying between 11.0 and 13.0 per cent of the total geographical area has been recorded in the Tala, Loharata, Mahmoodpur, Parsawan, Kherauna, Katara Phulkunwar, Dedhpasar, Mahaso, Loniapur, Raidaipur, Korari Girdhar Shah, Mochawa, Chaturbhujpur, Purabgaon, Ramgarh, Dehara, Himmatgarh, Bhusahari, Gaderi, Dhandhudhar, Kohara and Saidpur village panchayats of the Amethi Block. Very low percentage less than 11.0 per cent Avval land of the total geographical area has been observed in the Sarai Khema, Benipur, Raipur Phulwari, Rebha, Umapur Gannapatti, Darakha, Gangauli, Kakawa and Mahmadpur village panchayats of the Amethi Block.

It has been noted that in most of the village panchayats the area under the Avval land has been significantly reduced due to excessive doses of chemical fertilizers, over doses of canal irrigation, and application of insecticides and pesticides including excessive agricultural practices. The reduction in the Avval land quality has to be controlled and more areas under the Avval quality land has to be developed through soil testing and agricultural cropping has to be ensured as per the fertility status requirements of the soil. High yielding variety of seeds, and appropriate irrigation system with good quality water and suitable agricultural rotation in the cropping system will be helpful in controlling the reduction in the area under Avval land and also increasing the area under the Avval quality land. The appropriate and scientific agricultural cropping of the Avval land quality will be helpful in the restoration and proper conservation of the availability natural land resources of the Amethi Block of the District Amethi.

II. Doyam Land

The Doyam land has been classified as medium quality cultivable land in which multi-cropping including Rabi and Kharif season's crop are cropped. The land fertility, the soil characteristics, and the agro-climatic conditions are favourable for agricultural activities including intensive agricultural cropping. The distribution of the Doyam land in hectares and its percentage of the net cultivated land and Doyam land's percentage of the total geographical area of the village panchayats of the Amethi Block has been mentioned in Table 8.6. The proportionate distribution of the Doyam land to the net cultivated area of the village panchayat level in Amethi Block has been depicted in Map 8.76. It has been observed that the doyam land contributes more than 50 per cent land of the net cultivated area in the Loharata, Umapur Gana Patti, Nuwawan, Rebha and Mahmadpur village panchayats of the Amethi Block. Proportionately medium percentage between 45 to 50 per cent area under Doyam land of the net cultivated area has been recorded in Katara Maharani, Trilokpur, Bhaganpur, Tala, Mahmoodpur, Parsawan, Katara Phulkunwar, Kherauna, Dedhpasar, Mahaso, Dhandhudhar, Mochawa, Naraini, Nainaha, Ramgarh, Saraiya Duban, Kohara and Saidpur village panchayats of the Amethi Block. Low percentage varying 40 to 45 per cent area under Doyam land of the total net cultivated area has been observed in Kushi Tali, Darakha, Korari Girdhar Shah, Raidaipur, Katara Phulkunwar, Hathkila, Gangauli, Bhusahari, Himmatgarh, Gaderi, Purabgaon, Chaturbhujpur, Maharajpur, Darakha, Kakawa and Dehara village panchayats of the Amethi Block. Proportionately very low percentage area under Doyam land less than 40 per cent of the total net cultivated land has been observed in the Sarai Khema, Jangal Ram Nagar of the Amethi Block.

The proportionate distribution of the Doyam land capability of the total geographical area of the village panchayat has been depicted in Map 8.77. Proportionately very high percentage area under Doyam land more than 25 per cent of the total geographical area has been noted in the Katara Maharani, Benipur, Nuwawan, Loharata and Mahmadpur village panchayats of the Amethi Block. The medium percentage varying from 22 to 25 per cent area under Doyam land of the total geographical area has been observed in the Trilokpur, Ramdaipur, Bhaganpur, Tala, Kushi Tali, Umapur Ganapatti, Loniapur, Rebha, Hathkila, Dedhpasar, Mahaso, Bhusahari, Dhandhudhar, Purambgaon, Naraini, Darakha, Mochawa, Maharajpur, Saidpur, and Kohara village panchayats of Amethi Block. Low percentage distribution of Doyam land less than 19.0 per cent of the total geographical area has been noted in the village panchayats Sarai Khema, Jangal Ram Nagar, Korari Girdhar Shah, Gaderi and Agahar village panchayats of the Amethi Block.

It has been noted that the area under Doyam land quality has been recorded significantly than the Avval land quality in the various village panchayats of the Amethi Block. The significant reduction in the area under Doyam land has been great concern to the farmers and agricultural policy makers and planners of the grassroots level. The introduction of appropriate agricultural cropping including high yielding variety of seeds, application of the organic and inorganic amendments as per requirement of the soil to produce the agricultural crops and need based use of insecticides and pesticides including appropriate good quality water irrigation management with the views to ensure intensive agricultural cropping and multi-cropping system in the Rabi, Kharif and Jaid agricultural cropping seasons may be helpful in increasing the area under Avval land especially to increase the agricultural demand of the growing cattle and human population for the fuel wood, fodder, fibre, fruits, food grains and fisheries (F^6), wood and vegetables and also to improve the ecological imbalances and degraded environment of the area.

III. Soyam Land

The Soyam land is that cultivated land in which single and double agricultural cropping has been carried out and agricultural productivity has been reduced due to low fertility status of the soil and excessive over agricultural cropping and high doses of the chemical fertilizer and over irrigation has let to an increase in the area under Soyam land capability in the Amethi Block. The Soyam land capability distribution at village panchayat level of the total net cultivated area has been depicted in Map 8.78. It has been noted that proportionately very high percentage more than 20.0 per cent area under Soyam land quality of the total net cultivated area has been recorded in the Sarai Khema, Katara Phulkunwar, Korari Girdhar Shah and Gaderi village panchayats of Amethi Block. Medium percentage area under the Soyam land ranging between 18 and 20 per cent land of the total net cultivated area has been observed in the Trilokpur, Jangal Ram Nagar, Raipur Phulwari, Kherauna, Hathkila, Mochawa, Gangauli, Bhusahari, Dehara, Maharajpur, Saidpur and Kakawa village panchayats of Amethi Block. Very low percentage area under Soyam land ranging from 16 to 18 per cent of the total net cultivated area has been noted in the Tala, Kushi Tali, Mahmoodpur, Parsawan, Loniapur, Raidaipur, Dhandhudhar, Agahar, Chaturbhujpur, Ramgarh, Nainaha and Kohara village panchayats of Amethi Block. Very low percentage area under Soyam land varying less than 16 per cent area of the total net cultivated land has been observed in the Katara Maharani, Ramdaipur, Bhaganpur, Loharata, Darakha, Umapur Gana Patti, Nuwawan, Rebha, Mochawa, Dedhpasar, Purabgaon, Saraiya Duban, Naraini and Mahmadpur village panchayats of the Amethi Block of the district Amethi.

It has been observed that proportionately very high percentage area under the Soyam land more than 10 per cent of the total geographical area has been noted in Maharajpur, Bhusahari, Himmatgarh and Hathkila village panchayats of Amethi Block. Proportionate distribution of the area under Soyam land of the total geographical area of the village panchayats of the Amethi Block has been depicted in Map. 8.79.

Proportionately medium percentage of the area under the Soyam land varying between 9 and 10 per cent area of the total geographical area has been found in the Trilokpur, Ram Nagar, Katara Phulkunwar, Korari Girdhar Shah, Gangauli, Gaderi, Chaturbhujpur and Agara village panchayats of Amethi Block. Low percentage ranging between 8 and 9 per cent area under the Soyam land of the total geographical area has been recorded in the Sarai Khema, Raipur Phulwari, Kherauna, Tala, Kushi Tali, Darakha, Mochawa, Nainaha, Kakawa and Saidpur village panchayats of Amethi Block. It has been observed that less than 8 per cent land under the Soyam land of the total geographical area has been noted in the Katara Maharani, Ramdaipur, Bhaganpur, Mahmoodpur, Benipur, Loniapur, Loharata, Nuwawan, Raidaipur, Rebha, Dedhpasar, Mahaso, Dhandhudhar, Purabgaon, Ramgarh, Naraini, Saraiya Duban, Kohara and Mahmadpur village panchayats of Amethi Block.

It has been observed that the area under the Soyam land quality has been significantly distributed at village panchayat level in the Amethi Block. The area under the Soyam land can be brought under double cropping, multi-cropping or intensive agricultural cropping to increase the agricultural productivity of land through proper irrigation management, proper application of organic and inorganic manures, use of high yielding variety of seeds, testing of the soil to know the fertilizer requirements of the soil to produce the agricultural crops with the increase agricultural productivity. An attempt must be initiated to check the further degradation of the Soyam land and also to increase the fertility status of the soil classified under the Soyam land with the view to increase agricultural productivity of the land through multi and intensive agricultural cropping system in the various villages of the Amethi Block.

5. OTHER TYPES OF LAND CAPABILITY

The other types of the land capability class has been very less fertile cultivable and in which single cropping with difficulty has been carried out. The fertility status of the soil has been found poor and such lands are observed low fertile land with the low agricultural productivity due to soil limitation, hardship in the irrigation management. Due to excessive doses of the chemical fertilizer and over-irrigation and intensive agricultural cropping the land has lost its fertility status of the soil due to which the agricultural productivity of the land has been reduced. The other types of the land capability classes cultivable land has been deteriorated due to over cropping and mis-management, due to poor fertility status of the soil. The proportionate distribution of the other types of the land capability class area of the net cultivated area at the village panchayat level in the Amethi Block has been depicted in Map 8.80. The very high percentage of other types of the land more than 12 per cent land of the net cultivated area has been observed in the Sarai Khema, Jangal Ram Nagar, Ramdaipur, Gaderi, Agahar, Mahaso and Kakawa village panchayats of the Amethi Block. Medium percentage distribution of the other types of the land capability class verying between 10 and 12 per cent land of the total geographical area has been found in the Tala, Kushi Tali, Umapur Ganapatti, Darakha, Raipur Phulwari, Parsawan, Darakha, Korari Girdhar Shah Gangauli, Bhusahari, Himmatgarh, Hathkila, Chaturbhujpur, Maharajpur, Naraini, Nainaha, Dehara and Kohara village panchayats of the Amethi Block. Low percentage distribution of the other types of the land ranging between 8 and 10 per cent area of the total net cultivated land has been recorded in the Trilokpur Mahmoodpur, Dhandhudhar, Purabgaon, Ramgarh and Mahmadpur village panchayats of the Amethi Block. Very low percentage, distribution of other types of the land capability, less than 8 per cent land, to the Katara Maharani, Ramdaipur, Bhaganpur, Loharata Benipur, Kherauna, Katara Phulkunwar, Dedhpasar, Rebha, Loniapur, Mochawa, Nuwawan, Saidpur and Saraiya Duban village panchayats of the Amethi Block.

The proportionate distribution of the other types of the land capability class to the total geographical area of the village panchayats of Amethi has been depicted in Map 8.81. It has been observed that more than 8 per cent area under other types of the land capability class to the total geographical area has been found in the Sarai Khema, Jangal Ram Nagar, Kushi Tali, Tala, Darakha, Hathkila, Loniapur, Raipur Phulwari, Hathkila, Mahaso, Raidaipur, Korari Girdhar Shah, Gangauli, Mahaso, bhsahari, Himmatgarh, Dehara Nainaha, Gaderi, Chaturbhujpur, Maharajpur, Agahar, Naraini, Nainiha, Kohara and Kakawa village panchayats of the Amethi Block. Medium percentage distribution of the other types of the land capability ranging between 4 and 5 per cent area of the total geographical area has been noted in the Trilokpur, Mahmoodpur, Parsawan, Umapur Gana Patti, Nuwawan, Purabgaon, and Mahmadpur village panchayats of the Amethi Block. Proportionately low percentage distribution of other types of the land capability class with low fertility status of the soil varying between 3 and 4 per cent of the total geographical area has been recorded in the Katara Maharani, Bhaganpur, Loharata, Mochawa, Rebha, Dedhpasar, Dhandhudhar, Ramgarh and Saraiya Duban village panchayats of the Amethi Block. Very low percentage less than 3 per cent of other types of the land capability of the total geographical area has been observed in the Ramdaipur, Benipur, Kherauna, Katara Phulkunwar, Ramdaipur and Saidpur village panchayats of the Amethi Block.

It has been noted that the fertility status of the soil has been deterioraed in the intensive cultivated area and the agricultural productivity of the land has also been reduced significantly which has led to the crisis among the farmers of the Amethi and also cause ecological imbalance of the area and environmental degradation. It has become a great concern to the farmers that the cultivated area has been reduced and it has also lost the fertility status of the soil. The problem of salinity/alkalinity, calcareousness of the soil, water logged problem and loss of the fertility status of the soil due to over agricultural cropping, high doses of chemical fertilizer and excess irrigation and over use of insecticides and pesticides has been the main reasons of the deterioration of the fertility status of the soil. The land has been one of the most important natural resources/ infrastructure has to be properly innovated and possibilities should be explored for optimum utilization of such auspicious natural land resources. Proper multiple agricultural cropping, application of the organic and inorganic amendments including insecticides and pesticides as per head of the cultivated crops and suitability of the soil to produce the agricultural crops and proper irrigation management system and application of the bio-fertilizer and use of high yielding variety seeds and introduction of intensive and multiple agricultural cropping can control the deterioration of the fertility status of the soil and helpful in increasing the agricultural productivity of the land. Attempt should be made to reclaim the wastelands under agricultural cultivation activities, agro-forestry, social-forestry and horticultural activities in various village panchayats of the Amethi for the optimum exploitation of the available natural land resources and the natural infrastructure of the Amethi. The restoration of the degraded land and to check the degraded environment of the area and to increase the area under the net cropped area must be initiated at the war footing level. The conservation of the environment and optimum utilization of the available land resources will be an asset to meet the increasing demand of the growing cattle and human population and the maintain the ecological in balance of the area and to improve the environmental crisis of the Amethi.

5. CONCLUSION

The land capability has been defined as the capability of the land to produce the crop and other types of agricultural agro-forestry-social-forestry and horticultural activities. The term capability of the land has been perceived by the land users in their own terms. The land capability has been classified in terms of the cultivable land. As per the agricultural productivity and quality of the land the revenue charges has been fixed. The land capability classification has been classified historically in terms of Aana value land collection during the Mughal period. High Aana value land quality revenue charges has been high while for the low Aana value land quality of revenue charges has been low due to low fertility status of the soil and low agricultural productivity of the land. The revenue collection on the basis of land Aana value was collected during the zamindari and taluqdari tenures. The agricultural land adjacent to the settlement areas has been classified with higher Aana value land quality. The revenue authorities have classified the land quality based on the Aana value land quality having higher Aana value has been classified as Avval land and the lower value of the Aana value land has been classified as Soyam land quality. The revenue authorities have classified the land as Avval, Doyam and Soyam land quality and the other types of land capability has been classified for the very poor land quality and very poor fertility status of the soil.

The land quality as mentioned in the revenue records especially in the Khasara Register and the Khasara maps including for the collection of the Malguzari by the revenue authorities are fixed based on the Avval, Doyam, Soyam qualitiy of the land. The land capability classification has been classified in terms of the agricultural productivity of the land and the fertility status of the soil to produce the agricultural crops along with other factors related to the agricultural inputs and agro-climatic conditions.

It has been noticed that 1538 hectares land has been found under the Avval land while the area under the Doyam land has been recorded as 2707 hectares while 1067 hectare land has been observed as Soyam land in the Amethi Block.

It has been recorded that there are 11.61 per cent land under the Avval quality while 20.43 per cent land has been recorded under the Doyam land and 8.01 per cent land has been found under the Soyam land quality. The area under other types of the land quality has been recorded 6.17 per cent of the total geographical area of the Amethi Block. It has been observed that very high more than 16.2 per cent land under the Avval land quality has been found in village Katara Maharani and Hathkila village panchayats of the Amethi Block. Very low per cent of Avval land having 8.9 per cent of the total geographical area has been noticed in the Mahmadpur and 9.8 per cent area in Sarai Khema village panchayats of the Amethi Block. It has been observed that proportionately very high percentage 29.0 per cent area under Doyam land has been found in the Loharata and very low percentage having 16.1 per cent area under Doyam land of the total geographical area has been noticed in the village panchayat Sarai Khema of the Amethi Block. The highest percentage of the Soyam land having 11.5 per cent area of the total geographical area has been observed in the Hathkila and very low percentage area with 5.0 per cent area under the Soyam land has been recorded in the Loharata village panchayat of the Amethi Block. The highest percentage of the other types of land quality has been noticed in the Jangal Ram Nagar having 8.4 per cent area of the total geographical area and the very lowest percentage having 2.3 per cent land under the other types of the land quality has been observed in the village panchayat Benipur of the Amethi Block.

Thus, it has been noticed that the area under the Avval land quality enriched with the fertility status of the soil and highly agriculturally productive land has been significantly reduced and the area under the Doyam land and Soyam land quality has been significantly increased which has been found very difficult in terms of increasing agricultural productivity of the area. The reduction in the area under the Avval land quality has been due to deteriorated fertility status of the soil due to over agricultural intensive cropping and high doses of chemical fertilizers and over doses of canal irrigation water has led to an increase in the area under Soyam or poor quality land or an increase in the area under the wastelands. The reduction in the area under the Avval land has to be controlled and protected scientifically and the area in the Soyam and other types of land has to be made more fertile in terms of increasing agricultural productivity and through reclaiming the wastelands for agricultural agro-forestry and social-forestry programmes. The appropriate application of high yielding variety seeds proper application of organic and inorganic amendments, green manuring and the use of bio-fertilizer, rotation in the agricultural cropping system, proper irrigation management adoption of the latest agricultural scientific techniques should be adopted to increase the area under Avval land quality and adding the area in the multiple cropping and intensive agricultural cropping and agricultural cropping system should be initiated for the Kharif, Rabi, Jaid cropping seasons and commercial agricultural cropping should be promoted to with the view to increase agricultural productivity of the land, to maintain the ecological imbalance of the area and to improve the environmental degradation of the Amethi and to make the area and eco-friendly and to remaintain and improve the environmental crisis of the area. The fertility status of the soil should be maintained to increase the agricultural productivity of the land with the view to meet the ever increasing demand of the growing cattle and human population to the meet the need of the Fuel wood, Fodder, Food grain, Fruits, Fisheries, Fibre (F^6), wood and vegetables with the view to improve the ecological crisis and also to maintain the socio-economic crisis of the people of the Amethi.

9 Agro-Afforestation

1. INTRODUCTION

Agro-afforestation is a system of land use along with combined growing of agricultural crops with social-forestry, animal husbandry including the food grain, fodder, fuel wood, fibre, fruits and fisheries and vegetable. Generally, the common cultivators has been engaged in cultivating agricultural crops along with social-forestry and the domestication of the livestock, wood, fuel wood, fodder, fibre fruits and vegetables etc. to fulfil their day-to-day increasing demands. In response to the agro-ecological conditions of the area during 700 B.C. man changed himself from hunting and the collection of the food to the agricultural crop food production towards the development.

The propagation, promotion and development of the agro-forestry aims, its potential and positive interaction between ecological and economics, agricultural and forestry activities can be emphasized when proved technically and cropping and the increased agricultural production and productivity conservation are some of the important objectives of the agro-afforestation system.

The maintenance of the eco-system and the restoration of ecological and environmental degradation and to improve the socio-economic and ecological crisis, sustainable development of the available natural land resources have to be managed and maintained scientifically. The sustainable development comprises the preservation of environmental, equitable development, scientific and technological change is a dire necessity for obtaining self-sufficiency in food grains, fodder, fuel wood, fibre, fruits, fisheries, vegetables, wood and other necessary produce. An unbroken symbiotic relationship with man and nature has to be re-established. The development should aim at improving the eco-system and the environment for living cattle and human population providing then water, shelter, food and also reclaiming the wastelands for greening and interior rural area inhabitants. The higher standard of living and the increased agro-afforestation area and its productivity must be properly targeted and achieved with despoiling environmental conditions and the nature of its beauty, freshness and purity of the nature are so essential for the survival of down-trodden people surviving in the remote rural areas of the Amethi.

It has been noticed that there has been a shift in the policy during the recent past from sustained productivity to sustained development. The ecological, environmental, scientific, technological and socio-economics inputs are important for determining sustainable management systems. In the present scenario, it is necessary to link the eco-system processes with the multifaceted socio-economic activities centred for the sustainable development of an area.

There are different views and opinions regarding Agro-forestry systems like Agro-afforestation is not a perfect system because there has been both success and failure with the various agro-forestry system has provided viable alternative of land use pattern with a suitable environment and compatible social conditions agro-forestry system should be helpful in providing substantial output. A good agro-forestry-system should be able to solve ecological imbalances of the area, land and environmental degradation, and environmental crisis and socio-economic crisis, and common requirements of the people and it should be helpful in overcoming the physiological, ecological and environmental protection constraints that causes hindrance in the the development of the agro-forestry system for the sustainable development of the remote rural areas of the Amethi.

It has been realized that there is a tremendous pressure on the human population, which has resulted in deforestation and land degradation and also degradation of the valuable natural forest resources of the area. It has been an established fact that unless there is stabilization of population participatory management for the reclamation of the wastelands and agro-afforestation management for sustainable development to meet the increasing demand of the ever increasing cattle and human population for the fuel wood, fodder, fibre, food grain, fisheries, fruits and vegetable etc. and also to restore the degraded environment of the area.

2. AGRO-AFFORESTATION

The symbiotic relationship between man and nature and the symbiotic bond between the rural poor and forests especially among the Scheduled Castes and Scheduled Tribes has been observed closest through the years. The traditional rights of the down-trodden deprived rural poor multitudes to the minor forest produce, for the grasses and fallen dry wood for the fuel wood have kept the rural communities going through the several years. The relationship between man and nature now stands threatened. The agro-afforestation efforts rendered more difficult since the needs of the community and those of the individual are at variance with each other and it cannot be reconciled unless poverty amelioration programmes raise the level of living standard of those people surviving below the poverty line. So far there has been a slackening of the afforestation on the one hand and absence of support of the local communities for the protection and augmentation of the forest cover on the other. Agro-afforestation programmes has to be launched at larger extent through people's participation.

Thus, the agro-afforestation programmes must be initiated and launched as people's movement. The agro-afforestation programmes must be implemented by the people, for the people, and with the people, through leasing the forest and the non-forest land and the wastelands to the deprived rural poor. The Tree Patta Scheme must be reformulated and afforestation procedures should be simplified and the tree growers co-operatives should be promoted and the farmers should be encouraged to undertake free plantation on their own land and also the agricultural field boundaries. So far as the idea of leasing of wastelands is concerned, it has been realized that at the field level individuals must have vested interest in the tree plantation. The our programmes on agro-afforestation and social-forestry and under the poverty alleviation scheme, confirm themselves to provide wage employment to the deprived rural poor, the landless labourers and the unemployed people, in the ongoing agro-afforestation programmes and such programmes are planned properly and preciously because they fail to create such a vested interest among the people to plant trees, and to look after and maintain them with a view to enjoy use the fruit obtained from the agro-afforestation programmes in the remote rural area of the Amethi. Thus there is a tremendous pressure of cattle and human population which has led to deforestation and land degradation. The stabilization of population and people's participatory management for the reclamation of the wastelands and agro-afforestation management for sustainable development.

The magnitude of the problems of the wastelands, land degradation, deforestation and environmental degradation has also been given due priority and importance in the Constitution of India.

Article 48A of the Constitution of India enunciates that "The state shall endevour to protect and improve the environment and to safeguard the forest and wildlife of the country."

Thus the protection of our environment has become prime importance to our country. It was the vision and environmental zeal of the Indira Ji, that led to her for pioneering statement on Environment at Stockholm Conference with the subsequent establishment of Department of Environment. We must prevent further degradation of natural resources and the endowments, and life support system of land, water and vegetation.

The sustainable agro-afforestation and forest management and optimum conservation of natural resources and the orientation of the technological, institutional changes in a manner to attain and continued satisfaction of the human needs for the present and future generations. The sustainable development of the genetic resources, viz. agriculture, horticulture, social-forestry and fisheries. The optimum utilization of the natural resources must be environmentally non-degrading technologically appropriate, socially suitable and economically viable.

The sustainable development recognizes the utilization of natural resources which will change natural eco-system. The sustainable agro-afforestation management is the Agriculture-Social-Forestry, Agro-Forestry, Horticulture, Fisheries component of the sustainable development which aims at ensuring the values derived from the natural land resources to meet the existing requirements of the cattle and human population, while at the same time ensuring their continued ability and contribution to long term development needs of the common people living in the remote rural areas. The plants animal human life are closely interlocked with each other and the disturbance in any one provides an imbalance to the other elements of the nature and environment.

The conservation of the natural resources includes land and water resources which includes classification, enactment of law, micro-level area planning, extension and propagation and campaign for restoration and reclamation of degraded land and its development for the suitable agri-silvi-cultural techniques, and conservation of the degraded land for ensuring economic benefit for sustainable ecological development.

Intensive agro-afforestation management programmes that should treat afforestation in a holistic manner and seeks to involve the communities and the society in general in the preservation and the conservation of environment and forest has been deprived of financial and techno-scientific support system. The agro-afforestation schemes at village level **"Gram Panchayat Van Yojana"** to be worked out with the village panchayats to ensure appropriate utilization of the resources for successful implementation of the agro-afforestation programme with the people's participation.

The agro-afforestation programmes in the district Amethi has not been properly planned because it has been occasionally implemented without people's participation and common minimum requirements of the local people has not been given due importance for successful implementation, maintenance and management of the agro-afforestation programmes in the district Amethi. Various governments and the non-government departments, implementing agencies, viz. Agriculture-Horticulture-Social Forestry Department of the district Amethi has been engaged in the implementation of various agro-afforestation programmes, but these implementing agencies could not achieve the planned goals as per the vision and requirement of the local people and the degraded environment, land and forest cover of the Amethi, with the view to meet the ever increasing demands of the growing cattle and human population for the food grain, fuel wood, fodder, fibre, fisheries, fruits and vegetable etc. and also to maintain the ecological imbalances of the Amethi and to generate the degraded environment and restore the deforested land of the area.

3. AGRO-AFFORESTATION IN AMETHI

The agro-afforestation programme in the Amehti has not been properly planned and scientifically implemented without peoples participation because it requires joint venture and efforts of the implementing agencies, viz. Agriculture-Horticulture-Forest-Fisheries and with the continued involvement and sustained participation of the local people. The agro-afforestation programmes must be implemented for the people, by the people and with the people and demonstrated in the cultivators field in an intensive and integrated system adopting latest developed and innovated agricultural scientific-techniques. It has been noted that the eco-system of the Amethi has been degraded which has led to the ecological imbalances and environmental crisis and socio-economics backwardness of the Amethi. The flora and fauna, the wildlife and natural vegetation in Amethi has been significantly reduced in Amethi. The agro-afforestation is a system of land use along with the combined growing of agricultural crops with social-forestry, horticulture, animal husbandry including vegetables, fuel wood, fodder, fibre, fruits, fisheries, food grain etc. to fulfil the increasing demand of the growing cattle and human population of Amethi.

In fact the forest land in Amethi is insignificant and the forest land has been degraded and the forest land in the revenue record has been very insignificantly noted in the existing revenue records which indicates that the area under the forest land without forest cover has been changed in the revenue records.

The natural vegetation in Amethi has been very insignificantly distributed. The area under bushes, natural forest plants, trees etc. has been very nominal in comparison to the other land use categories. It has been noticed that there are 302 hectares area under the natural vegetation in the Amethi Block. The area under the natural vegetation contributes 2.27 per cent land of the total geographical area of the Amethi Block. The area under the natural vegetation also includes the reserved forest land and the traditional forest areas with the Dhak and other jungle tree but such forest lands are rarely distributed in the Amethi Block. The area under the natural vegetation has been reduced due to over-exploitation of the forest for fuel wood, fodder fruits, wood etc. and subsequent replacement of the tree plantation on the forest land has not been carried which led to the degradation of the forest level.

The social-forestry programmes has been carried out by the District level forest department. The social-forestry plantation programmes has been carried out in the community land, i.e. village panchayat common land. The afforestation programmes have been carried out through plantation of fuel wood, fodder and wood trees and also to reclaim the wastelands including the usar land. The social forestry plantation programmes has also been implemented along with rail, road, canal and chakroads in the village panchayats. The area under the social-forestry has been recorded 284 hectares land in the Amethi Block. The area under the social-forestry shares 2.14 per cent land of the total geographical area of the Amethi Block.

The total area under the agro-afforestation and its percentage to the total geographical area has been mentioned in Table 9.7.

Table 9.7: Agro-Afforestation (Amethi Block)

Sl. No.	*Name of the Village Panchayat*	*Area in Hectares*					*%age of total geographical area*				
		Agro Forestry	*Horticulture*	*Social Forestry*	*Natural Vegetation*	*Agro-afforestation*	*Agro Forestry*	*Horticulture*	*Social Forestry*	*Natural Vegetation*	*Agro-afforestation*
1	*2*	*3*	*4*	*5*	*6*	*7*	*8*	*9*	*10*	*11*	*12*
1	Katra Maharani	3	7	2	2	14	2.2	5.2	1.4	1.4	10.44
2	Sarai Khema	10	25	5	10	50	1.7	4.4	2.6	1.7	8.97
3	Trilokpur	5	20	10	7	42	1.5	6.2	3.2	2.2	13.1
4	Ramdaipur	9	10	5	1	19	2.8	9.5	4.7	0.9	18.0
5	Jangal Ram Nagar	20	60	25	3	108	1.6	5.0	2.0	0.2	9.0
6	Benipur	3	10	5	2	20	2.3	7.8	3.9	1.5	15.6
7	Parsawan	10	35	5	4	54	3.1	11.0	2.5	1.2	17.0
8	Mahmoodpur	10	35	5	10	60	3.1	10.9	1.5	3.1	18.7
9	Raipur Fulwari	5	25	10	5	45	1.7	8.7	3.5	1.7	15.7
10	Umapur Gana Patti	10	22	5	12	49	3.0	6.7	1.5	3.6	14.9
11	Bhaganpur	5	15	5	4	29	3.3	9.9	3.3	2.6	19.2
12	Tala	5	15	5	4	29	2.7	8.3	2.7	2.1	15.9
13	Kushi tali	5	20	10	13	48	1.4	5.8	2.9	3.9	14.0
14	Loharta	2	5	2	1	10	3.9	9.8	3.9	1.9	19.0
15	Darkha	10	25	9	1	45	3.3	8.3	3.0	0.3	15.0
16	Mochwa	4	20	5	1	30	2.4	12.0	3.1	0.6	18.0
17	Nuwawan	10	45	10	0	75	2.1	9.8	2.1	2.1	16.4
18	Loniapur	3	5	1	6	15	2.7	4.5	0.9	5.5	13.7
19	Kherauna	4	15	5	2	26	2.3	8.8	2.9	1.1	15.3
20	Katra Phulkunwar	4	15	5	5	29	2.2	8.5	2.8	2.8	16.5
21	Hathkila	5	20	5	8	38	2.3	9.2	2.3	3.7	17.5
22	Rebha	2	15	1	1	19	2.0	15.0	2.0	1.0	19.0
23	Raidaipur	10	20	5	5	40	3.8	7.6	1.9	1.9	15.3
24	Korari Girdhar Shah	10	22	15	12	59	2.6	5.7	3.9	3.1	15.4
25	Dedhpasar	7	22	10	10	49	2.2	7.0	3.1	3.1	15.6
26	Mahaso	1	10	2	1	14	1.1	11.6	2.3	1.1	16.2
27	Gangauli	5	35	10	10	60	1.3	9.2	1.1	2.6	15.7
28	Dhadhudhar	2	15	3	3	23	1.5	11.2	2.2	2.2	17.0
29	Gaderi	3	30	5	6	48	1.1	11.1	1.8	2.2	17.7
30	Bhusahari	8	20	5	10	43	3.4	8.5	2.1	4.2	18.2
31	Purabgaon	5	20	7	4	36	2.2	9.0	3.1	1.8	16.2
32	Chaturbhujpur	3	10	5	3	21	2.5	8.3	4.1	2.5	17.5
33	Mahrajpur	5	10	3	14	31	3.0	6.0	1.8	8.5	18.9
34	Aghar	5	15	7	14	41	2.3	6.9	3.2	6.4	18.8
35	Saidpur	3	20	2	8	33	1.6	11.1	1.1	4.4	18.3
36	Kakwa	10	25	5	14	54	2.7	6.7	1.3	3.8	14.6
37	Saraiya Duban	5	25	10	9	49	1.6	8.3	3.3	3.0	16.3
38	Kohra	15	35	10	21	81	2.6	6.1	1.7	3.6	14.1
39	Mahmadpur	5	40	5	12	62	1.2	8.9	1.2	2.6	13.9
40	Naraini	5	30	10	15	60	0.8	5.1	1.7	2.5	10.2
41	Nainaha	5	20	5	3	33	2.7	10.9	2.7	1.6	18.1
42	Ramgarh	11	40	10	13	75	1.6	6.0	1.5	1.9	11.3
43	Himmatgarh	10	35	5	10	60	2.9	10.2	1.4	2.9	17.6
44	Dehara	3	20	5	3	31	1.2	8.5	2.1	1.2	13.1
	Amethi Block	**280**	**983**	**284**	**302**	**1857**	**2.11**	**7.41**	**2.14**	**2.27**	**14.01**

The agro-forestry includes such land which has been under cultivation but along with cultivated land boundary or the Medium and inside in different rows as per the suitability of the farmers social-forestry, horticulture plants are planted and the agricultural cropping during the Rabi, Kharif and Jaid cropping seasons are also simultaneously carried out under the multi-agriculture and intensive agriculture farming. The agro-forestry mean that if the agricultural activities and the social-forestry and horticultural trees plantation programmes are carried out in a combined agricultural production system has been termed as agro-forestry system with the view to obtain food grain, wood, fuel wood, fodder, fibre, fruits etc. to meet the growing demand of the increasing cattle and human population. It has been recorded that there are 280 hectares land under the agro-forestry programmes in the Amethi Block. The area under the agro-forestry programmes contributes 2.11 per cent area of the total geographical area of the Amethi Block. The area under the agro-forestry has been very insignificantly distributed in the various village panchayats of the Amethi Block.

The area under the horticulture included the Bagwani land, i.e. Bagh, horticulture and vegetable land and the garden land but such lands has been recorded as area under the Bagh in the revenue records. The area under the Bagh has been recorded for the individual land owners/farmers/cultivators but not the community land or common land or the village panchayat land lying in the village panchayats. It has been observed that the area under the horticulture land has been significantly distributed in the various village panchayats of the Amethi Block. There are 983 hectares land under the horticulture including the Bagh and vegetables and horticulture-cum-vegetable field in the Amethi Block. The area under the horticulture land contributes 7.41 per cent area to the total geographical area of the Amethi Block. Thus the area under fruit plants, vegetable field and combined agricultural cropping during the agricultural cropping seasons has to be promoted and propagated to ensure intensive agriculture cropping and also to propagate agricultural-scientific farming in the various village panchayats of the Amethi Block for sustainable integrated development.

The total area under the agro-forestry, horticulture, social-forestry and natural vegetation has been mentioned under the agro-afforestation at village panchayat level in Amethi Block has been depicted in Table. 9.7. It has been recorded that total area under the agro-afforestation including agro-forestry, horticulture, social-forestry and natural vegetation has been recorded 1857 hectares in the Amethi Block. The total area under the agro-afforestation contributes 14.01 per cent land of the total geographical area of the Amethi Block. It is surprising to note that to maintain the ecological imbalances of an area 33 per cent forest cover has been earmarked to maintain the eco-system of the area but the agro-forestry, social-forestry, horticulture and natural vegetation total area contributes only 14.01 per cent against the 33 per cent forest land to ensure ecological balance and to maintain the ecosystem of the Amethi. Thus, approximately more than 19 per cent land has to be brought under the agro-afforestation area through reclaiming the wastelands which has significantly increased more than 33 per cent in most of the village panchayats of the Amethi Block. The constant increase in the area under the wastelands, insignificant forest land and very nominal area under agro-afforestation has also developed hardship to the people of the Amethi and developed environmental crisis along with the socio-economic crisis in the various village panchayats of the Amethi Block.

4. AGRO-AFFORESTATION IN VILLAGE PANCHAYATS

The intensive agriculture cropping and multi-cropping along with agro-forestry, horticulture, social-forestry and natural vegetation jointly termed as agro-afforested land has been comparatively lower than the total area under wastelands. The fuel wood, fodder, fibre, fruits, food grains, fisheries, wood and vegetables combined cropping has been nominally cultivated in the various village panchayats of the Amethi Block. Lowest, only one hectares agro-forestry land has been noted in the Mahaso village panchayat followed by 2 hectares area under agro forestry has been noted in the Dhandhudhar, Rebha and Loharata village panchayats while highest 20 hectares land has been found Jangal Ram Nagar followed by 15 hectares land under agro-forestry area in Loharata village panchayat of the Amethi Block.

It has been noticed that the highest 60 hectares land in Jangal Ram Nagar, followed by 45 hectares land in Nuwawan village panchayat has been recorded under horticulture area in Amethi Block. Lowest area under horticulture land 5 hectares land in Loniapur and 7 hectares land in Katara Maharani village panchayats has been recorded in the Amethi Block.

It has been observed that highest 25 hectares land in Jangal Ram Nagar and 15 hectares land under social-forestry area has been recorded while lowest area with one hectare land in Loniapur and Rebha village panchayats has been found in the Amethi Block under the social-forestry plantation programmes.

The area under the natural vegetation has been found very insignificantly distributed at village panchayat level in Amethi. It has been noticed that highest 21 hectares land under natural vegetation has been found in Kohara village panchayat and the lowest area having one hectare land under the natural vegetation has been found in the Ramdaipur, Loharata, Darakha, Mochawa, Rebha and Mahaso village panchayats of the Amethi Block.

The total agro-afforestation area has been found very high 108 hectares land in the Jangal Ram Nagar followed by 81 hectares land in the Kohara village panchayat while lowest area under agro-afforestation has been recorded 14 hectares in the village panchayat Katara Maharani of the Amethi Block.

The proportionate distribution of the area under agro-forestry, social-forestry, horticulture, natural vegetation and total agro-afforested area of the total geographical area has been mentioned in Table 9.7. The proportionate distribution of the Agro forestry, social forestry, natural vegetation, horticulture and total agro-afforested land of the total geographical area at the village panchayat level has been analysed.

I. Agro Forestry

The proportionate distribution of the area under the agro-forestry village panchayat level to the total geographical area has been depicted in Map 9.82. Very high percentage of area under agro-forestry more than 3 per cent to the total geographical area of the village panchayats has been recorded in the Bhaganpur, Loharata, Darakha, Mahmoodpur, Parsawan, Raidaipur, Bhusahari, village panchayats of the Amethi Block. Medium percentage ranging between 2 and 3 per cent area under agro-forestry to the total geographical area has been recorded in the Katara Maharani, Benipur, Kherauna, Katara Phulkunwar, Hathkila, Dedhpasar, Loniapur, Umapur Gana Patti, Nuwawan, Mochawa, Tala, Korari Girdhar Shah, Purabgaon, Himmatgarh, Nainaha, Maharajpur, Chaturbhujpur, Nainaha, Kohara, Saidpur and Kakawa village panchayats of the Amethi. Low percentage between 1 and 2 per cent area under agro-forestry to the total geographical area has been observed in the Trilokpur, Sarai Khema, Jangal Ram Nagar, Raipur Phulwari, Rebha, Mahaso, Gangauli, Kushitali, Gaderi, Ramgarh, Dehara,

Agahar, Saraiya Duban and Mahmadpur village panchayats of the Amethi Block. It has been noticed that less than one per cent area under the agro-forestry to the total geographical area has been noticed in the Naraini village panchayat of the Amethi Block which has been recorded very insignificant proportionate area under the agro-forestry at the village panchayat level in the Amethi Block.

Thus, it has become pertinent to increase the area under agro-forestry through reclaiming the available wastelands natural resources in Amethi which has adversely affected more than one-third area in most of the village panchayats of the Amethi Block. Introduction of the tissue culture raised high yield variety plants related to the fuel wood, fodder, fibre, fruits, wood etc. has to be propagated which can provide the results within the short duration to the farmers. Proper irrigation management system, green manuring, application of the organic and inorganic amendments and the bio-fertilizer and requisite application of the insecticides and pesticides may be helpful in an increase in the area under agro-forestry system. The combined or mixed or multi-agricultural cropping must be initiated with the view to increase in the area under agro-forestry to meet the increasing demand of the ever growing of cattle and human population for the wood, fuel wood, fodder, fibre, fruits, fisheries, food grains (F^6) and vegetables etc. and to regenerate the degraded environment of the area and also be conserve the available natural wasteland resources to maintain the eco-system and ecological imbalance of the various village panchayats of the Amethi Block.

II. Horticulture

The horticulture land has been recorded in the revenue records of the Amethi. The area under Bagh has been recorded in the revenue records including village khasara maps. Khasara register and other revenue records. The Bagh has been mentioned as one of the land use category in the revenue records. The Bagh has also been recorded in the Khatauni of the individual land owners. The traditional Bagh land has been left very nominal except very few pockets of Bagh area where traditional fruit plants like Mango, Mahua, Neem, Pipal and other trees are still surviving in the traditional Bagh. Very few farmers have also developed horticulture land through plantation of high yielding fruit plants and other trees as well including wood, fuel wood, fodder etc.

The proportionate distribution of the horticultural are the total geographical area at the village panchayat level in Amethi Block has been shown in Map 9.83. Very high percentage of the horticulture land of the total geographical area more than 11.0 per cent of the total geographical area has been observed in Kakawa, Mochawa, Dhandhudhar, Rebha, Mahaso and Saidpur village panchayats of the Amethi Block. The medium percentage distribution of the horticulture land varying between 10 and 11 per cent land of the total geographical area has been noted in the Parsawan, Mahmoodpur, Loharata, Himmatgarh and Nainaha village panchayats of the Amethi. Proportionately low percentage distribution of the horticulture land between 9 and 10 per cent land of the total geographical area has been recorded in the Ramdaipur, Bhaganpur, Nuwawan, Gangauli, Hathkila and Purabgaon village panchayats of the Amethi Block. It has been observed that very low percentage less than 9 per cent horticulture land of the total geographical area has been noted in the Katara Maharani, Sarai Khema, Trilokpur, Jangal Ram Nagar, Benipur, Tala, Kushi Tali, Umapur Gana Patti, Darakha, Raipur Phulwari, Kherauna, Katara Phulkunwar Dedhpasar, Raidaipur, Loniapur, Korari Girdhar Shah, Bhusahari, Dehara, Ramgarh, Maharajpur, Naraini, Chaturbhujpur, Agahar, Saraiya Duban, Kohara, Mahmadpur village panchayats of the Amethi Block of the district Amethi.

It has been noticed historically that sufficient area under Bagh has been available in most of the village panchayats for the use of fodder, fruits, wood and fuel wood plants but the trees are disappeared through deforestation but the area under the Bagh is still recorded in the records of the Tehsils including Khatauni of the individual land owners. The excessive deforestation of the Bagh plant has brought severe hardship to the down-trodden section of the village poors especially for the fruits, fuel wood, fodder, grazing, wood etc. and it has also created environmental crisis and the degradation of the environment. Thus, an attempt must be made to develop the bagh and also to reclaim the available natural wasteland resources for the fuel wood, fodder, fibre, fruits, wood plantation to larger extent with the people movement and peoples participation high yield variety and tissue culture raised horticultural plants must be scientifically propagated and planted to a larger extent. Proper soil treatment, application of organic and inorganic amendments, proper irrigation water arrangements etc. must be horticulturally, scientifically promoted and demonstrated at grassroots level, so that the growing demand of the increasing cattle and human population can be fulfilled and the environmental crisis of the area can be restored and the ecological imbalances of the area can be improved and also to improve the socio-economic crisis of the peoples of the Amethi.

III. Social-Forestry

The proportionate distribution of the area under social-forestry to the total geographical area at the village panchayat level has been depicted in Map 9.84.

The social forestry programmes have been implemented by the Department of the Social-Forestry and various other organizations have also been engaged in the implementation of social forestry plantation programmes. The social forestry plantation programmes have been implemented mostly in the community land, common land and the village panchayat land and also along with the rail, road, canal, chakroad sides. The improper management of social-forestry programmes without people's participation could not result appropriately and most of the social forestry plantation programmes could not achieve the requisite targets, unsuccessful social-forestry plantation programmes have led towards reduction in the social-forestry land and also an increase in the area under wastelands.

The proportionate distribution of the area under social forestry to the total geographical area at the village panchayat level reveals that high percentage more than 3 per cent area under social forestry has been recorded in the Trilokpur, Ramdaipur, Bhaganpur, Benipur, Loharata, Raipur, Phulwari, Dedhpasar, Korari Girdhar Shah, Agahar, Chaturbhujpur, Purabgaon and Saraiya Duban village panchayats of the Amethi. Medium percentage distribution of the social forestry land between 2.5 and 3 per cent to the total geographical area has been found in the Tala, Kushi Tali, Sarai Khema, Kherauna, Katara Phulkunwar, Hathkila, Nuwawan, Bhusahari, Dehara, Nainaha and Maharajpur village panchayats of the Amethi. Low percentage between 1.5 and 2 per cent area under the social-forestry to the total geographical area has been observed in the Jangal Ram Nagar, Mahmoodpur, Parsawan, Rebha, Raidaipur, Umapur Gana Patti, Kohara, Ramgarh, Naraini and Gaderi village panchayats of the Amethi Block. It has been noticed that very less than 1.5 per cent area under social forestry programmes to the total geographical area has been recorded in the Katara Maharani, Gangauli, Himmatgarh, Saidpur, Kakawa, Saraiya Duban and Mahmoodpur village panchayats of the Amethi.

The reclamation of the wastelands must be started at war-footing level. Tissue culture raised high yielding tree plants related to fuel wood, fodder, fibre, fruit plants etc. must be propagated and demonstrated at grassroots level, to a larger extent. The social forestry plantation programmes should be implemented through peoples participation and tree plantation should be carried out as people's movement. The need based social forestry plantation programmes should be implemented in the community land and also the individual lands to fulfil the ever increasing demand of growing cattle and human population for the fodder, grazing, fuel wood, fibre, fruits, wood etc. and to conserve the available natural wastelands resources and also to ensure the degraded environment of the area and to improve the ecological imbalances of the area and also to make the environmental conditions of the village eco-friendly with the people, by the people and for the people of the Amethi.

IV. Natural Vegetation

The natural vegetation includes the flora, fauna, wild life and traditional jungle with local trees like Dhak and other plants, bushes and thorny bushes and grasses. Historically and traditionally the local level jungles and forest cover had been available in the most of the villages of Amethi but the traditional jungles has been deforested and degraded and at present such jungles has been vanished. Although natural vegetation and insignificant jungles has been found in the various villages of the Amethi. The village Jangal Ram Nagar has been named due to excessive jungles and traditional jungles with local flora and fauna but at present the jungles of the Jangal Ram Nagar has been disappeared but only the village panchayat Jangal Ram Nagar and revenue village Jangal Ram Nagar exists for the name only in the revenue records and as a local level village development unit of the Amethi Block. The area under the natural vegetation has been very significantly distributed at village panchayat level in Amethi Block. The proportionate distribution of the natural vegetation to the total geographical area at the village panchayat level of the Amethi Block has been depicted in Map 9.85. It has been observed that high percentage more than 6.0 per cent area under the natural vegetation of the total geographical area has been noted in the Agahar and Maharpur village panchayats of the Amethi Block. Medium percentage varying between 4.0 and 6.0 per cent area under natural vegetation has been observed in the Saidpur, Loniapur and Bhusahari village panchayats of the Amethi Block. Low percentage between 2 and 4 per cent area under the natural vegetation of the total geographical area has been noted in Trilokpur, Bhaganpur, Tala, Kushi Tali, Umapur Gana Patti, Mahmoodpur, Katara Phulkunwar, Hathkila, Dedhpasar, Nuwawan, Korari Girdhar Shah, Dhandhudhar, Gaderi, Gangauli, Himmatgarh, Chaturbhujpur, Saraia Duban, Kohara, Mahmadpur, Naraini and Kakawa natural vegetation of the Amethi Block. Very low percentage distribution of the natural vegetation and less than 2 per cent of the total geographical area has been observed in the Katara Maharani, Ramdaipur, Jangal Ram Nagar, Sarai Khema, Benpur, Parsawan, Raipur Phulwari, Kherauna, Raidaipur, Rebha, Mahaso, Dehara, Ramgarh, Nainaha, Purabgaon, Loharata, Darakha and Mochawa village panchayats of the Amethi Block. The constant decrease in the area under the natural vegetation has caused ecological crisis and environmental degradation of the area. An attempt must be ensured to control the deforestation and degraded of the natural vegetation. Large scale programmes should be launched to restore the area under the natural vegetation and regeneration of the flora, fauna and wild life should be implemented. The reclamation of the degraded land should be implemented with the people's participation and fuel wood, fodder, wood, fibre, fruits, grasses, flora, fauna, etc. should be planted in the reclaimed wastelands to maintain the natural vegetation cover of the various village panchayats of the Amethi. The traditional jungles and one-third natural vegetation and afforestation land must be developed to maintain the ecological imbalance of the area and improve the eco-system of the Amethi.

V. Total Agro-Afforestation

The total agro-afforestation area includes the area under the agro-forestry, social forestry, natural vegetation and horticultural lands. The total agro-afforestation area in the Amethi has been noted 14.01 per cent land to the total geographical area of the Amethi Block which has been very less than the one-third area under the forest cover required to maintain the ecological imbalances and also to maintain the environmental crisis of the various villages of the Amethi because more than one-third area of the Amethi suffers from the chronic problems of the wastelands. The proportionate distribution of the total afforested area of the total geographical area at the village panchayat level of the Amethi Block has been depicted in Map 9.86. It has been recorded that very high percentage more than 15.0 per cent area under total agro-afforested land to the total geographical area has been noted in the Bhaganpur, Tala, Loharata, Benipur, Mahmoodpur, Parsawan, Kherauna, Raipur Phulwari, Katara Phulkunwar, Hathkila, Dedhpasar, Rebha Raidaipur, Dhandhudhar, Korari Girdhar Shah, Gangauli, Mahaso, Bhusahari, Himmatgarh, Nainaha, Purabgaon and Saidpur village panchayats of Amethi Block. Medium percentage area under the total agro-afforested land between 12 and 15 per cent area of the total geographical area has been noted in the Trilokpur, Kushi Tali, Darakha, Umapur Gana Patti, Loniapur, Kohara, Kakawa and Mahmadpur village panchayats of Amethi Block. Low percentage distribution of the total agro-afforested area varying between 9 and 12 per cent land to the total geographical area has been found in the Katara Maharani, Jangal Ram Nagar, Naraini, Ramgarh and Dehara village panchayats of Amethi Block. It has been observed that very low percentage area under the total agro-afforested area, less than 9 per cent area has been recorded in the Sarai Khema village panchayat of Amethi Block.

The total agro-afforestation area at village level in Amethi Block has been significantly reduced due to over-grazing, over-exploitation of the forest, deforestation and changes in the land use pattern to meet the day-to-day increasing requirements of the local people who are directly or indirectly dependent on the forest produce especially for the fuel wood, fodder, fruits, wood, fibre etc. The large scale deforestation and land degradation has cause an increase in the area under the wastelands. The area under the wastelands has been increased upto more than one-third area of the total geographical area. Therefore, an attempt must be made to reclaim the wastelands for agro-afforestation especially for the agro-forestry, social-forestry, horticultural plants and the plantation of the natural vegetation to protect the environmental conditions of the area and also to meet the increasing demand of the growing cattle and human population for the fuel wood, fodder, fibre, food grains, fruits, fisheries, wood and vegetables etc. and also to maintain the ecological imbalances of the area and also to improve the socio-economic crisis of the people of the Amethi.

5. CONCLUSION

The agro-afforestation is a system of land use along with the combined growing of

Agricultural crops with social forestry. Animal husbandry including the food grain, fodder, fuel wood, fibre, fruits, fisheries, wood and vegetables etc. to fulfil the increasing demand of growing cattle and human population. The development of the agro-forestry aims its potential and positive interaction between ecological, economics, agricultural and forestry activities can be emphasized when proved scientifically and technically. The intensive agricultural cropping and the increased agricultural production and productivity, sustainable land management and ecological conservation are the some of the important objectives of the agro-afforestation system. The maintenance of the ecosystem and the restoration of the ecological and environmental degradation and to improve socio-economic and ecological crisis the sustainable development of the available natural land resources have to be managed scientifically.

There are different views and opinions regarding Agro-forestry system. Agro-Afforestation is not a perfect system because there has been both success and failure with the various agro-forestry system. No doubt agro-forestry system has provided viable alternatives or land use pattern with a suitable environment and compatible social conditions, the agro-forestry system should be helpful in providing substantial output.

The agro-afforestation programmes must be initiated and launched as a people's movement. The agro-afforestation programmes should be implemented by the people, for the people, and with the people through leasing the forest and the non-forest land and the wastelands to the deprived rural poor multitudes on the Tree Patta System of agro-afforestation programmes. The agro-afforestation programmes are necessary because there is a tremendous pressure of the growing cattle and human population which has led to deforestation and land degradation. The magnitude of the problems of the wastelands, deforestation, land degradation and deterioration of the environmental conditions has also been given due priority and importance in the Constitution of India.

The appropriate utilization and conservation of the natural infrastructural resources which includes classification, enactment, extension and propagation and campaign for the restoration of the degraded land and its development for the suitable agri-silvicultural, techniques and conservation of the degraded lands for ensuring the economic benefit for the sustainable ecological development.

The intensive agro-afforestation management programmes that should treat afforestation in a holistic manner and seeks to involve the communities and society in general in the preservation and the conservation of the environment and forest has been deprived of the techno-scientific financial support system. The agro-afforestation programmes at village level "Panchayat Van Yojana" to be worked out with the village panchayats to ensure appropriate utilization of the resources for the successful implementation of the agro-afforestation programmes with the peoples participation with the view to meet the increasing demand of the fuel wood, fodder, fibre, fruits, food grains, fisheries (F^6), wood, vegetables etc. It has been noticed that there are 14.01 per cent area under the total agro-afforested land of the total geographical area of the Amethi Block which comprises agro-forestry area 2.11 per cent, horticulture land 7.41 per cent, social forestry land 2.14 per cent and area under the natural vegetation 2.27 per cent of the total geographical area of the Amethi Block.

The very low percentage 1.1 per cent area under the agro-forestry land has been found in Gaderi and Mahaso village panchayats while highest percentage 3.9 per cent land in the agro-forestry has been observed in the village panchayat Loharata of the Amethi Block. Thus, it has been observed that the area under agro-forestry has been significantly distributed at village panchayat level in Amethi.

It has been observed that very low percentage 4.4 per cent area under the horticulture land has been found in the Sarai Khema village panchayat and the highest 15 per cent area under the horticulture land of the total geographical area has been found in the Rebha village panchayat of the Amethi Block.

The highest area under the social forestry having 4.7 per cent of the total geographical area has been observed in the Ramdai Pur village panchayat while lowest area under social-forestry having 0.9 per cent land of the total geographical area has been found in village panchayat Loniapur of the Amethi Block.

It has been recorded that very low percentage with 0.2 per cent area under natural vegetation of the total geographical area has been found in the Jangal Ram Nagar and 0.3 per cent land under natural vegetation in the Darakha village panchayat of the Amethi Block. The highest percentage area having 8.5 per cent under the natural vegetation has been found in the Maharajpur followed by 6.4 per cent land in Agahar village panchayat of the Amethi Block.

It has been observed that the highest percentage 19.2 per cent area under total agro-afforestation land of the total geographical area has been found in the Bhaganpur followed by 19 per cent area under the total agro-afforested land has been seen in Rebha, and Loharata village panchayats of the Amethi Block. The lowest agro-afforested are. having 8.9 per cent in the Sarai Khema village panchayat followed by 9.0 per cent area under total agro afforested area of the total geographical area in the Jangal Ram Nagar village panchayat of the Amethi Block of the district Amethi.

Thus, the total agro-afforested land including agro-forestry, social-forestry, horticulture and natural vegetation land has been very insignificantly distributed at village panchayat level in Amethi Block. The area under the agro-afforested land has also been insignificantly reduced due to diversion in the land use pattern and due to over-grazing, and over-exploitation of the natural forest resources and land degradation which has led to an increase in the area under wastelands. The area under wastelands has also been significantly increased and it shares more than one-third area to the total geographical area of the various village panchayats of the Amethi Block. The significant growth in the area under wastelands and decrease in the area under agro-afforested land including natural vegetation and forest land has caused environmental degradation and ecological imbalances of the area and created environmental crisis.

The growth in the wastelands and decrease in the area under agro-afforestation and excessive deforestation has to be halted and the reclamation of the wastelands and an increase in the area under agro-afforested land has to be technically, scientifically launched at the war-footing level in the village panchayats of the Amethi Block. The propagation of techno-scientific reclamation measures of the wastelands must be demonstrated at the grassroots level. The arrangement of the irrigation water system, tissue-culture raised high yielding variety of the tree, plants of different varieties as per the local needs with the people's participation must be started. The appropriate varieties of the fodder, fuel wood, fibre, fruits, wood etc. with salt tolerant capacity plants should be scientifically developed and demonstrated at village level. The techno-scientific methods of the wastelands reclamation procedures and plantation techniques and it management system and use of organic and inorganic amendments, application of the insecticides and pesticides etc. has to be appropriately demonstrated with the view to utilize and conserve the available natural wasteland resources and also to increase the area under agro-forestry, social-forestry, horticulture and natural vegetation area with the view to utilize the natural land resources and also regenerate the degraded environmental conditions of the area. The reclamation of the wastelands and an increase in the area under agro-afforested land may lead towards conservation of the degraded natural land resources and environmental conditions of the area. The reclamation of the wastelands and to increase the area under agro-afforestation have become an important task to fulfil the increasing demand of ever growing cattle and human population for the fuel wood, fodder, fiber, food grains, fisheries, fruits, wood, vegetables, etc. and also to regenerate the degraded environment and ecological imbalances of the areas and also to improve the environmental crisis and to maintain the socio-economic crisis of the down-trodden deprived rural poor people surviving in the remote rural areas of the Amethi.

10 Conclusions

1. INTRODUCTION

The sustainable economic development has been realized and commendable advances have been made in the diverse field of rural development programmes, would in reality be incomplete, if such rural development programmes fail to translate into better and dignified life for the majority of people who live in the remote rural area because the benefit of development must reach to the down-trodden deprived rural resourceless multitudes of the society.

To obtain the equal and sustained development proper planning and programme implementation is necessary. The planning is an organized process by which the people achieve their development goals or the planning is a means to achieve the development. The policy makers and planners should not only be technocrat to provide projections but they should study and innovate the available resources/infrastructure at village level before the formulation of village development planning.

The development is pursued with genuine concern for human values. The development means the evolution of the people to acquire better perception and best control over its natural and human infrastructure at National-State-District-Block-Village level perspectives.

The existing process of development is neither true, nor good, nor beautiful but it does not mean that everything is wrong. Really the development lies in the development of Science and Technology and its application by the common people.

The formulation of the planning process and the implementation of integrated rural development programmes are an information intensive task. The available natural resources/infrastructural information system at grassroots level, i.e. village level-block level-district level are not properly arranged while such available infrastructural information needs to be integrated for the development planning process.

The innovation of available natural, social, economic and institutional infrastructure at village level and developing village information system at grassroots level has become a challenging task for the policy makers and planners to formulate need based village development planning and also to assess the progress of on-going implementation of various rural development programmes being implemented at village level/grassroots level. The people's participatory planning process and implementation of various rural development programmes through the peoples involvement may be an asset for successful implementation of the programmes for realizing the aspiration of the people for better and dignified quality of life.

The development is after all a multifaceted phenomenon, which can neither be understood nor promoted unless it is seen from the interdisciplinary and holistic perspectives. Moreover, the development is basically a human enterprise.

There are no such study carried out for the innovation of infrastructure and developing village information system at grassroots level some studies have been carried out to study the planning and development studies at macro and meso level while at the grassroots level insignificant studies have been carried out. Natural Resources Data Management System (NRDMS) programmes have been evolved to develop spatial data management technologies for integrated rural development planning and its demonstration at local level in the different districts of the country.

It has been felt that an integrated and faster development is required to bridge the gap between rural to urban and rurban. An appropriate faster development of rural areas is required to fulfil the increasing demand of ever growing cattle and human population. The formulation and implementation of rural development programmes at village level is an intensive, innovative, informative and infrastructure (Natural and Human) oriented activity. Participatory planning and development should be initiated and sustained through the participation and involvement of the rural people at grassroots level. The real meaning of democracy is important to the people and self-government is the main soul of the rural development and to establish village panchayat for self-governance, i.e. for the people, by the people and with the people, to ensure participatory planning and successful implementation of various rural development programmes for self-reliance.

In view of the above an attempt has been made in this study to innovate the available natural resources/infrastructure at village level in Amethi. To study the physiographic, soil, climate, land use, waste lands, land capability and agro-afforestation are available resources at village level in Amethi Block. To develop village information system of the available natural infrastructure resources has been used at village panchayat level in Amethi Block. To analyze and develop the geo-referencing of the available natural infrastructure village information system based on the village khasra map has been used for all 44 village panchayats of Amethi Block, District Amethi.

The newly created District Amethi of Uttar Pradesh has been the study area while Amethi Block has been selected to study the innovation of available natural and human infrastructure at village panchayat level. There are five Tehsils, viz. Salon, Tiloi, Gauriganj, Musafirkhana and Amethi of the district while there are 44 village panchayats in the Amethi Block which have been taken to study the innovation of available natural and human infrastructure and developing the village information system at grassroots level, i.e. village panchayat level in Amethi.

The topography of the district is irregular in shape but fairly compact tract of undulating land. In general the slope is from north-west to south-east. The land of the district is fairly flat and gently undulating, the highest points being the crown of the watersheds.

Geologically the district forms the part of the Gangetic plain which is the recent origin according to the geological chronology and reveals ordinary Gangetic alluvium containing the nodules of calcareousness or Kankar pan. Geologically, there is no noteworthy feature in the physiognomy of the district. It is characterized by unrelieved monotony of alluvial deposits which is completely shroud the old land surface to a depth of several thousand metres. The Kankarili and ravine land is found in the district. Older alluvial soils, saline, alkaline and saline-alkaline soils are found in the different parts of the district.

The Gazetteer of the district published in 1903 A.D. throws some light on the historical background of the area. It has been observed that the main land owning families of the past were the Rajputs of the different clans. These Rajputs were owning 76.16 per cent land of the total area of the district. The Rajkumars alone held over one-fourth of the land of the district while their kins-men Bachhgotis and Rajwars owned 11.4 per cent and 3.4 per cent land respectively. The Rajkumars were main proprietors of the land.

Religiously, historically and archaeologically the District Amethi had an important key role in the overall development of the area. There are Hindu's and Muslim's religious important places in district. There are no archaeological existence in the district except the forts of Rajas and Maharajas.

The drainage system of the district plays an important role in the integrated development of the district. The southern part of the district is drained by the Sai River while northern part of the district is drained by the Gomati River. There are large number of marshes and irregular inland water bodies in the district. Most of the water-bodies do not serve any useful purpose and the ponds cannot be developed for fish farming.

The various categories of land use and different types of wastelands, land capability and area under agro-afforestation has been taken to innovate and analyse the natural resources/infrastructure and developing village information system at village panchayat level in Amethi Block of the District Amethi.

The agro-climatic conditions of the district, viz. Kharif-Rabi-Jaid cropping seasons are appropriate farming seasons for the agricultural practices in the district. The annual rainfall in the district is received within a short period of four months. Rainfall during the other months is scanty and sporadic. The seasonal and monthly variability of rainfall is rather more significant than the annual variability from the point of view of the agricultural practices.

The available land revenue records field observation, fieldwork and direct field survey has been the basic source of collecting the available natural infrastructure information and development of the village information system at village panchayat level in Amethi Block of the District Amethi.

There are older alluvial soils, saline, alkaline and saline-alkaline soils, found in district. The bangar soil, loamy soil, clayey loam, clayey soil, khadar soil and saline and alkaline soils found in the district. Depending upon the fertility status of the soil—Avval, Doyam and Soyam land—capabilities has been classified in the district by the revenue authorities for the fixation of revenue charges.

The district has an almost inhabited human occupance for over three millennia and centuries of plough and pastoral culture has reduced the natural vegetation except in the pockets of low-lying areas and some river banks with a moderate and suitable climatic conditions and fertile soil of the area, except the patches of Dhak, Mahua, Jamun, Mango and other trees including Dhak etc.

Thus, it has been observed that the available natural resources or infrastructure of the area is well enriched except the saline-alkaline, soils, water logged land and ravine land. The latest scientific and technological application for agricultural, horticultural, social forestry and agro-forestry programmes through people participation and involvement may be an asset to meet the ever increasing demand of growing cattle and human population for fuel wood, fodder, fibre, fruits, food grains, fisheries (F^6) and vegetables for integrated sustainable development of Amethi.

2. CLIMATE

The District Amethi experiences semi-arid climate characterized by hot summer, cold winter and rainy season. The hot summer months are May and June, while severe cold conditions prevail during the months of December and January. By the end of June monsoon starts and lasts till the months of September.

During the month of May and June the extreme mean minimum temperature of 44.1°C and 38.9°C has been observed. In the month of April the temperature is also quite hot having mean temperature of 38.1°C. The extreme maximum temperature has been recorded as 45.6°C. The mean minimum temperature during the month of December and January has been 7.7°C and 7.9°C respectively. The lowest minimum temperature has been recorded as 2.2°C and 7.9°C.

The least annual normal rainfall of 885.7 mm and the highest of 1100.2 mm has been recorded in the district. The average annual rainfall maximum per cent of co-efficient of variations is as high as 60.4 per cent. From the agricultural points of view effective rainfall should be high enough to charge the soil profile with enough moisture after meeting the evaporation losses. It has been observed that the effective monsoon in the district commencing in the beginning of July and ends towards the end of September. The district experience wet spell as well as critical dry spell in a different characteristics.

It has been recorded that the minimum, maximum and normal rainfall of the specific duration and the probabilities of the normal rainfall values being equal or exceeded as well as the expected minimum precipitation (rainfall) with their return period has been analyzed. It has been observed that 80 per cent of the annual rainfall during the month of July to September has been recorded.

During the month of April and May atmosphere of the district is dry. The mean relative humidity in the morning during April and May has been recorded 33 per cent and 37 per cent respectively.

The mean cloud amount varies during the months of monsoon from 4 to 6 oktas in the morning and 3 to 6 oktas in the evening. Approximately 90 per cent of the clouds during the monsoon months are also low level clouds. There is a slight increase in the cloud cover during the month of January and February.

There is quite moderate mean wind speed in the district. During the month of June about 12 kmph wind speed has been recorded in Amethi. During the monsoon season wind speed ranges from 8 to 10 kmph while during the winter season the mean wind speed varies from 4 to 6 kmph.

Hail storms are very rare in Amethi. Even if at all hail storm occurs, it is only during the months of March, April and May.

It has been observed that there is occurrence of dust storm during the pre-monsoon season from the month of April to the month of June.

The minimum evaporation during the winter has been recorded about 2.2 to 4.0 mm per day. The rate of evaporation during the summer season has been found high. The evaporation status of the district ranges from 9 to 11 mm per day in Amethi.

The mean Potential Evapotranspiration has been recorded as 185 mm during the hot summer months of May and June. During the monsoon months Potential Evapo-transpiration has been found much less and it ranges between 116 to 126 mm. During the winter evapotranspiration varies between 47 to 73 mm. The annual Potential Evapo-transpiration has been observed as 1379 mm in Amethi.

The sunshine has been received on an average of 8 hours during the winter season. It has been observed that during the summer season 10-11 hours sunshine has been recorded. The sunshine during the monsoon season has been recorded from 5 to 6 hours per day. The recorded sunshine during the different season has been very appropriate and suitable for agricultural practices especially for the Kharif, Rabi and Jaid agricultural cropping season in Amethi of the District Amethi.

It has been observed that the climatic conditions of Semi Arid Zone in Amethi of the District Amethi has been an appropriate and suitable for agricultural practices in the district provided that the occurrence of climatic situations are normal because the most of the farmers of the area are engaged in the traditional agriculture farming.

The occurrence of the rainfall during the rainy season and very hot summer during May and June and very cold season during December and January and the moderate hot cold weather in the area are suitable and feasible for the agricultural practices and the agricultural farmers are also immuned with the such agro-climatic conditions. The mild weather and semi-arid agro-climatic conditions of the district has forced the farmers for adoption of single, double and multiple-agriculture cropping in a traditional farming practices with a medium agricultural productivity and also the agro-forestry and social-forestry programmes to produce fuel wood, fodder, fibre, fruits, food grains, fisheries (F^6) and vegetable to meet the increasing demand of growing cattle and human population.

In spite of the above if there is occurrence of very high rainfall or dry critical spell, and very high evapotranspiration in the climatic conditions may adversely affect the agricultural practices, but in view of these worst climatic conditions farmers of the district are not having any techno-scientific know how to combact the worst agro-climatic conditions like dry land agriculture farming or immediate farming after the flood or utilization of water logged lands in the district.

The techno-scientific awareness and latest knowledge like Israel the farmers of the area can also develop agricultural agro-forestry and social-forestry practices in the district to meet the increasing demand of growing cattle and human population.

Thus the agro-climatic conditions of the Amethi and the District Amethi are suitable for moderate agricultural practices. The agro-climatic conditions available in the area as natural infrastructure are feasible, suitable, viable and acceptable to the cultivators single, double cropping and also for the multiple cropping as per the farmers awareness. The techno-scientific knowledge and awareness of the farmers and exploitation of existing seasonally agro-climatic conditions the farmers may develop the agricultural-agro-forestry and social-forestry programmes which can fulfil the increasing demand of growing population for fuel wood, fodder, fibre, food grains, fruits, fisheries (F^6) and wood, vegetable and it can also generate employment and improve the socio-economic conditions of the deprived rural multitudes surviving in the remote rural areas of the Amethi of the District Amethi.

3. SOILS

It has been observed that the soils of the district are formed through the alluvial deposits carried out by the river drainage system. The soil structure of the district has been divided into ravines of the Gomati river, flat lands of Gomati, Sai river uplands, Sai river low lands and Sai river flat lands.

The younger alluvial and older alluvial tracts and saline and alkaline soils having hard kankar layers are found in the district Amethi.

The soil sample surveys has been carried out to assess the fertility status of the soil by the Soil Survey Organisation of the Sharda Sahayak Command Area Development Authority, Department of Agriculture, Government of Uttar Pradesh and the collected soil samples has been tested and analyzed by the Water Technology Centre of the Indian Agricultural Research Institute, New Delhi.

The physical and chemical soil characteristics of the soil has been analyzed. The physical characteristics of the soil has been explained through mechanical composition of the soil water retentivity, soil depth and salinity and alkalinity problems of the soils.

The chemical characteristics of the soil has been analyzed based on the cation exchange capacity of the soil and salinity and alkalinity problems of the soils.

The physical and chemical characteristics of the soil has been carried out at Amethi, Raipur Phulwari, Ramdaipur and Trilokpur villages of the Amethi Block. The textural classes of the soil has been found silt clay loam, silt loam, silty clay loam and silty clay in the different villages of the Amethi. The field capacity, wilting point of the soil, *p*H value of the soil, Electrical Conductivity Exchangeable Sodium percentage and cation exchange capacity of the soil has been analyzed which can be maintained and managed to increase the fertility status of the soil. The textural class of the soil and moisture of the soil, *p*H value of the soil and Electrical Conductivity value of the soil, Exchangeable sodium percentage and cation exchange capacity of the soil analysis indicates the higher level of soil sodicity in the Amethi. The high *p*H value of the soil and Exchangeable Sodium percentage has been found at different layers of the depth of the soil.

The soil samples has been collected, tested and analyzed at different depths, layers of the soil profile in the various villages of the Amethi. The macronutrients and micronutrient values of the soil has also been analyzed.

The mechanical composition and chemical properties of the top 30 cm soil has been analyzed in the selected villages of the Amethi. The mechanical composition of the soil including sand, silt and clay percentage has been explained and the textural classes and *p*H value of the soil and other physical and chemical properties of the soil for the selected villages at 0.30 cm depth, 30-50 cm depth 50-90 cm depth, and 130-150 cm depth soil texture and other properties of the soil have been explained to maintain the fertility status of the soil. The macronutrient and micronutrient values of the soil has been analyzed through the nutrition index of the soil. The nutrition index of the soil ranges from 1.8 (very poor) to 4.2 (very high) value of the soil has been found in the various villages of the Amethi.

The macronutrients (Nitrogen, Phosphorus and Potash) and micronutrients (Zinc, Copper, Iron and Maganese) content of the soil in the various sample locations at Amethi, Raipur Phulwari, Ramdaipur and Trilokpur villages has been analyzed. It has been observed that higher value of the macronutrients and micronutrients content at different layers of the soils has been found in Amethi. The land capability classification has been carried out and four different types of land capability classes has been explained based on soil texture and content of the soil and fertility classes of the soil.

Land irrigability classes has been classified in Amethi through A, B, C, D irrigability classes of the soil based on the soil properties. Agricultural production is possible through adopting adequate ameliorative measures to maintain the fertility status of the soil.

The reclamation of the usar land has also been carried out in different villages of Amethi. Proper bunding, levelling, leaching, flushing, good quality water irrigation management, appropriate application of organic and inorganic amendments and cultivation of salt tolerance high yielding variety agricultural crops, trees, grasses, vegetable and multi-cropping through techno-scientific management of the reclamation of usar land to increase the fertility status of the soil and to enrich the macronutrient and micronutrient content of the soil for cultivation of the crop, with the view to conserve the environmental conditions of the area and to utilize the available natural resources of the Amethi, and to maintain the eco-system on the one hand and to meet the growing demand of growing cattle and human population of the area for fuel wood, fodder, fibre, fruits, fisheries, food grains (F^6) wood, and vegetables through which socio-economic conditions of the deprived resourceless rural poor can be improved for sustainable eco-friendly integrated development of the Amethi and adjoining areas of the other districts.

4. WATER RESORUCES

It is known fact that the water is the basic source which affects the survival of plants and also affects the agricultural crop production of the area. It has been observed that there is abundant ground water resources which could be harnessed for facilitating the irrigation purpose for the agricultural crops through installation of shallow and deep tube-wells. The canal irrigation is another source of irrigation system.

The major natural drainage systems of the district are the tributaries of the Gomati and Sai river, Chamraura and Kunwar rivers. The Gomati river flows from north-west border of the district covering Shukul Bazar, Jagdishpur and Musafirkhana blocks of the district. The Sai river flows from south-east portion of the district covering Salon Tehsil of the district Amethi. The most areas of the district are low-lying and there are numerous lakes, ponds, water logged land and marshy land. Chamraura river and Kunwar river are small tributaries. The natural resources of the district are abundant and sufficient provided that appropriate techno-scientific planning systems are evolved to tap the available natural water resources of the district.

The irrigation network of the district is another source of the drainage network to facilitate the irrigation of agricultural land. Sharada Sahayak Command Area Irrigation Project of the Government of Uttar Pradesh has constructed irrigation network especially Inhauna, Singhpur and Amethi distribution tributaries and other minors of the Sharada Sahayak Irrigation Project in the district.

The ground water and tanks has been the major source of irrigation, prior to introduction of canal irrigation system. The extensive development of the available surface water resources and the ground water resources contributes to the irrigation up to a limited extent. It has been observed that the ground water resources, provides irrigation of 20 per cent of the net cultivated area of the district and 29.5 per cent of the net irrigated area. The existing ground water potential and the increased exploitation of ground water resources has been considered attentively for appropriate exploitation of the available natural ground water resources of the district.

The Gomati, Sai, Chamraura and Kunwar rivers the main natural drainage system of the district. Due to excessive rain, low-lying area the water logging, flood etc. are major problems of district. It has been observed that about 40 per cent of the district has been found flood-prone area and about 50 per cent area of the district has been classified and identified as poor drainage system of the area. The extensive canal network, resulting from the rise of sub-surface water table except in the few patches of the district. The occurrence of heavy rainfall caused large scale water logging. There are variant approaches for the design of drainage system having conventional practices because the slope up to 0.5 per cent on the flat field. Constant drainage design and development has been attempted. The tolerance of the crops of surface poundings varies in their proportionate tolerance to the water logging conditions of the district. The surface drainage requires the removal of excess rain water and irrigated water in the cropping land. The development and drainage design and construction should be carried out within the limited days to avoid the crop loss in the field.

The ground water infrastructure has been one of the most important source of irrigation because more than 55 per cent area of the district are irrigated through the ground water source of irrigation through open-wells, tube-wells, dug-wells, bore-wells which are uniformly distributed all over the district. Appropriate progressive development of ground water resources through private tube-wells, government tube-wells and other sources of irrigation has been implemented in the district.

The movement and occurrence of the ground water resources of the area is dependant on the sub-surface formation of the underground water and hydrological characteristics. The lithology of the soil and sub-surface soil comprises sandy clay of varying thickness of sand and gravel of the tectonic to the recent age. The acquifer zone extends to the depths greater than 100 m below the ground surface. The transmissivity values lies in the range of 1000 to 2000 m^2/day. The specific yield of the formations in the ground water has been in the range of 10 to 18 per cent in the district.

The ground water fluctuation varies from one village to another based on the occurrence of the rain fall in the area. The rainfall has been principal source of ground recharge. The seasonal fluctuations of the water table reflect primarily the effect of the monsoon rainfall on the ground water, the annual ground water fluctuations represent the net effect of the recharge and draft of the ground water in the district.

The average annual draft of the ground water depends on the number and yield of wells operational in the area and the capacities of the water pumps. It has been observed that about 60 cm of water is required for irrigation of the one hectare land. The irrigation potential may be expressed in terms of the corresponding volume of the ground water extracted for each type of ground structure in the respective types of wells of any area in the district. The annual draft of the ground water structure ranges from 660 million cubic metres to 1260 million cubic metres in the district.

It has been recorded that ground water recharge from rainfall is 20 per cent, due to seepage from unlined canals 1.8 to 2.5 cumecs/million square metres of the wetted area and return seepage from irrigation fields under canal irrigation is 35 per cent of water delivered at outlet and the return flow from irrigation under ground water irrigation is 30 per cent of water delivered at the outlet in the district. It has also been recorded that there is 16 cumecs recharge from the large distributaries and 9 cumecs recharge from the other distributaries of the canal branches of the district.

The net change in the ground water table indicated a marginal rainfall in the ground water table of the district. The total ground water inflow or outflow is not known but can be estimated that the change in the depth 0.1 m. corresponds to a net drop in the ground water storage of 55.3 MCM. The ground water outflow has been estimated as 1045 million cu.m in the district.

The ground water natural resources has been aportioned into the static and dynamic water resources. The depth of the zone of water fluctuations varies between 3.9 to 10.0 m. The specific field of aquifer in the zone of water table fluctuations of the area has been assumed to be uniform at 12 per cent in the district. The estimated average recharge potential for the normal rainfall has been estimated up to 4066 MCM. The average ground water outflow has been about 1150 MCM as per the occurrence of pre-monsoon and post-monsoon seasons.

The seepage water losses based on the water perimeter has been estimated about 13 per cent of the design discharge capacity. The other large distributaries of the canal irrigation system have their design discharge capacity varying from 5.7 to 14.2 cumecs in the district. The gross water supply to the remaining canal distributaries and minors (less than 100 cusec discharge capacity) has the standard. Value of 7 per cent has been adopted to estimate the seepage losses. The water losses based on the water loss rates of 0.76 cumecs and 1.52 cumecs per million sq.metres has been estimated for the monsoon and non-monsoon seasons respectively. The comparative estimated water seepage losses has been about 27 per cent in the district varying from 5.7 to 14.2 cumecs.

It has been observed that the different irrigation water distribution system, i.e. 10 per cent in the canal branches network, 13 per cent in the large canal network and 7 per cent in the small level irrigation distributaries and 30 per cent in the water course has been the water conveyance efficiency of about 51 per cent in the district. The water requirements for the irrigation varies from time to time and cropping pattern. The consumpting use of water by the various crops effective rainfall and water requirements for the specific purposes including land preparation, leaching and flushing of the saline/ alkaline soil, green manuring and other agricultural practices. The consumption use of irrigation water for crop or its potential evaporation is function of the stage of crop growth and the evaporative demand of the atmosphere and climatic conditions.

The conjunctive water use based on the estimates of the water available in the different situations—rainfall, canal water supplies, ground water and moisture storage at the various points of time. The irrigation water distribution system comprises water courses, cultivated field channels and field water application network. The water management is the basic responsibility of the farmers and various measures must be adopted for the appropriate and optimum utilization of available water resources of the district. The diversion of the cultivated field existing into rectangular plots and consolidation of the land ownership on the basis of suitable exchange between different owners of the field has been essential for increasing the water management efficiency at the level of the cropping field.

The water seepage from the field channels are influenced by the soil permeability, wetted area of the channel and distribution of the irrigation water at the cropping field level. The losses of the water through seepage in the water course are function of the duration of the water flow. The optimum length of the lined section of water course has been estimated on the basis of net water saving. The saving from the losses of the water seepage through construction of lining of length of drainage network from it needs to end. The 50 per cent lining of the upstream section of water course, approximately 80 per cent water seepage losses can be controlled.

The contour map of the outlet of the area depicting contours at 30 cm intervals and indicating the field boundaries are important for planning layout of water course and field channels. The channels season and the bed slope of the channel must be fixed from the canal outlet. Attempt should be carried out to run the water course in partial cutting and bank filling and maintaining water surface elevation in the field channel at least 15 cm from the ground level to permit the gravity of the irrigation in the cultivated land.

The proper lining material for the field channels and water course must be carried out through local masonry work including bricks, flat stones, stones, lateriate blocks, and cement. The joints of bricks, stones, lateriate blocks are constructed in cement or lime mortar and pointed and the plastering of the inside of the field channels has been optional. The field channel structures are constructed for the water diversion to control soil erosion and crossing or the obstructions in the field channels.

The field drains has been constructed at the lower ends of the crop fields to facilitate the escape of excess run off water reaching the field due to rains and excess irrigation water of the crop field. The construction of field drains must join the outlet of the irrigation drain The subsequent green maintenance of the drains by removal of the weeds and clearing the ridges and bunds which restricts entry of water into the drain has been essential to ensure frequent and efficient surface drainage network of the district.

Thus, it is evident that the average annual rain fall in the district has been sufficient to carryout the agricultural practices, the surface water through river systems, lacks, ponds and accumated water in the low-lying areas are enough to fulfil the water requirements of the area. The ground water resources are sufficiently available in the area which is utilized through dug-wells, tube-wells and bore-wells to irrigate the cropping land. The drainage discharge capacity and water seepage losses has to be controlled to avoid the water loss and check the soil erosion of the area. Appropriate irrigation water management through techno-scientific interventions are necessary and proper practical demonstration to save water and management of available water resources of the area must be ensured and water resources should be stored and safeguarded to utilize during the drought period and also during the pre-monsoon and post-monsoon time. Since the water is life, therefore, it should be properly utilized maintained as per the need of the cattle and human population plants and agricultural cropping requirements of the farmers to meet the increasing demand of the growing population for the conservations of the available natural water resources and to improve the eco-system and environmental conditions of the area and to ensure sustainable Eco-Socio-Economic sustainable development of the Amethi.

5. FOREST

It is known fact that the forest is an important natural resource which is a source of survival for cattle and human. Forest plays an important role in maintaining the Eco-System and improving the environmental conditions of the area. It has been observed that a centuries ago the district had been covered with extensive forests, flora, fauna and birds. The *jhils* and *tals* were famous for the birds sanctuary. Due to ever increasing cattle and human population, urbanization, the forest cover of the area had been destroyed and turned into cultivated lands and developed for the settlements. The patches of the forest cover has been observed along with on river banks of the district.

During the period of National campaign on "Grow More Food" following the Second World War the forests were recklessly cut down. During the early fifties extensive afforestation measures had been initiated by the Government, and Dhak, Khair, Babool, Shisham, Neem, Arjun, Kanji, Seras, Eucalyptus, Mango, Mahua, Jamun trees etc. were planted in the district. The major impact of deforestation and ecological degradation was because half of our nation depends for its fuel wood, fodder on the common lands.

Due to the extensive jungles in the Amethi, village panchayat Ram Nagar was named as Jangal Ram Nagar covering one revenue village and 48 purwas, tolas, and hamlet villages. The village Jangal Ram Nagar is the largest village of Amethi in terms of geographical area and population. The area under forest has been 0.44 per cent of the total geographical area of the district. The per capita forest area of the district has been found 0.0021 hectare which is lower than the Rae Bareli district and higher than the neighbouring Sultanpur district. Musafirkhana range and Salon forest range has proportionately higher area under forest than other parts of the district.

The district was covered with the extensive forest which had been destroyed and turned into cultivated lands. During the early fifteen various afforestation programmes had been launched and several varieties of trees like—Dhak, Khair, Babool, Shisham, Neem, Arjun, Kanji, Siras, Eucalyptus, Jamun, Mahua etc. were planted in the district with the view to meet the increasing demand of growing cattle and human population. The tiger, wild buffalows, used to be found in the jungles along the bank rivers. The larger carnivore has disappeared and even the wolf and hyena have become very rare in the area. The Indian Antelope has been protected species in the district. The main game birds found in the district are of several varieties of ducks, partridges, pigeons, geese, coots, jacana, moorhen, snipe and sand piper are also met in the area. During the cold weather season plenty of water fowl, snipe, quail are found in the district.

Afforestation has been an important plan for the integrated development of an area. The afforestation approach has been distinguished into 'Productive' and 'Social' Forestry. The social forestry has been considered as an asset for the rural economy, preventing soil erosion and sedimentation of reservoirs and increasing the soil moisture retentivity.

Social forestry programme has been launched along the road, rail road, small category forests, panchayat land, banjar land, and along with canal banks. The selection of need based appropriate tree species play an important role for the success of afforestation schemes.

The suitable species for the social-forestry are mahua, jamun, mango, kanji, eucalyptus, casurina, shisham, arjun and jetropha and various other tree species suitable for the different agro-climatic conditions have been promoted, propagated and planted by the various implementing agencies.

Thus, it can be concluded that afforestation or social forestry programmes are basic requirements of the cattle and human population of the district. The successful afforestation programmes are helpful in the reclamation of waste lands, conserving the environmental degradation, improving the eco-system of the area. The afforestation programme is an asset to obtain fuel wood, fuel energy, fodder, fibre, fruits, wood and wood for the rural cottage industry and wood for the agricultural implements and household needs of the people. Afforestation programme also generates employment to the needy resourceless poor people and it has also been helpful in increasing the income of the deprived resourceless, rural poor multitudes surviving in the remote rural areas of the district.

Lastly, it can be said that the natural resource jungle has saved the land and can improve the Agro-climatic conditions and helpful for the rainfall in the area. Therefore, the jungle (Forest) is more important than *Jameen* (Land) and *Jal* (Water) but to save the earth, cattle and human population along with Jal, Jangal and Jameen plays an important role in the development of rural economy and improving the socio-economic conditions of the rural poor and forest is an asset for maintaining the Ecological imbalance and conserving the environmental degradation of the area and also improving the environmental crisis of the various villages of the Amethi.

6. LAND USE

The land is a free gift of the nature like Jal, Jungle and Jameen (water, forest and land). The land has much importance and significance among the natural resources. The land provides space for economic and other activities whether they are related to primary, secondary or tentiary sector of economy. Everything we use, can be ultimately innovated on the land.

"Life without land is unconceivable"

The nature has gifted the land with unique life sustaining potential which has made it invaluable and indispensable for the various ecological needs.

Agricultural census and National Bureau of Soil Survey and Land Use Planning, Government of India considered forest land, land not available for cultivation, other uncultivated land excluding fallows, fallow land and net cultivated land which has been re-classified as per the needs and suitability of revenue system.

Village level revenue record system including village Khasara Register, Khasara Map, Land Khetauni, and other land revenue records has been the basic source of the land use system and natural infrastructure of the various village panchayats of Amethi.

The land use pattern of the district Amethi indicates that the forest area has been found 0.4 per cent, and permanent pastures and grazing land has been noted 0.3 per cent to the total geographical area of the district. The current fallow land has been recorded 8.8 per cent and the fallow land has been noted 9.7 per cent of the total area of the district. The culturable wastelands has been observed 3.6 per cent of the total area of the district. The area under the miscellaneous trees etc. has been noted 1.9 per cent of the total area of the district. The land not available for cultivation has been observed 21.0 per cent of the total area of the district. The barren and uncultivated land shares 10.9 per cent of the total area of the district. The net cultivated area of the district has been recorded 53.8 per cent of the total geographical area of the district Amethi.

There are 44 village panchayats and 82 revenue villages and 377 purwas/tolas/hamlet villages in the Amethi Block. The land not available for cultivation has been noted 10.66 per cent and the barren and uncultivated land has been observed 3.47 per cent of the total geographical area of the Amethi Block. The net cultivated area of the Amethi Block has been recorded 6109 hectares which contributes 47.90 per cent land of the total geographical area of the Amethi Block. The area under the culturable waste lands contributes 24.87 per cent of the total geographical area of the Amethi Block. The area under permanent pasture contributes 1.57 per cent and other uncultivated land excluding fallows shares 2.17 per cent to the total reporting area of the Amethi Block. The area under miscellaneous trees shares 2.45 per cent and the area under the fallows other than the current fallow land has been noted 4.53 per cent and the area under the current fallow to the total geographical area of the village panchayat Ramdaipur, Mahso and Chaturbhujpur has been noted while very low percentage less than 8 per cent are under the land not available for cultivation has been recorded in the Trilokpur, Jangal Ram Nagar, Village panchayats of the Amethi Block.

Very low percentage less than 2 per cent area under the barren and uncultivated land has been observed in the Ramdaipur, Raipur Phulwari, Kushi Tali, Dedhpasar, Bhusahari, Maharajpur and Himmatgarh village panchayats while very high percentage more than 6.0 per cent has been recorded in the village panchayat Rebha of the Amethi Block.

Proportionate distribution of the net cultivated area having high percentage more than 50.0 per cent has been observed in the Trilokpur, Jangal Ram Nagar, Loharata, Darakha, Nuwanwa, Bhusahari and Dehra village panchayats while very high more than 55.0 per cent net cultivated area has been noted in the Hathkila and Maharajpur village panchayats. The very low percentage less than 45 per cent area under net cultivated land has been observed in the Sarai Khema, Umapur Gana Patti, Loniapur, Kherauna, Katara Phulkunwar, Raidaipur, Korari Girdhar Shah, Dedhpasar, Gaderi, Saidpur, Kakawa, Saraiya Duban and Ramgarh village panchayats of the Amethi Block has been recorded 1.96 per cent to the total geographical area of the Amethi Block.

It has already been analyzed that there are 44 village panchayats and 82 revenue villages and 377 purwas/tolas/hamlet villages in the Amethi Block. It has been found very difficult to analyze the natural resources including land use pattern categorically for all 82 revenue villages and 377 purwas/tolas/hamlet villages adopting the revenue khasara maps as a base map for area specific description of the land use pattern. It is known fact that Village Panchayats ? Block Panchayats ? District Panchayats are the three Tier system of strengthening the Panchayati Raj Institutions. At the grassroots level village panchayats has been taken as lowest/grassroots level rural development units. In view of this village panchayats of the Amethi has been taken as a base to analyse the area specific distribution of natural resources including land use pattern and other socio-economic-institutional infrastructures and development planning in Amethi.

It has been observed that proportionately very high percentage more than 20 per cent of the Amethi Block.

Very high percentage of the culturable wastelands more than 30 per cent has been observed in the Darakha, Katara Phulkunwar, Mochawa, Dedhpasar and Raidaipur village panchayats while very low per centage less than 15 per cent of the area under the culturable wastelands has been recorded in the Ramdaipur village panchayat of the Amethi Block.

It has been observed that the very high percentage more than 4.0 per cent area under the permanent pastures and the grazing land has been noted in the village panchayat Tala and Loniapur while very low percentage less than 1.0 per cent has been recorded in the village panchayat Raipur Phulwari, Umapur Gana Patti, Kushi Tali, Darakha, Hathkila, Purabgaon, Chaturbhujpur, Maharajpur and Dehara of the Amethi Block.

Other uncultivated land excluding fallows has been found very high more than 4 per cent to the total geographical area in the Kherauna, Rebha and Saraiya Duban village panchayats while very low percentage area less than 1.0 per cent has been observed in the Jangal Ram Nagar, Kushi Tali, Dedhpasar and Himmatgarh village panchayats of the Amethi.

It has been noted that the area under the miscellaneous trees etc. has been found very high more than 6 per cent in the Jangal Ram Nagar and Agahar Village Panchayat while very low more than 1.0 per cent area under the miscellaneous trees etc. has been recorded in the Darakha and Mochawa village panchayats of the Amethi Block.

It has been observed that the very high percentage 10.0 area under the fallows other than the current fallow land has been recorded in the Jangal Ram Nagar village panchayat while very low percentage less than 2.0 per cent area under the fallows other than current fallow land has been noted in the village panchayat Bhusahari of the Amethi Block.

Proportionate distribution of the area under the current fallow land more than 6.0 per cent area has been noted in the village panchayat Bhaganpur while very low percentage less than 1.0 per cent area under the current fallow land has been recorded in the Trilokpur, Jangal Ram Nagar, Raipur Phulwari, Kushi Tali, Kherauna, Hathkila and Himmatgarh village panchayats of the Block Amethi of the district Amethi.

Thus, it has been observed that there is significant decrease in the area under the net cultivated area or the net sown area in most of the village panchayats of the Amethi Block. Subsequently, there is increase in the area under the culturable wastelands and the area under the uncultivated land in most of the village panchayats of the Amethi Block of the District.

Misconception of over irrigation, excess doses of fertilizer and insecticides and pesticides, careless over cultivation etc. has led to deterioration in the fertility status of the soil of the area. Appropriate scientific testing of the soil and as per the soil requirement the application of organic and inorganic amendments and use of NPK fertilizer and insecticides and pesticides etc. may be helpful in controlling the deterioration of the soil fertility status of the soil.

Thus, it has become an important task to the cultivators and agricultural policy makers and planners including the various agricultural implementing agencies to control the deterioration of the fertility status of the soil and to apply organic and inorganic amendments as the requirements of the soil. To ensure proper drainage system of the water logged areas especially to drain out the excess accumulated water. To yield high salt tolerant varieties of the agricultural crops which can easily grow in the salt affected land. To ensure high yielding varieties of the crops.

Proper techno-scientific demonstration of the reclamation of wastelands must be ensured with the continued participation of the farmers. Appropriate scientific techniques should be demonstrated to the farmers for agricultural activities including the systematic cropping system and crop rotation system. The requisite agro-forestry, horticulture and social-forestry programmes should be practically demonstrated to the farmers as per the fertility status of the soil to meet the increasing demand of the ever growing cattle and human population for the fuel wood, fodder, fibre, fruits, fisheries, food grains (F^6), wood and vegetables. The excessive and over exploitation of the available land, Agri-Horti-Social Forestry has to increase to check land degradation and control deforestation and to maintain the eco-system and to improve environmental conditions of the various villages of the Amethi.

7. WASTELANDS

The wastelands of the Amethi contributes approximately one-third area of the total geographical area. The unutilized leftout uncultivated land of the area has also been an important natural wasteland resources of the Amethi. The appropriate utilization and conservation of the available wasteland resources has become an important task to the policy makers and planners at the grassroots level as well as the district level.

Therefore, it is imperative that the rehabilitation process be taken up on a war-footing through mobilizing every conceivable natural wasteland resources and adopting imaginative strategies of the reclamation of wastelands natural infrastructural resources. Thus everyone has to strive to create a true consciousness amongst the entire people of the Amethi. The urgency of the problems of the wastelands and its reclamation process must be taken as a challenge to meet the increasing demand of the growing cattle and human population for fuel wood, fodder, fibre, fisheries, food grains (F^6), wood, vegetables etc. and to restore the ecological imbalance and environmental degradation of the area and to improve environment crisis of the Amethi.

The wastelands has been used for the common land usually for the less fertile soil which failed to yield crops and to give any return to the medieval farmers. The common man may define the wastelands as the land lying uninhibited and uncultivated and the such land is no longer serving any purpose. The land which has been left out of cultivation for one to two years to maintain the fertility status of the soil, has been termed as fallow land and if such land is left out of cultivation for two to five years has been termed as old fallow land and if the uncultivated land is left out of cultivation has been termed as banjar land.

The land which has lost its agricultural production capability due to various reasons and due to improper agricultural practices, mismanagement and in which the cost of production is higher than the profit occurred out of it, such types of the land may be termed as wastelands.

The wastelands of the Amethi has been classified as Kankarili land, Banjar land, usar land, old fallow land, fallow land, waterlogged land and other types of wastelands.

It has been noted that 33.11 per cent area of the Amethi has been lying as wasteland and uncultivated out of which the Kankarili land shares 1.87 per cent, Banjar land 4.67 per cent noted in the Saraiya Duban village panchayat of the Amethi Block.

Proportionately very high percentage fallow land 3 per cent area of the total geographical area has been noted in the village Dhandhudhar, Maharajpur and 3.1 per cent in the Parasawan village panchayats recorded in the Gangauli and Benipur Village Panchayats of the Amethi Block.

It has been observed that very high 13.3 per cent other types of wastelands have been noted in the village panchayats Raipur Phulwari and very low 1.2 per cent land to the total geographical area has been found in the village panchayats Maharajpur of the Amethi Block.

The total wastelands of the various village panchayats of the Amethi Block has been found very high in the Ramgarh, Ramdaipur, Sarai Khema, Trilokpur, Parsawan, Raipur Phulwari, Umapur Gangapatti, Kushi Tali, Loharata, Darakha, Mochawa and Kherauna village panchayats of Amethi Block.

Thus, it has been observed that the various types of wastelands contribute significant proportion at village panchayat level to the total geographical area of the village panchayats of the Amethi. The significant availability of wastelands in the Amethi Block is usar land 7.45 per cent and water logged land 7.5 per cent. The old fallow land contributes 2.9 per cent, fallow land 2.82 per cent and other types of the wastelands shares 7.51 per cent of the total geographical area of the Amethi Block.

The proportionate distribution of the Kankarili land at village panchayat level varies upto 0.4 per cent to the total geographical area in Tala and 3.3 per cent in the Kohara village panchayats of the Amethi Block.

Highest percentage of the Banjar land has been recorded 9.5 per cent in the Ramgarh and the lowest percentage of the Banjar land 1.8 per cent of the total geographical area has been noted in the Maharajpur village panchayat of the Amethi Block.

Proportionately very high percentage of usar land has been found 11.2 per cent in the Dhandhudhar and very low 3.0 per cent of the total geographical area of the village panchayat Maharajpur of the Amethi Block.

Very high percentage of the water logged 19.0 per cent has been found in Saidpur and very low 3.2 per cent of the total geographical area in the Agahar village panchayat of the Amethi Block.

It has been recorded that very high 6.0 per cent of old fallow land has been found in the village Jangal Ram Nagar and very low 0.3 per cent old fallow land of the total geographical area has been in various village panchayats of Amethi Block has been of great concern to the village people and the agricultural policy makers and planners at the grassroots level.

Thus, it has become an important task to initiate the reclamation of wastelands at war-footing level. The reclamation of the wastelands should be ensured as per the suitability and feasibility of the wastelands, to meet the increasing demand of the growing cattle and human population for the fuel wood, food grain, fodder, fibre, fruits, fisheries (F^6), wood and vegetables. The techno-scientific reclamation measures of the wastelands may be helpful in the conservation of available natural wastelands resources. The reclamation of the wastelands may be an asset to conserve the environmental degradation and maintain the ecological imbalances of the Amethi. The reclamation of the wastelands will be an asset in maintaining the eco-system of the area and regenerating the environmental conditions of the area and also to maintain the environmental crisis as well as socio-economic crisis of the people of the Amethi.

8. LAND CAPABILITY

The land capability has been defined as the capability of the land to produce the crop and other types of agricultural agro-forestry-social-forestry and horticultural activities. The term capability of the land has been perceived by the land users in their own terms. The land capability has been classified in terms of the cultivable land. As per the agricultural productivity and quality of the land the revenue charges has been fixed. The land capability classification has been classified historically in terms of Aana value land collection during the period of the Mughal period. High Aana value land quality revenue charges has been high while for the low Aana value land quality of revenue charges has been low due to low fertility status of the soil and low agricultural productivity of the land. The revenue collection on the basis of land Aana value was collected during the zamindari and taluqdari tenures. The agricultural land adjacent to the settlement areas has been classified with higher Aana value land quality. The revenue authorities have classified the land quality based on the Aana value land quality having higher Aana value has been classified as Avval land and the lower value of the Aana value land has been classified as Soyam land quality. The revenue authorities have classified the land as Avval, Doyam and Soyam land quality and the other types of land capability has been classified for the very poor land quality and very poor fertility status of the soil.

The land quality as mentioned in the revenue records especially in the Khasara Register and the Khasara maps including for the collection of the Malguzari by the revenue authorities are fixed based on the Avval, Doyam, Soyam qualitiy of the land. The land capability classification has been classified in terms of the agricultural productivity of the land and the fertility status of the soil to produce the agricultural crops along with other factors related to the agricultural inputs and agro-climatic conditions.

It has been noticed that 1538 hectares land has been found under the Avval land while the area under the Doyam land has been recorded as 2707 hectares land while 1067 hectare land has been observed as Soyam land in the Amethi Block.

It has been recorded that there are 11.61 per cent land under the Avval quality while 20.43 per cent land has been recorded under the Doyam land and 8.01 per cent land has been found under the Soyam land quality. The area under the other types of the land quality has been recorded 6.17 per cent to the total geographical area of the Amethi Block. It has been observed that very high more than 16.2 per cent land under the Avval land quality has been found in village Katara Maharani and Hathkila village panchayats of the Amethi Block. Very low per cent of Avval land having 8.9 per cent to the total geographical area has been noticed in the Mahmadpur and 9.8 per cent area in Sarai Khema village panchayats of the Amethi Block. It has been observed that proportionately very high percentage 29.0 per cent area under Doyam land has been found in the Loharata and very low percentage having 16.1 per cent area under Doyam land of the total geographical area has been noticed in the village panchayat Sarai Khema of the Amethi Block. The highest percentage of the Soyam land having 11.5 per cent area to the total geographical area has been observed in the Hathkila and very low percentage area with 5.0 per cent area under the Soyam land has been recorded in the Loharata village panchayat of the Amethi Block. The highest percentage of the other types of land quality has been noticed in the Jangal Ram Nagar having 8.4 per cent area of the total geographical area and the very lowest percentage having 2.3 per cent land under the other types of the land quality has been observed in the village panchayat Benipur of the Amethi Block.

Thus, it has been noticed that the area under the Avval land quality enriched with the fertility status of the soil and highly agriculturally productive land has been significantly reduced and the area under the Doyam land and Soyam land quality has been significantly increased which has been found very difficult in terms of increasing agricultural productivity of the area. The reduction in the area under the Avval land quality has been due to deteriorated fertility status of the soil due to over agricultural intensive cropping and high doses of chemical fertilizers and over doses of canal irrigation water has led to an increase in the area under Soyam or poor quality land or an increase in the area under the wastelands. The reduction in the area under the Avval land has to be controlled and protected scientifically and the area in the Soyam and other types of land has to be made more fertile in terms of increasing agricultural productivity and through reclaiming the wastelands for agricultural agro-forestry and social-forestry programmes. The appropriate application of high yielding variety seeds proper application of organic and inorganic amendments, green manuring and the use of bio-fertilizer, rotation in the agricultural cropping system, proper irrigation management adoption of the latest agricultural scientific techniques should be adopted to increase the area under Avval land quality and adding the area in the multiple cropping and intensive agricultural cropping and agricultural cropping system should be initiated for the Kharif, Rabi, Jaid cropping seasons and commercial agricultural cropping should be promoted to with the view to increase agricultural productivity of the land, to maintain the ecological imbalance of the area and to improve the environmental degradation of the Amethi and to make the area and eco-friendly and to re-maintain and improve the environmental crisis of the area. The fertility status of the soil should be maintained to increase the agricultural productivity of the land with the view to meet the ever increasing demand of the growing cattle and human population to the meet the need of the Fuel wood, fodder, food grain, fruits, fisheries, fibre (F^6), wood and vegetables with the view to improve the ecological crisis and also to maintain the socio-economic crisis of the people of the Amethi.

9. AGRO-AFFORESTATION

The agro-afforestation is a system of land use along with the combined growing of Agricultural crops with social forestry. Animal husbandry including the food grain, fodder, fuel wood, fibre, fruits, fisheries, wood and vegetables etc. to fulfil the increasing demand of growing cattle and human population. The development of the agro-forestry aims its potential and positive interaction between ecological, economics, agricultural and forestry activities can be emphasized when proved scientifically and technically. The intensive agricultural cropping and the increased agricultural production and productivity, sustainable land management and ecological conservation are the some of the important objectives of the agro-afforestation system. The maintenance of the ecosystem and the restoration of the ecological and environmental degradation and to improve socio-economic and ecological crisis and the sustainable development of the available natural land resources have to be managed scientifically.

There are different views and opinions regarding Agro-forestry system. Agro-Afforestation is not a perfect system because there has been both success and failure with the various agro-forestry system. No doubt agro-forestry system has provided viable alternatives or land use pattern with a suitable environment and compatible social conditions, the agro-forestry system should be helpful in providing substantial output.

The agro-afforestation programmes must be initiated and launched as a people movement. The agro-afforestation programmes should be implemented by the people, for the people, and with the people through leasing the forest and the non-forest land and the wastelands to the deprived rural poor multitudes on the Tree Patta System of agro-afforestation programmes. The agro-afforestation programmes are necessary because there is a tremendous pressure of the growing cattle and human population which has led to deforestation and land degradation. The magnitude of the problems of the wastelands, deforestation, land degradation and deterioration of the environmental conditions has also been given due priority and importance in the Constitution of India.

The appropriate utilization and conservation of the natural infrastructural resources which includes classification, enactment, extension and propagation and campaign for the restoration of the degraded land and its development for the suitable agri-silvicultural, techniques and conservation of the degraded lands for ensuring the economic benefit for the sustainable ecological development.

The intensive agro-afforestation management programmes that should treat afforestation in a holistic manner and seeks to involve the communities and society in general in the preservation and the conservation of the environment and forest has been deprived of the techno-scientific financial support system. The agro-afforestation programmes at village level "Panchayat Van Yojana" to be worked out with the village panchayats to ensure appropriate utilization of the resources for the successful implementation of the agro-afforestation programmes with the people's participation with the view to meet the increasing demand of the fuel wood, fodder, fibre, fruits, food grains, fisheries (F^6), wood, and vegetables. It has been noticed that there are 14.01 per cent area under the total agro-afforested land of the total geographical area of the Amethi Block which comprises agro-forestry area 2.11 per cent, horticulture land 7.41 per cent, social forestry land 2.14 per cent and area under the natural vegetation 2.27 per cent.

The very low percentage 1.1 per cent area under the agro-forestry land has been found in Gaderi and Mahaso village panchayats while highest percentage 3.9 per cent land in the agro-forestry has been observed in the village panchayat Loharata of the Amethi Block. Thus, it has been observed that the area under agro-forestry has been significantly distributed at village panchayat level in Amethi.

It has been observed that very low percentage 4.4 per cent area under the horticulture land has been found in the Sarai Khema village panchayat and the highest 15.0 per cent area under the horticulture land of the total geographical area has been found in the Rebha village panchayat of the Amethi Block.

The highest area under the social forestry having 4.7 per cent of the total geographical area has been observed in the Ramdaipur village panchayat while lowest area under social-forestry having 0.9 per cent land of the total geographical area has been found in village panchayat Loniapur of the Amethi Block.

It has been recorded that very low percentage with 0.2 per cent area under natural vegetation of the total geographical area has been found in the Jangal Ram Nagar and 0.3 per cent land under natural vegetation in the Darakha village panchayat of the Amethi Block. The highest percentage area having 8.5 per cent area under the natural vegetation has been found in the Maharajpur followed by the 6.4 per cent land in Agahar village panchayat of the Amethi Block.

It has been observed that the highest percentage 19.2 per cent area under total agro-afforestation land of the total geographical area has been found in the Bhaganpur followed by 19.0 per cent area under the total agro-afforested land has been seen in Rebha, and Loharata village panchayats of the Amethi Block. The lowest agro-afforested area having

8.9 per cent in the Sarai Khema village panchayat followed by 9.0 per cent area under total agro afforested area of the total geographical area in the Jangal Ram Nagar village panchayat of the Amethi Block of the district Amethi.

Thus, the total agro-afforested land including agro-forestry, social-forestry, horticulture and natural vegetation land has been very insignificantly distributed at village panchayat level in Amethi Block. The area under the agro-afforested land has also been insignificantly reduced due to diversion in the land use pattern and due to over-grazing, and over-exploitation of the natural forest resources and land degradation which has led to an increase in the area under wastelands. The area under wastelands has also been significantly increased and it shares more than one-third area of the total geographical area of the various village panchayats of the Amethi Block. The significant growth in the area under wastelands and decrease in the area under agro-afforested land including natural vegetation and forest land has caused environmental degradation and ecological imbalances of the area and environment crisis.

Thus, it is evident that Amethi has been enriched through historical and religious infrastructure including the archaeological existence in the form of the Forts of the Rajas and Maharajas. The agro-climatic conditions of the Amethi is appropriate for the agricultural multi-cropping including Agri-Horti-Social-Forestry programmes. The soil structure of the area forms the alluvial deposits of the rivers which is suitable for agricultural activities except the usarization of the land, i.e. salinity and alkalinity problems of the soils and water logging which has caused large scale land degration in Amethi. The surface water and underground water including the natural drainage system of the Amethi has appropriate for agricultural activities except flooding and water logging due to excess rain water and over irrigation water through canal irrigation system.

The forest of the Amethi has been deforested and destroyed and the Jungles of the Amethi have been recklessly cut down. The afforestation programme in Amethi has not been properly implemented due to which the deforestation and land degradation has been increased in Amethi. The land use pattern of the Amethi has been appropriately distributed at village panchayat level except significant increase in the area under wastelands especially Banjar land, water logged land, usar land, old fallow land and fallow lands which has led to environment crisis and socio-economic crisis of the people of the Amethi. Avval, Doyam and Soyam land capability classes have been classified by the revenue authorities of the district. It has been noticed that Doyam land has comparatively higher area than the Soyam and Avval land quality. Therefore, an attempt has to be ensured to increase the area under Avval land to ensure multiple agricultural cropping system to increase the agricultural productivity of the area. The area under agro-forestry, social-forestry, has been significantly distributed at village panchayat level in Amethi. The severe land degradation and extreme deforestation have led to environmental degradation and ecological imbalances of the area.

The significant growth in the wastelands and decrease in the area under agro-afforestation and excessive deforestation has to be halted and the reclamation of the wastelands and an increase in the area under agro-afforested land has to be technically, scientifically launched at the war-footing level in the village panchayats of the Amethi Block. The propagation of techno-scientific reclamation measures of the wastelands must be demonstrated at the grassroots level. The arrangement of the irrigation water system, tissue-culture raised high yielding variety of the tree, plants of different varieties as per the local needs with the people participation must be started. The appropriate varieties of the fodder, fuel wood, fibre, fruits, wood etc. with salt tolerant capacity plants should be scientifically developed and demonstrated at village level. The techno-scientific methods of the wastelands reclamation procedures and plantation techniques and its management system and use of organic and inorganic amendments, application of the insecticides and pesticides have to be appropriately demonstrated with the view to utilize and conserve the available natural wasteland resources and also to increase the area under agro-forestry, social-forestry, horticulture and natural vegetation area with the view to utilize the natural land resources and also regenerate the degraded environmental conditions of the area. The reclamation of the wastelands and an increase in the area under agro-afforested land may lead towards conservation of the degraded natural land resources and environmental conditions of the area. The reclamation of the wastelands and to increase the area under agro-afforestation have become an important task to fulfil the increasing demand of ever growing cattle and human population for the fuel wood, fodder, fibre, food grains, fisheries, fruits, wood, and vegetables, etc. and also to regenerate the degraded environment and ecological imbalances of the areas and also to improve the environmental crisis and to maintain the socio-economic crisis of the down-trodden deprived rural poor people surviving in the remote rural areas of the Amethi.

References

I. BOOKS

Babarata, K.C., "Planning for Forest Resources and Bio-Diversity Management", Concept Publishing Co. Pvt. Ltd., New Delhi.

Bhattacharya, J.P., "Study of Soil Conservation for Agricultural Land", New Delhi, 1964.

Bland Fort, H.F., "The Climates and Weather of India", Cylon and Burma (London), 1889.

Browrigg, F.W., "Settlement Report", District Sultanpur, 1898.

Chauhan, D.S., "Studies in Utilization of Agricultural Land", Aggarwal Publication, Agra, 1966.

Ganguli, H.N., "Restoration of Soil Structures with Gypsum", Bikaner Gypsum Ltd., Bikaner, 1954.

Govinda Rajan, S.V. and Gopal Rao, H.G., "Studies on Soils of India", New Dehli, 1978.

Krishnan, M.S., "Geology of India and Burman", Madras, 1960.

Kuldeep, K.P., "The Climates of Continents", Oxford University Press, Oxford, 1927.

Leather, J.W., "Investigation of Usar Land in United Province", Government Press, Allahabad, 1914.

Mishra, R.P. (Ed.), "Changing Deceptions of Development Problems", Singapore, Maruze in Asia, 1983.

Mishra, R.P., "Regional Planning", 1992, Concept Publishing Co. Pvt. Ltd., New Delhi.

Mohammad Ali, "Dynamics of Agricultural Development in India", Concept Publishing Co., New Delhi, 1978.

Nevill, H.R., "District Gazetteer of the District Sultanpur", Allahabad, 1903.

Rae Chaudhury, S.P., "Land and Soil", National Book Trust, New Delhi, 1966.

Rao, K.L. "India's Water Wealth: its assessment, Uses and Projections", Orient Longman, New Delhi, 1979.

Shafi, M., "Land Utilization in Eastern Uttar Pradesh", Aligarh Muslim University, Aligarh, 1960.

Shame, K.B. and Rae Chaudhary, S.P., "Rating of Soils of India", IARI, New Delhi, 1960.

Singh A.K. *et al.*, "Forest Resource Economy and Environment", Concept Publishing Co., New Delhi, 1987.

Srivastava, K.G., "Agro Forestry", Central Book House, Rajpur, M.P., 1995.

Tajwani, K.G., "Agro-Forestry in India", Concept Publishing Co., New Delhi, 2001.

Wadia, D.N., "Geology of India", London, 1960.

Yadav, H.R., "Agro-Afforestation Management on Wastelands", Concept Publishing Co. Pvt. Ltd., New Delhi, 2011.

Yadav, H.R., "Dimension of Wastelands Development", Concept Publishing Co., New Delhi, 1988.

Yadav, H.R., "Genesis and Utilization of Wastelands", Concept Publishing Co., New Delhi, 1986.
Yadav, H.R., "Reclaiming Wastelands", Concept Publishing Co., New Delhi, 1990.
Yadav, H.R., "Village Development Planning", Concept Publishing Co. Pvt. Ltd., New Delhi, 2009.
Yadav, H.R., "Wastelands Diagnosis and Treatment", Concept Publishing Co., New Delhi, 1987.

II. ARTICLES

Abrol, I.P., "Reclamation of Wastelands and World Food Prospects", Trans, Int. Cong. Soil Science, New Delhi, 1982.
Aggarwal, L.I. and Mehrotra, C.L., "Soil Survey and Soil Work in U.P." Vol. I, Allahabad, 1952.
Aggarwal, L.I., "Soil Survey and Soil Work in U.P.", Vol. II, Allahabad, 1952.
Aggarwal, L.I., "Soil Survey and Soil Work in U.P.", Vol. III, Allahabad, 1954.
Aggarwal, L.I., "Soil Survey and Soil Work in U.P.", Vol. IV, Allahabad, 1954.
Aggarwal, R.R. and Mehrotra, C.L., "Quality of Water Irrigation Waters in U.P." *India J. Science*, 31, (1966).
Bland Ford, H.F., "Hotwinds of Northern India", Mem., Indian Met, Deptt., Vol. 6, 1986.
Bland Food, H.F., "The Rainfall of India", Mem., Indian Met. Deptt., Vol. 3, 1986-88.
Choudhary, Kamala, "Greening of India", *Indian Express*, 21 December, 1982.
Clark, K.G.T., "The Vicissitudes of Summer Rainfall of the Indo-Gangetic Plain and the Valley", *Geography*, 18 (Pt. IV) 1932.
Dhawan, C.L., "Land Reclamation", Central Board of Irrigation and Power, Publication No. 43, Vol. III, 1.
Eliot, J., "The Hot Winds of Northern India", Indian Met., Mem. 6 (Pt.) (III), 1980.
Hart, D.R.D. and Stands, M.W., "Sustainable Land Use System Research and Development" in International Workshop on Sustainable Land Use System Research, Feb. 12-16, p. 13.
Hill, S.P., "Variations of Rainfall in Northern India: Indian Met., Mem. 1 (Pt. III) 1897.
Kanwar, J.S., "Reclaiming Retriorated Soils", *Fertilizer News*, 6 (21) 1961.
Leather, J.W., "Agriculture Soils" in Agriculture, Ledger (3) 1898.
Mathur, R.K., "Economics of Agro-Forestry for Wastelands Reclamation", Agriculture Ledger, 1983, (175-87).
Rae Chaudhary, S.P., "Extension of Cultivated Areas for Increased Agricultural Production", Producing of Symposium of Science and India's Food Problem, ICAR, New Delhi, 1971.
Rae Chaudhary, S.P., "Final Report of the All India Soil Survey Scheme": *Bulletina*, India Council of Agricultural Research, 1952, 13.
Stamp, L.D., "The Land Utilization Survey of Britain", *Geographic Journal*, London, and (8), 1931.
St. Duster, S.A., "Intervening Opportunities: A Theory Relating Mobilizing and Distance", *American Sociological Review*-Vol.-II, Dec. 840, pp. 845-67.
Taylor, C.C., "Weed Control", *Indian Farming*, 9 (4) 1948.
Tewari, D.N., "Technology for Sustainable Development", National Seminar on "Technology for Sustainable Development, March, 12-18, Guru Ghasi Das University, 1980, p. 17.
Wadia, D.N. and Auden, J.B., "Geology Structure of Northern India", Mem. Geological Survey of India, 78 (1939).

III. GOVERNMENT AND U.N. PUBLICATIONS

Agricultural Census of Uttar Pradesh, 1990-71, Agricultural Statistics of Uttar Pradesh, 1971.
District Census Hand Book, Sultanpur District, 2001.
F.A.O. Frame Work for Land Evaluation, *Soil Bulletin No.* 1, 1976
I.C.A.R., Final Report of All India Soil Survey Scheme, 1958.

I.A.R.I., Resources Analysis of the District Sultanpur, 1989.
I.A.R.I., Resources Analysis of the District Rae Bareli, 1989
I.C.A.R. Hand Book of Agriculture, 1969.
Land Revenue Records (Khasara Register and Khasara Maps) Amethi Tehsil, Distt. Amethi.
National Thematic Maps Organisation (NATMO), Govt. of India.
State Soil Conservation Research, Annual Reprot, U.P. p. 61.
Statistical Book, District Sultanpur, 2001.
Survey of India, Topographical Sheets No. 63F/11, 63F/10, 63F/12, 63F/14, 63F/15, 63F/16, 63J/3, 63J/4, 63J/5, 63J/8, 63J/11, 63J/12, 63K/3, 63K/9.
U.S.D.A., "A Manual on Conservation of Soil and Water", Hand Book of Professional Agriculture Workers, Oxford Publishing Co., 1984.
The Imperial Gazetteer of India, Vol. XXIII, Reprinted, Today and Tomorrow Printers and Publishers, New Delhi, 1974, 130-37.

IV. UNPUBLISHED PH.D. THESES

Shafi, M., "Land utilization in eastern Uttar Pradesh", Ph.D. Thesis, Aligarh Muslim University, Aligarh, 1960, 40-41.

Coloured Maps

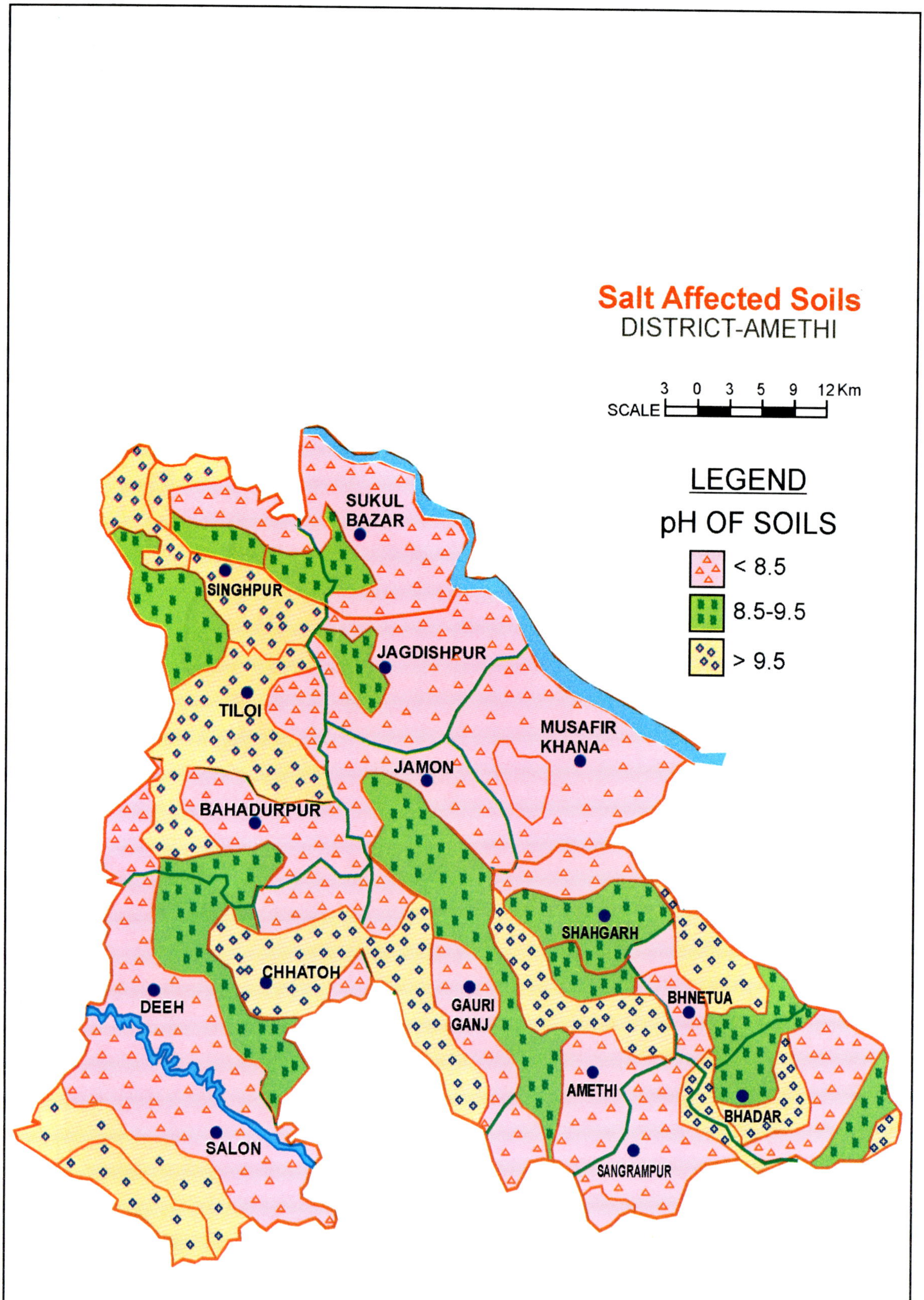

Map 3.12: Salt Affected Soils

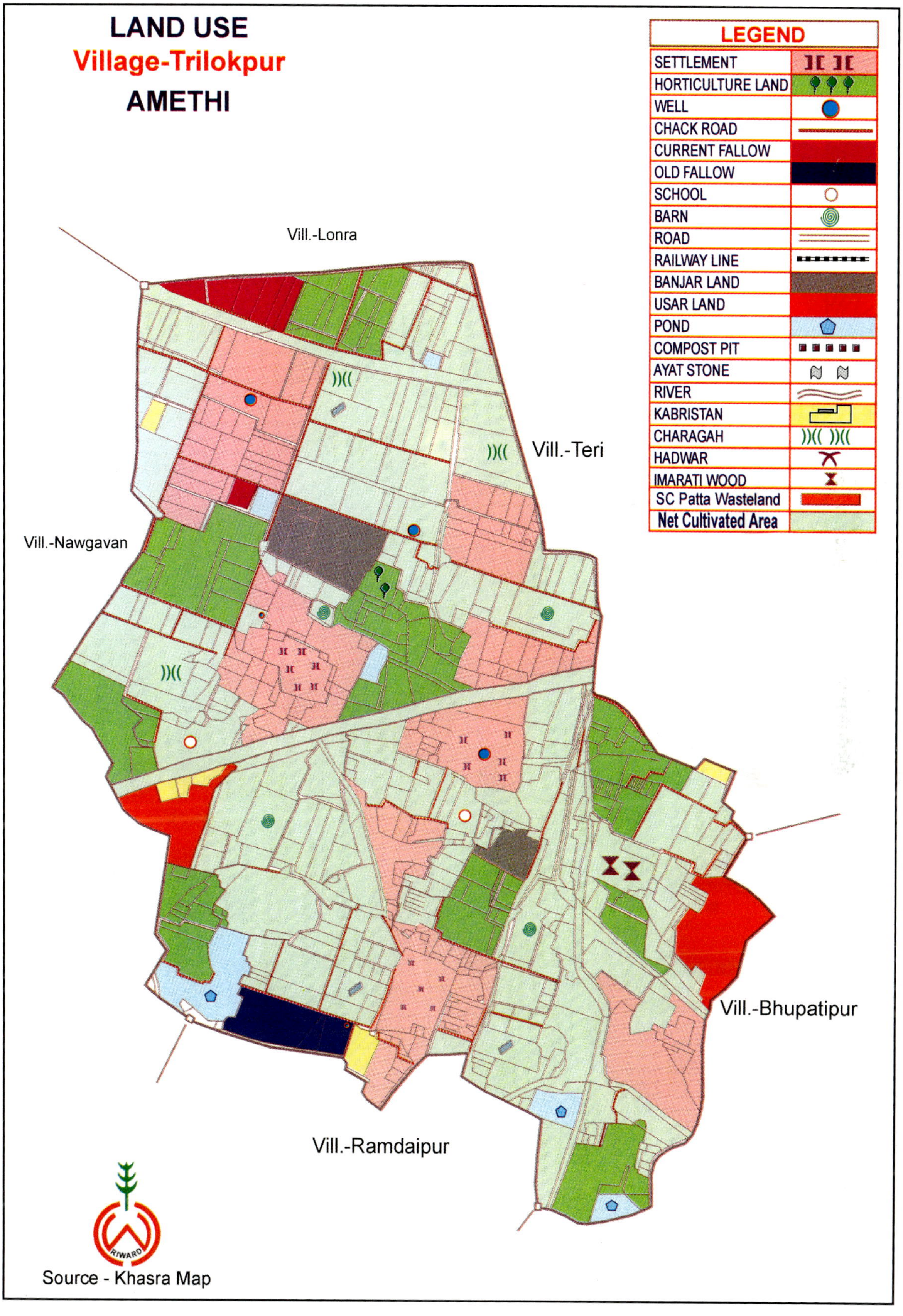

Map 6.17: Land Use of Village Panchayat Trilokpur

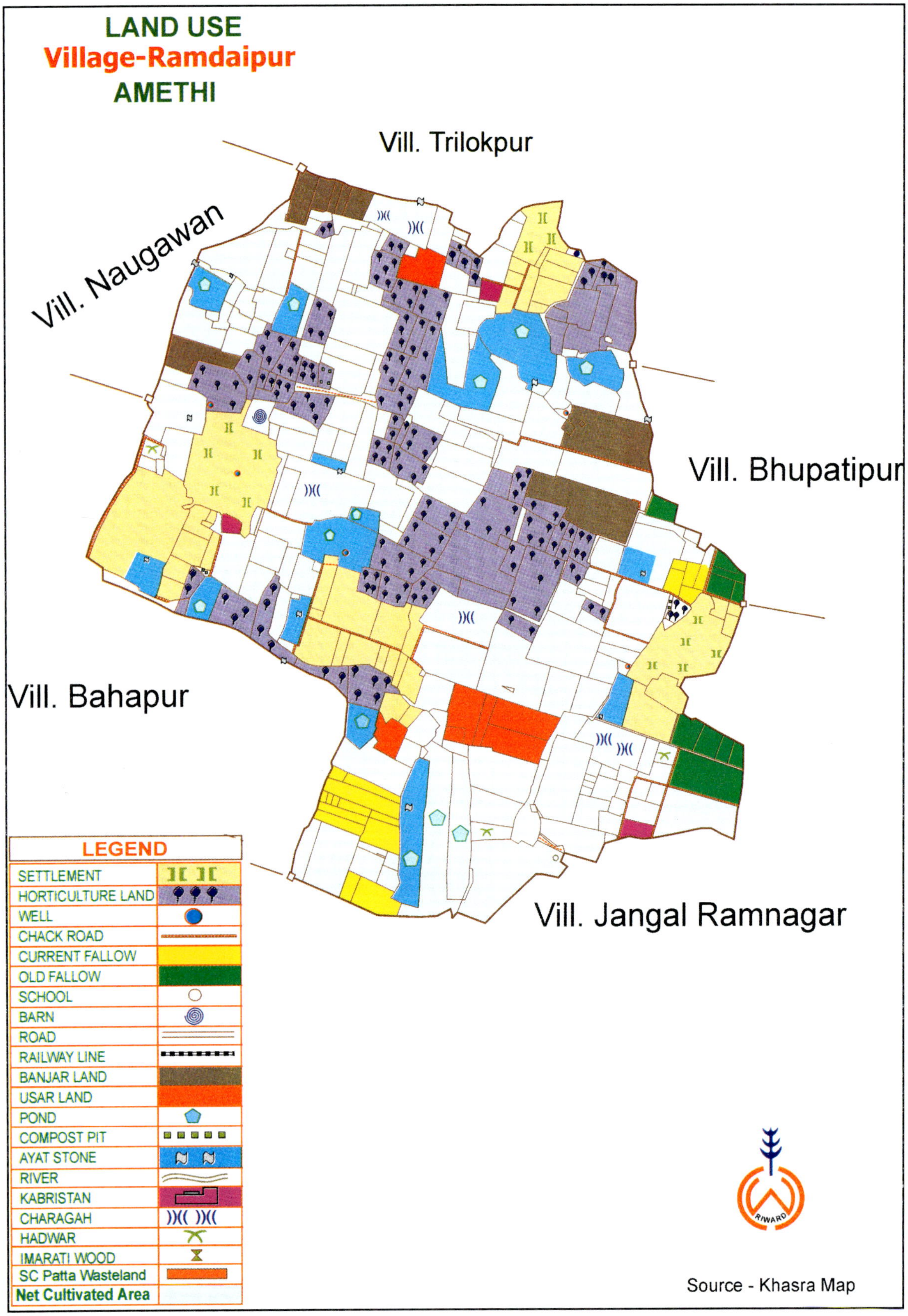

Map 6.18: Land Use of Village Panchayat Ramdaipur

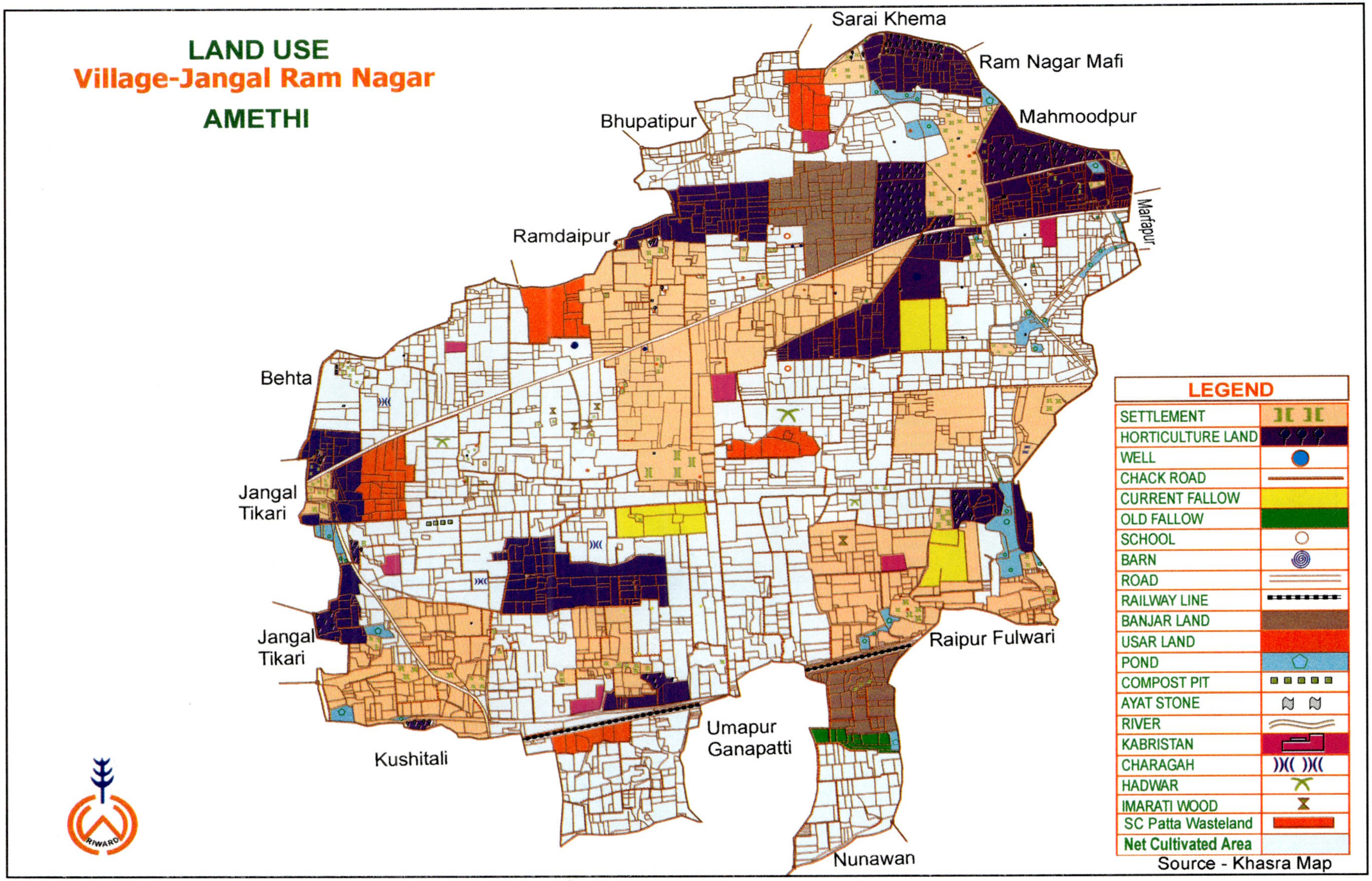

Map 6.19: Land Use of Village Panchayat Jangal Ram Nagar

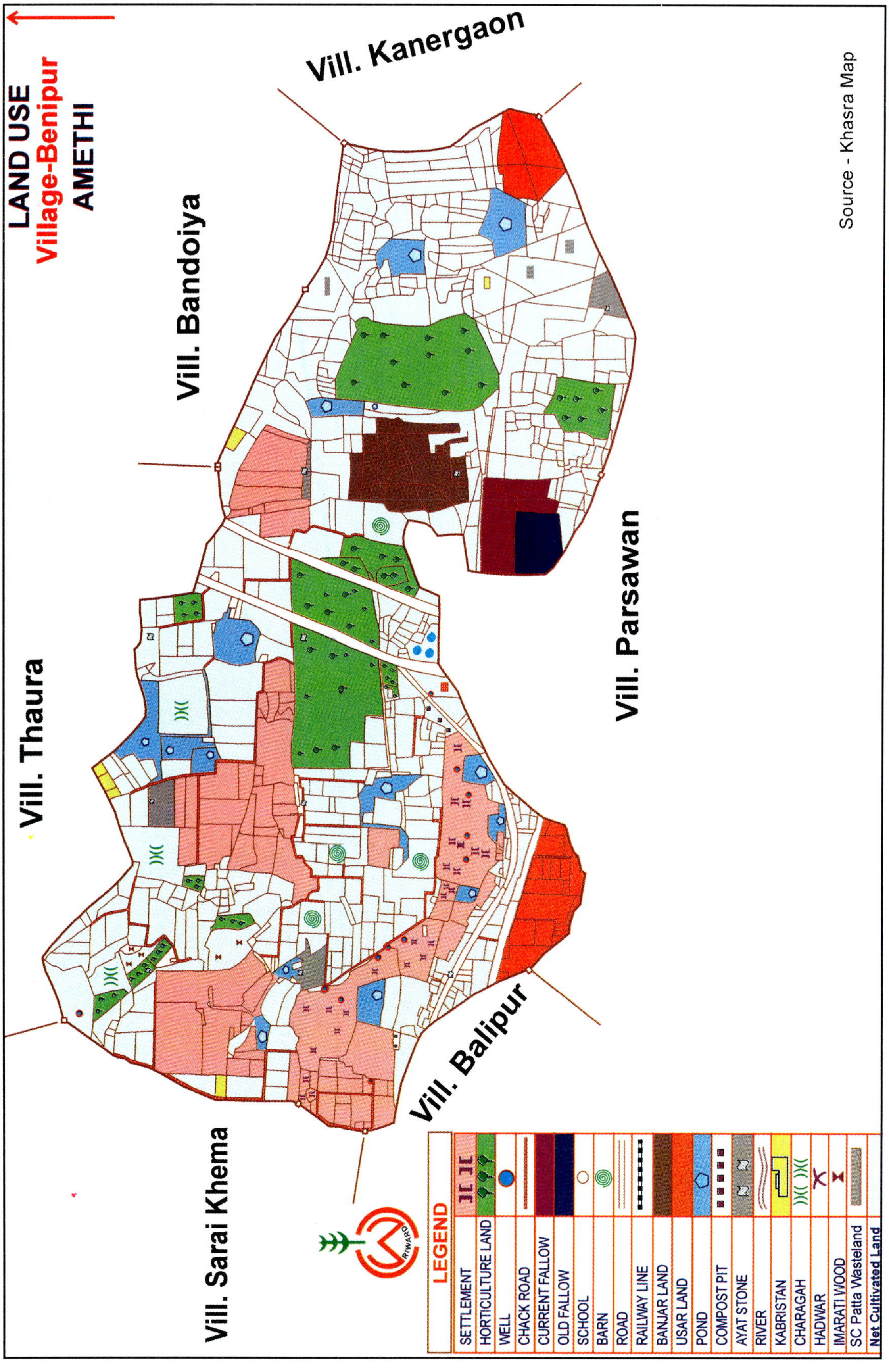

Map 6.20: Land Use of Village Panchayat Benipur

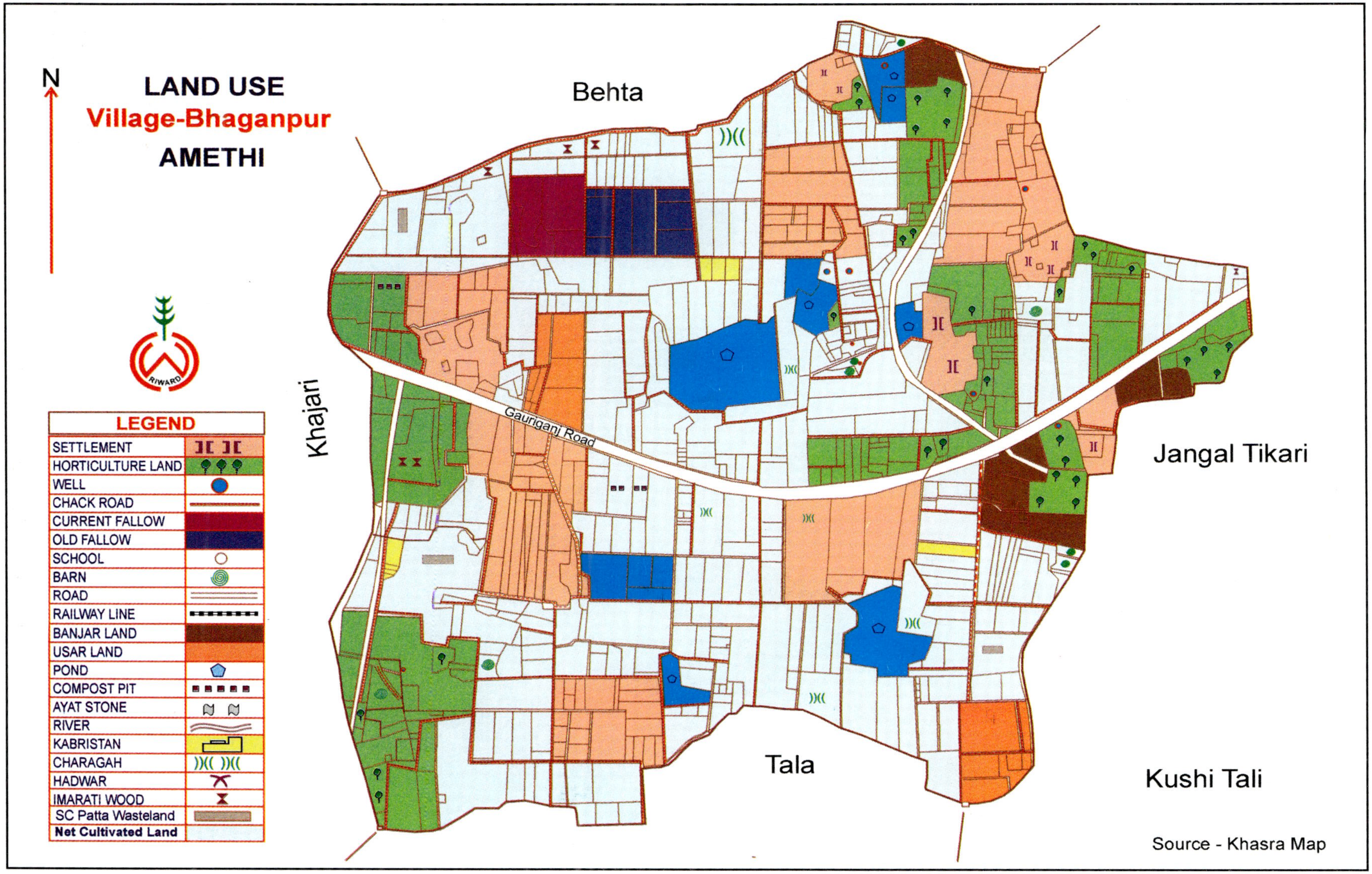

Map 6.25: Land Use of Village Panchayat Bhaganpur

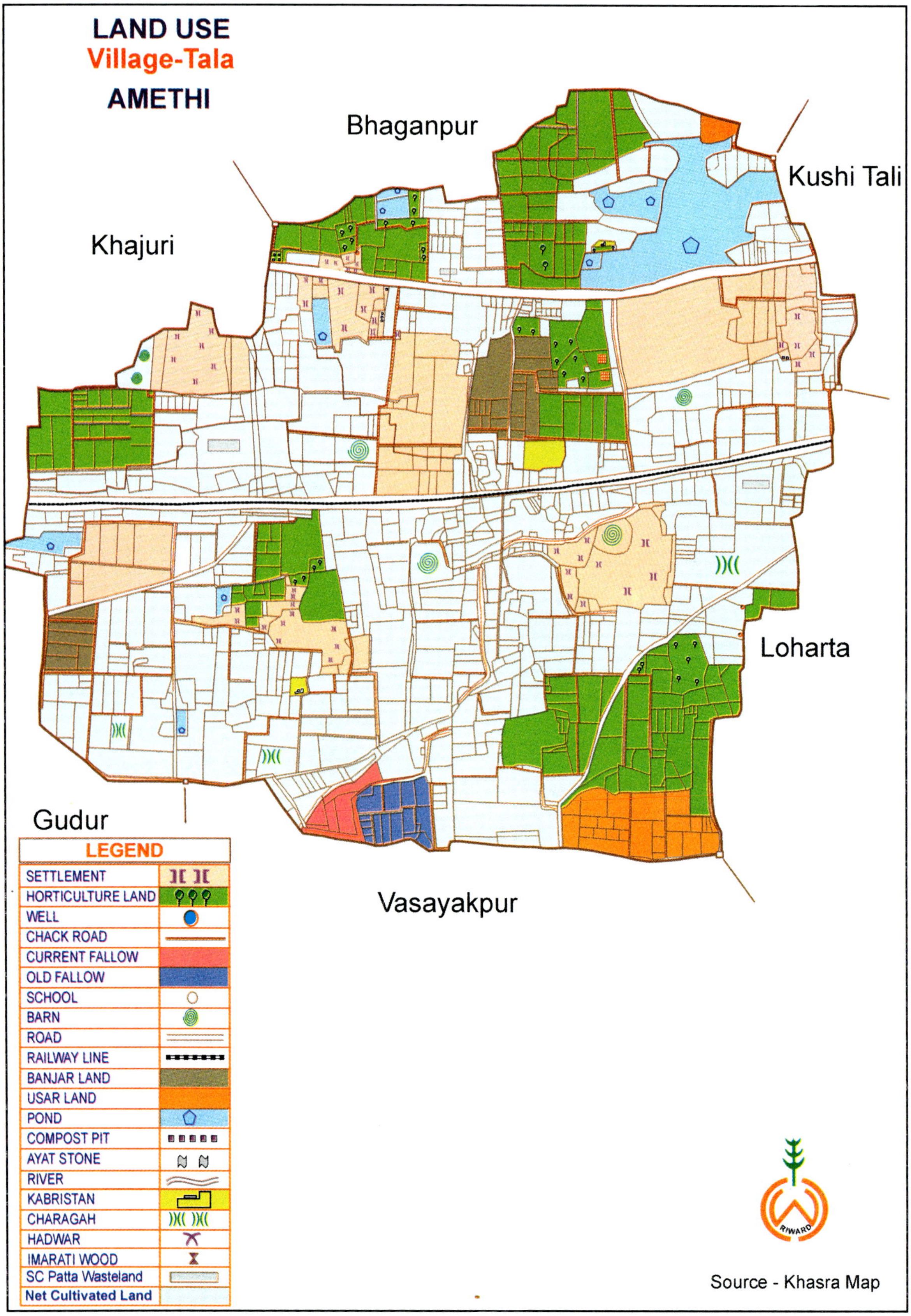

Map 6.26: Land Use of Village Panchayat Tala

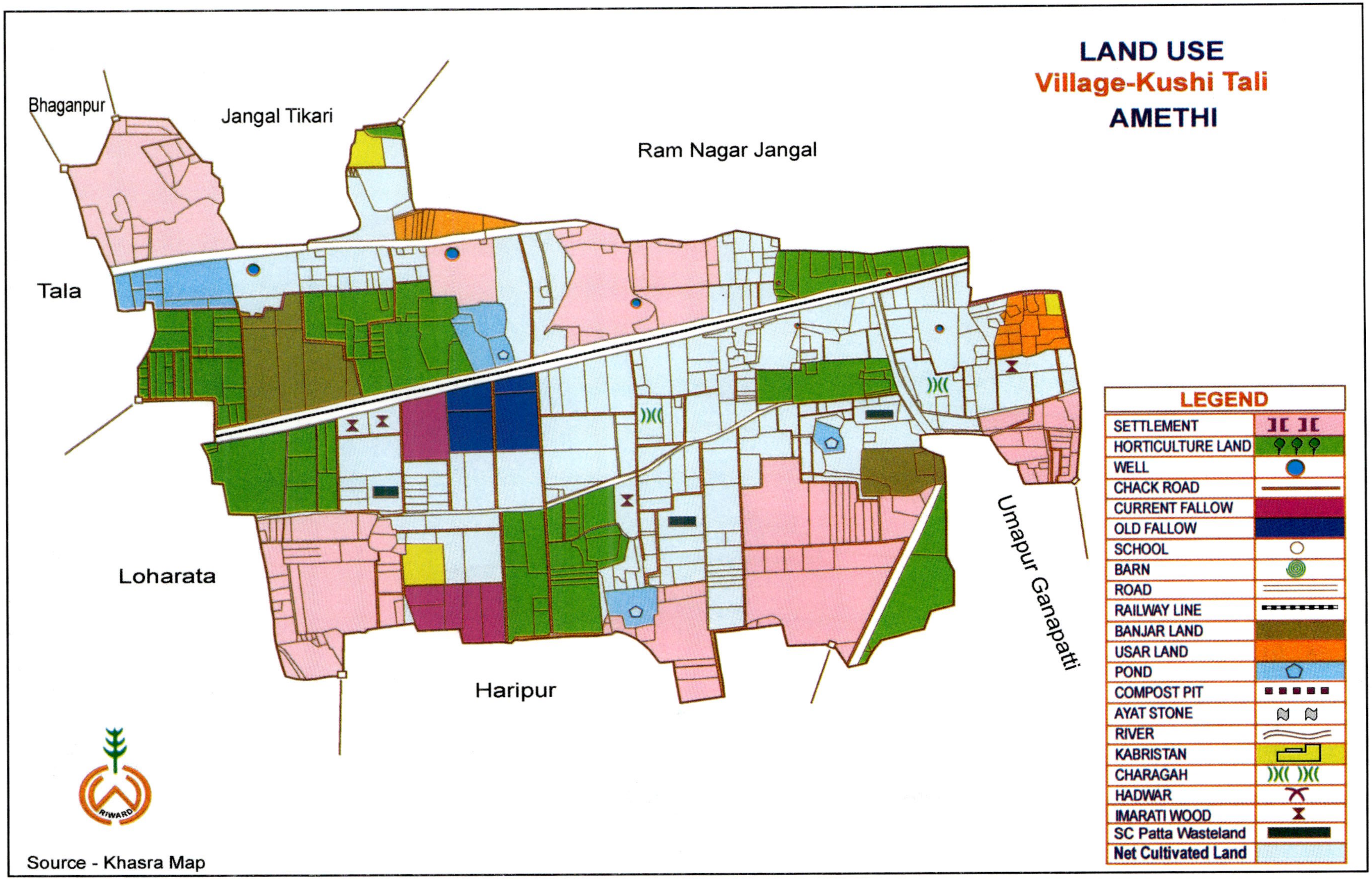

Map 6.27: Land Use of Village Panchayat Kushitali

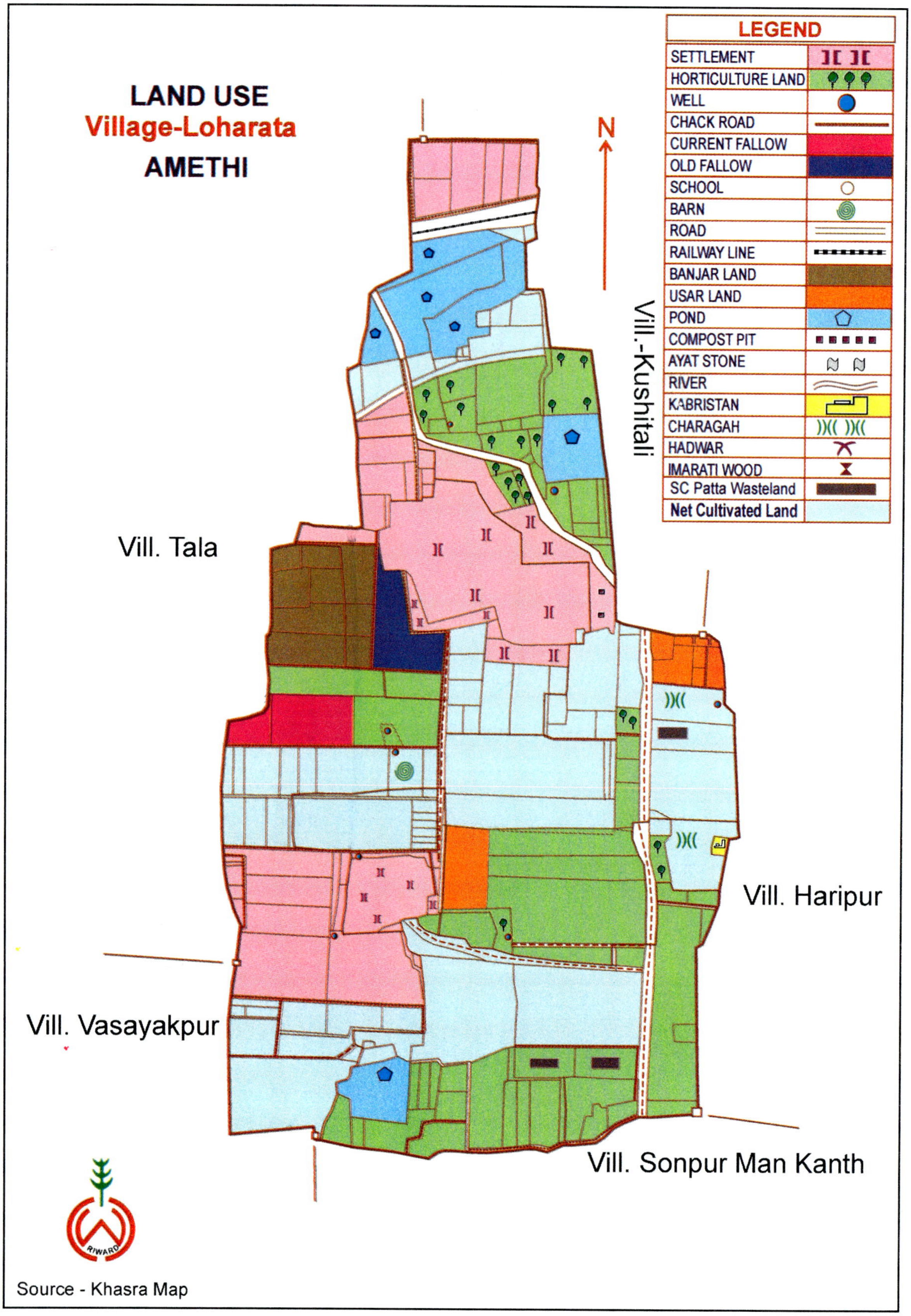

Map 6.28: Land Use of Village Panchayat Loharta

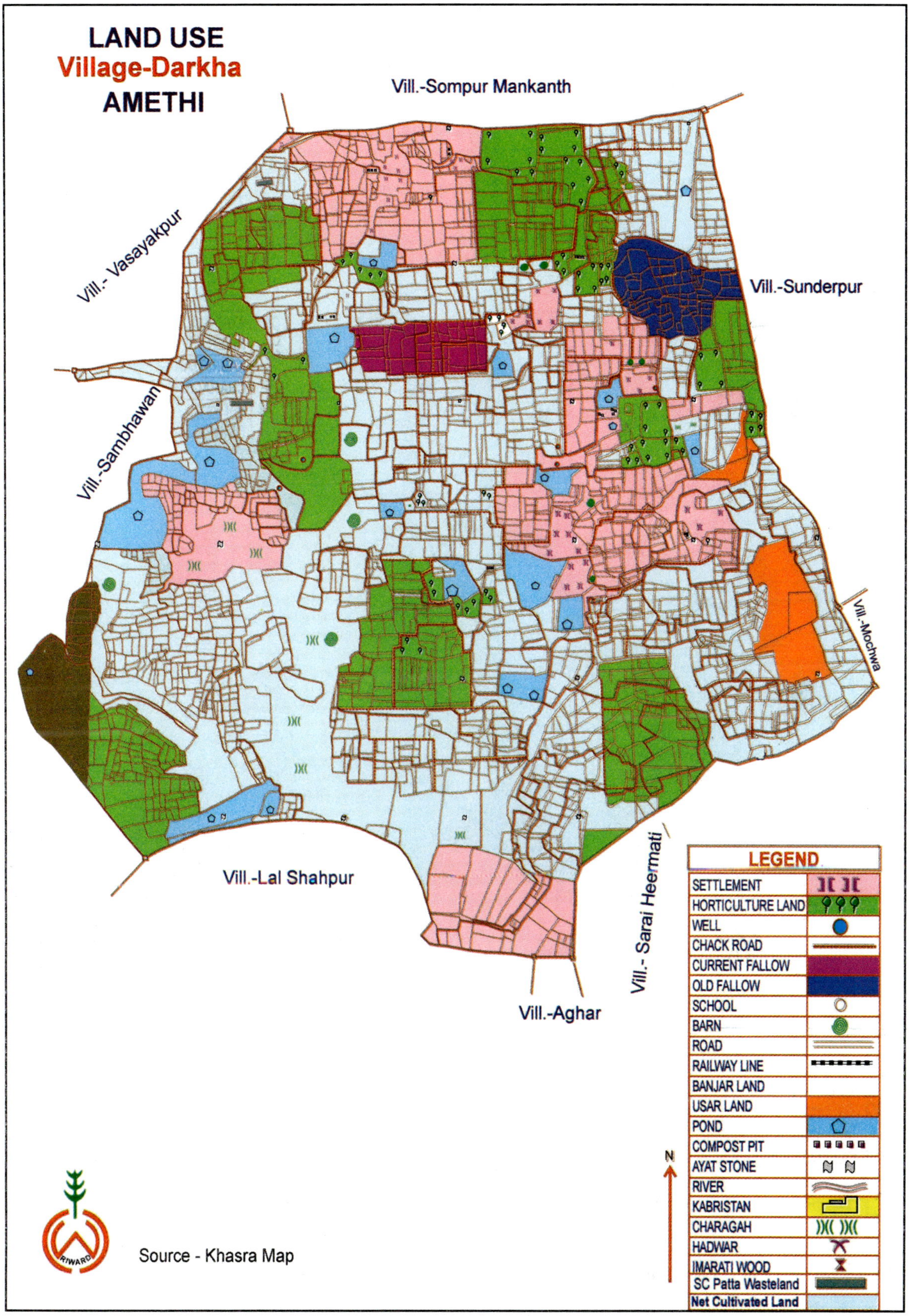

Map 6.29: Land Use of Village Panchayat Darakha

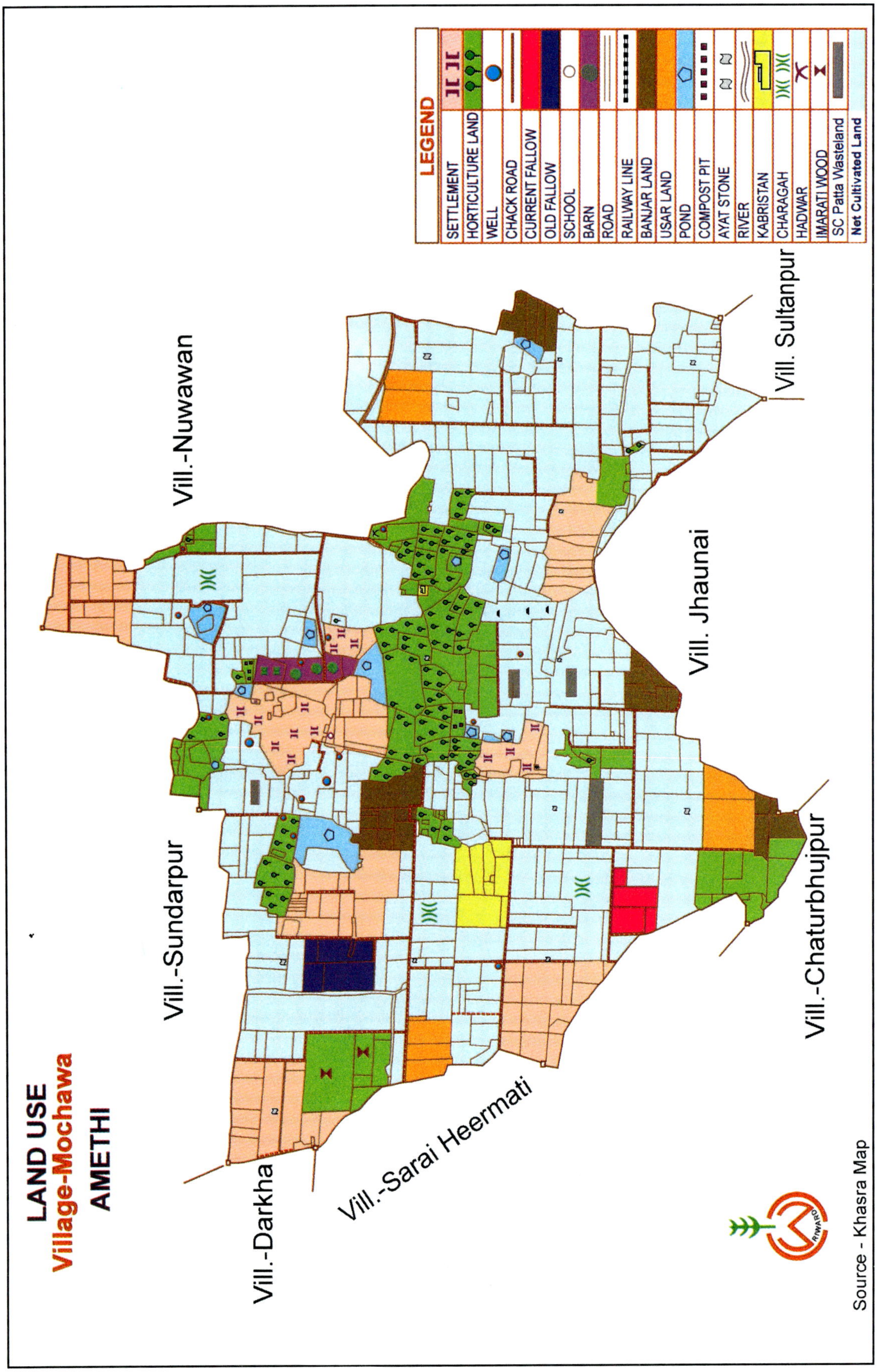

Map 6.30: Land Use of Village Panchayat Mochawa

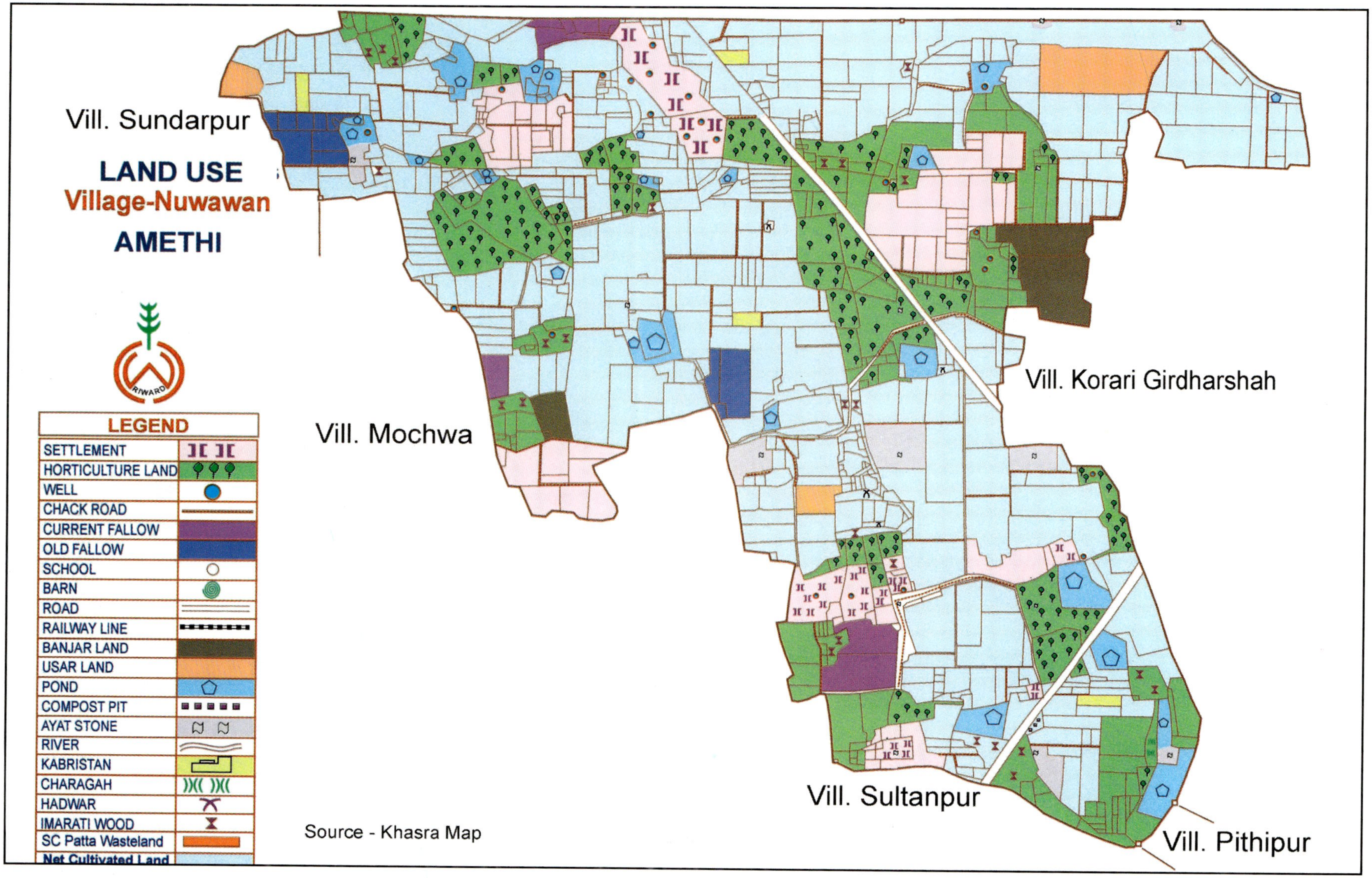

Map 6.31: Land Use of Village Panchayat Nuwawan

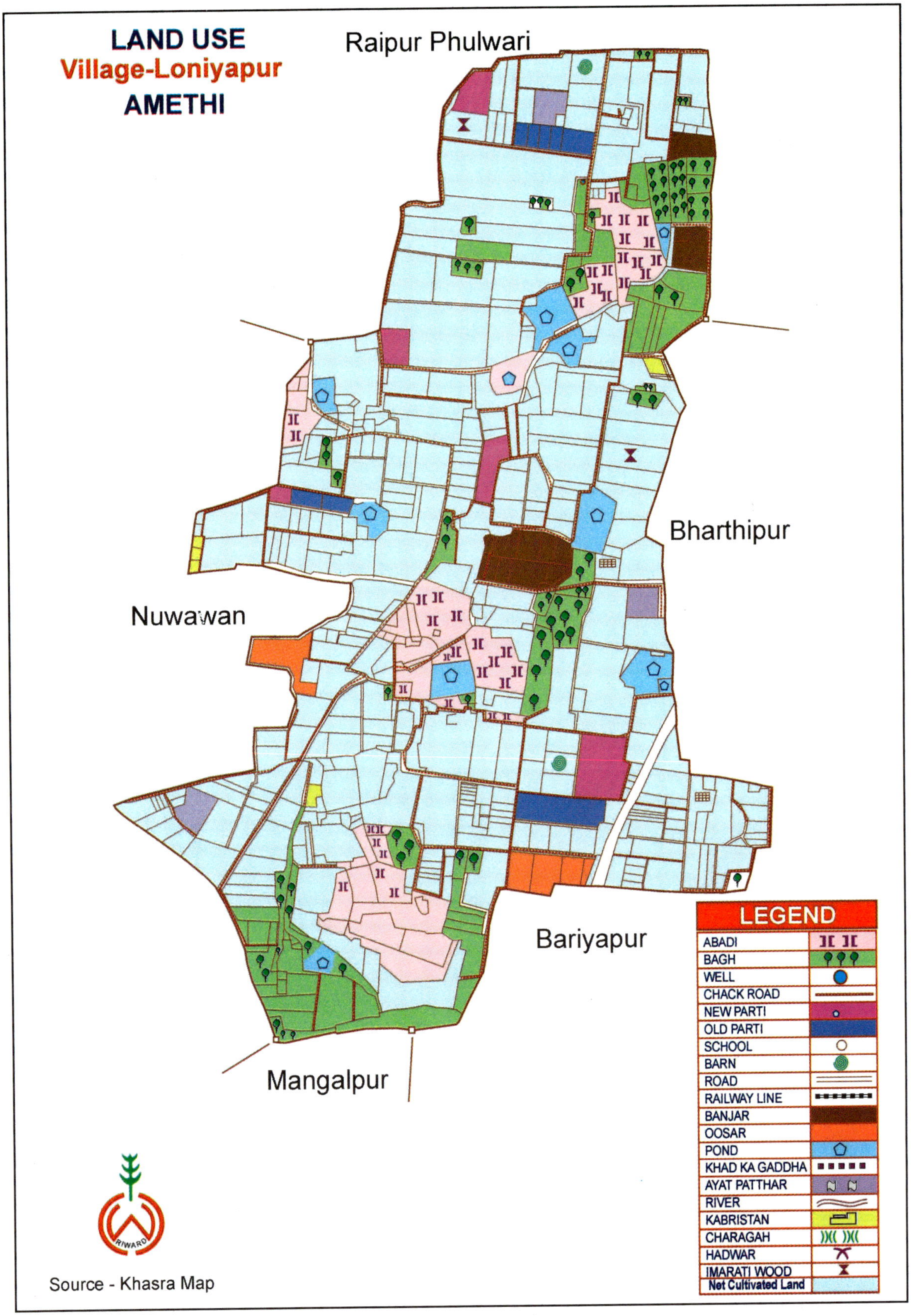

Map 6.32: Land Use of Village Panchayat Loniapur

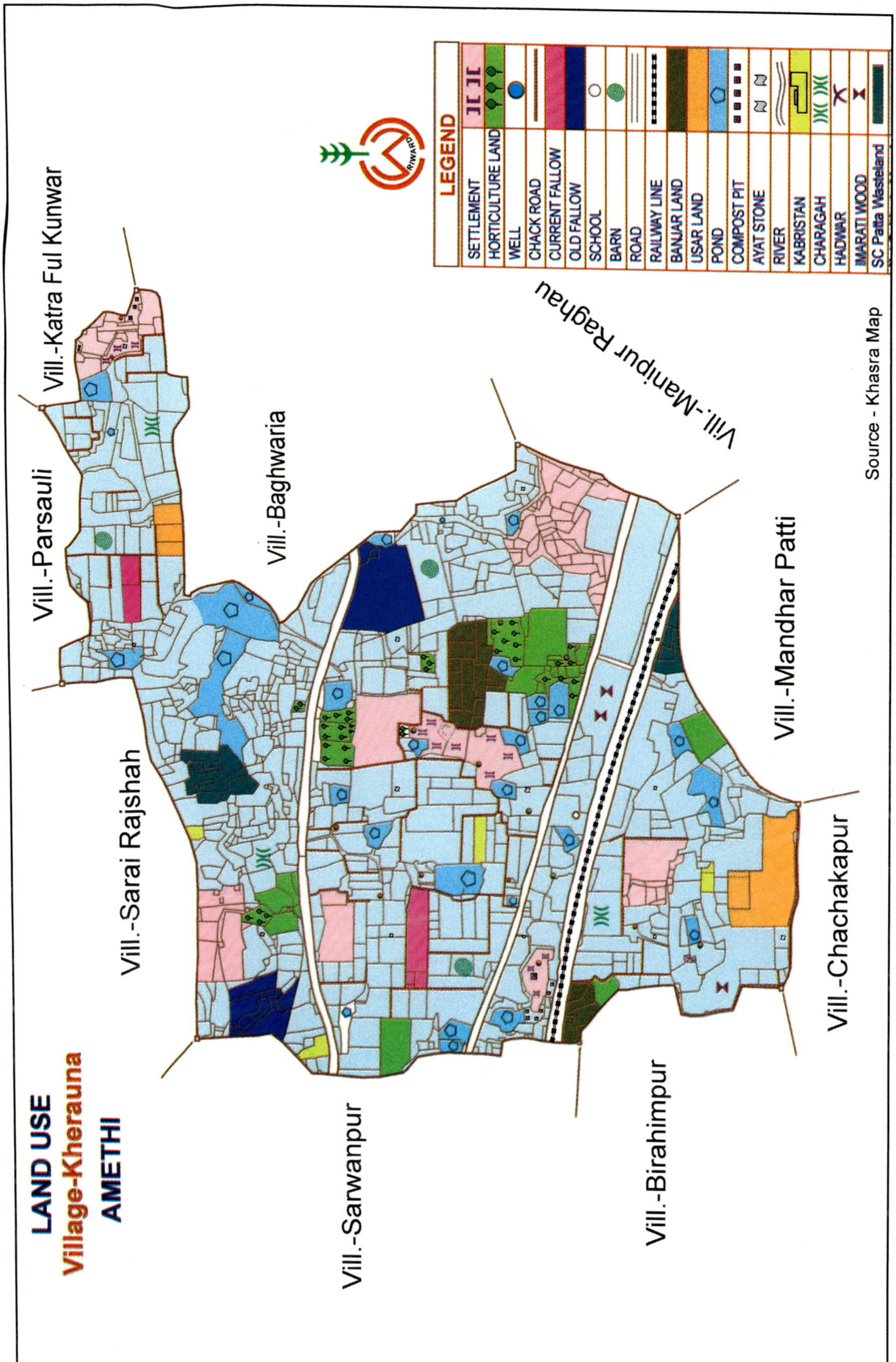

Map 6.33: Land Use of Village Panchayat Kherauna

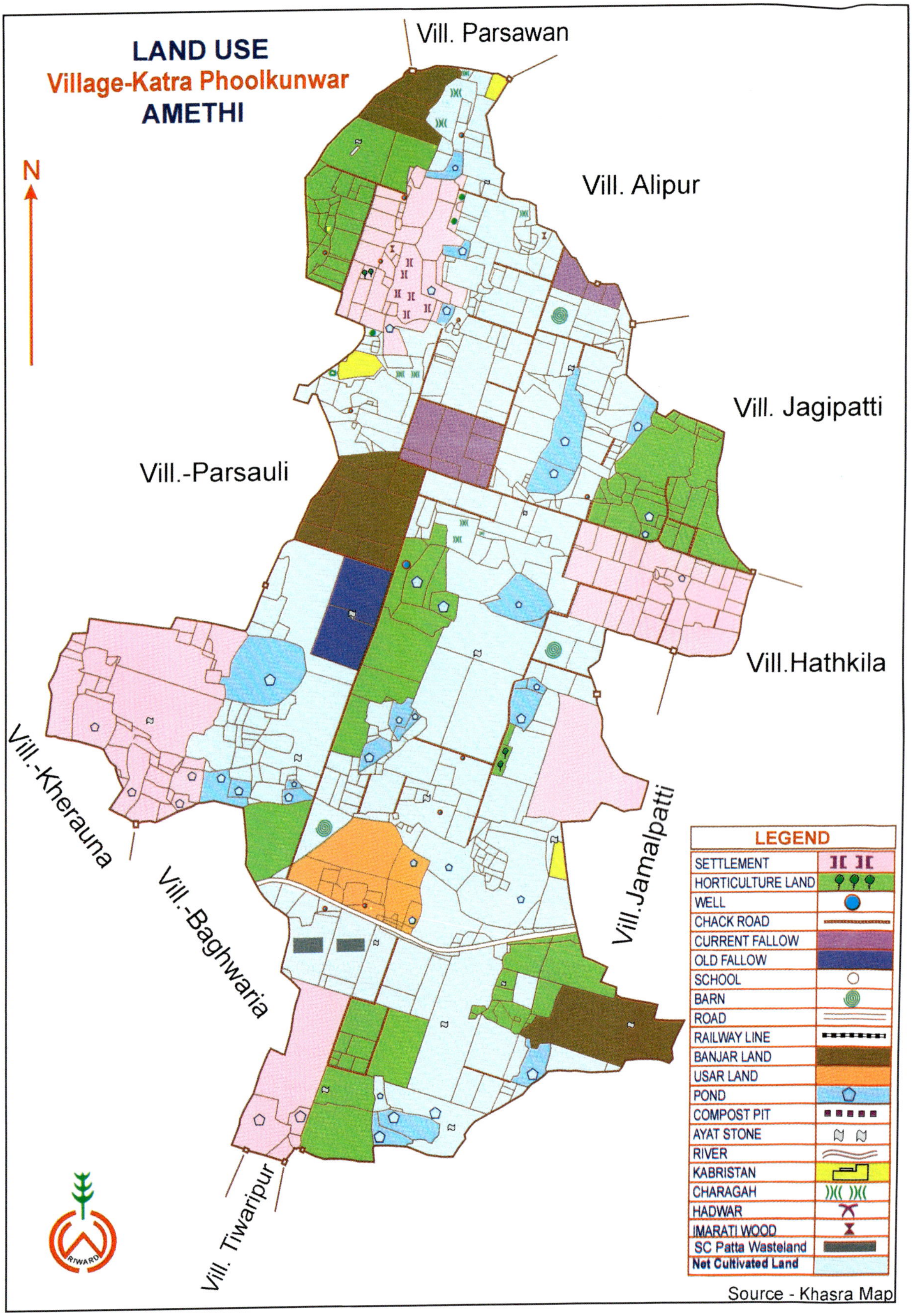

Map 6.34: Land Use of Village Panchayat Katra Phoolkunwar

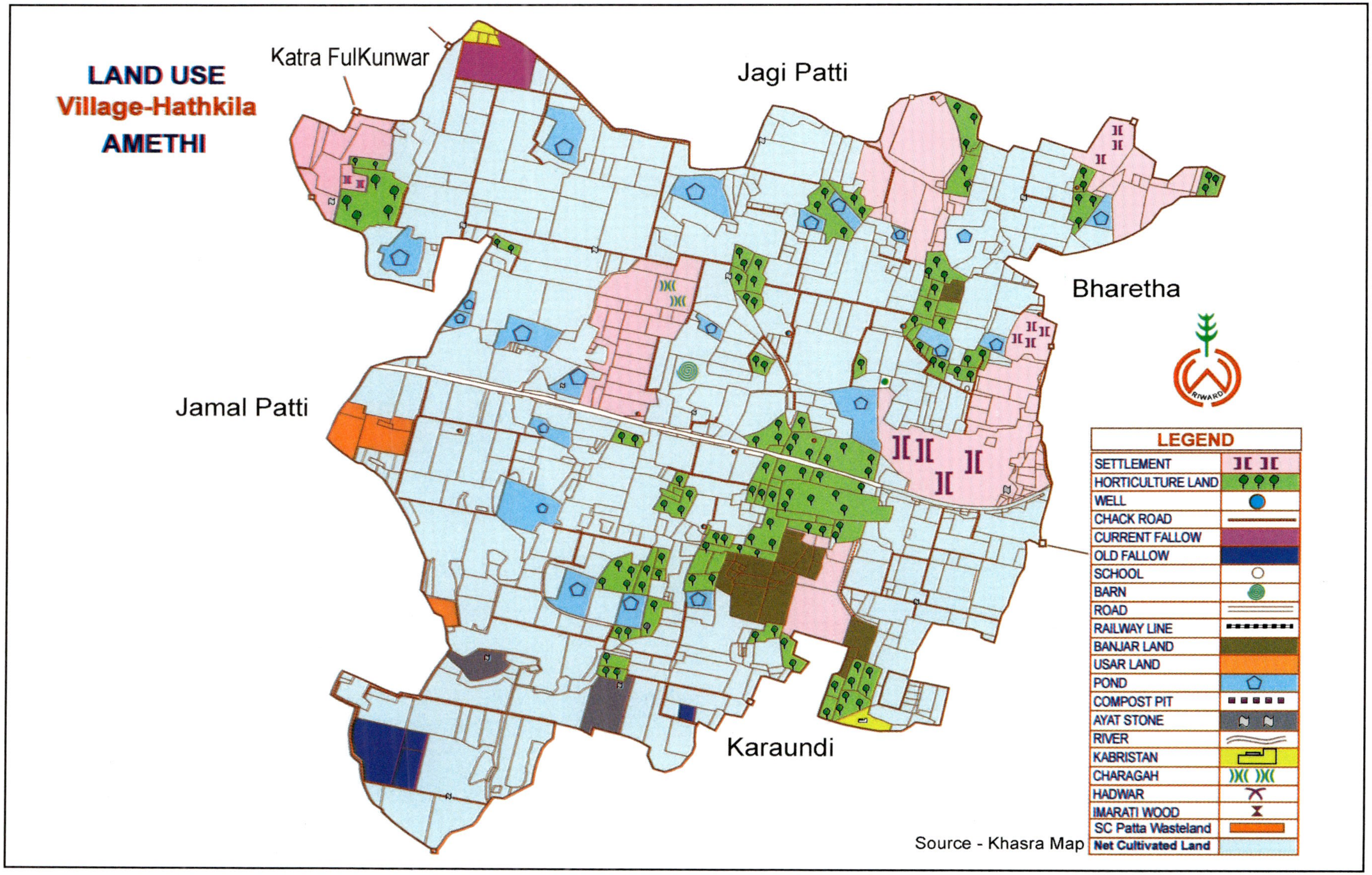

Map 6.35: Land Use of Village Panchayat Hathkila

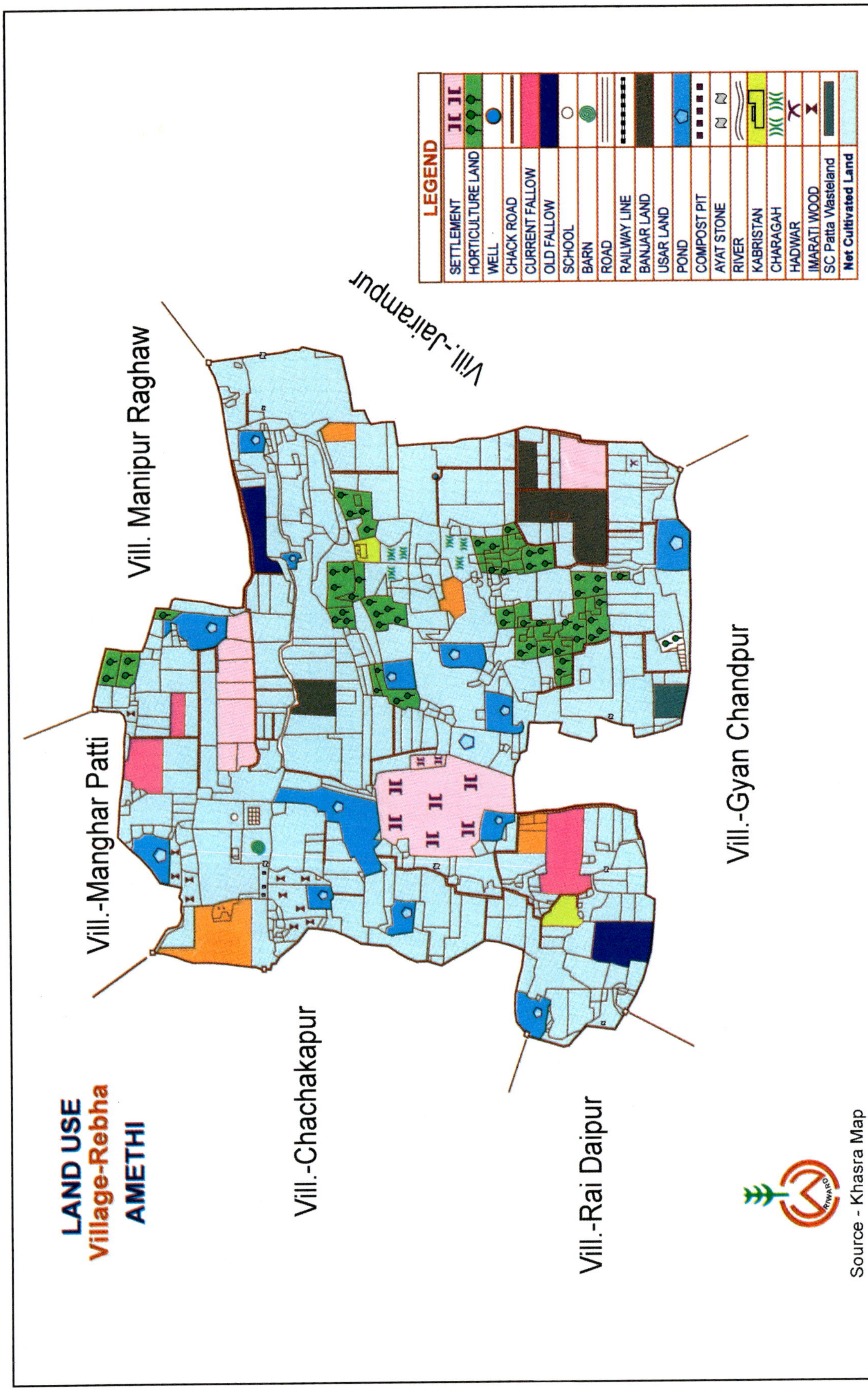

Map 6.36: Land Use of Village Panchayat Rebha

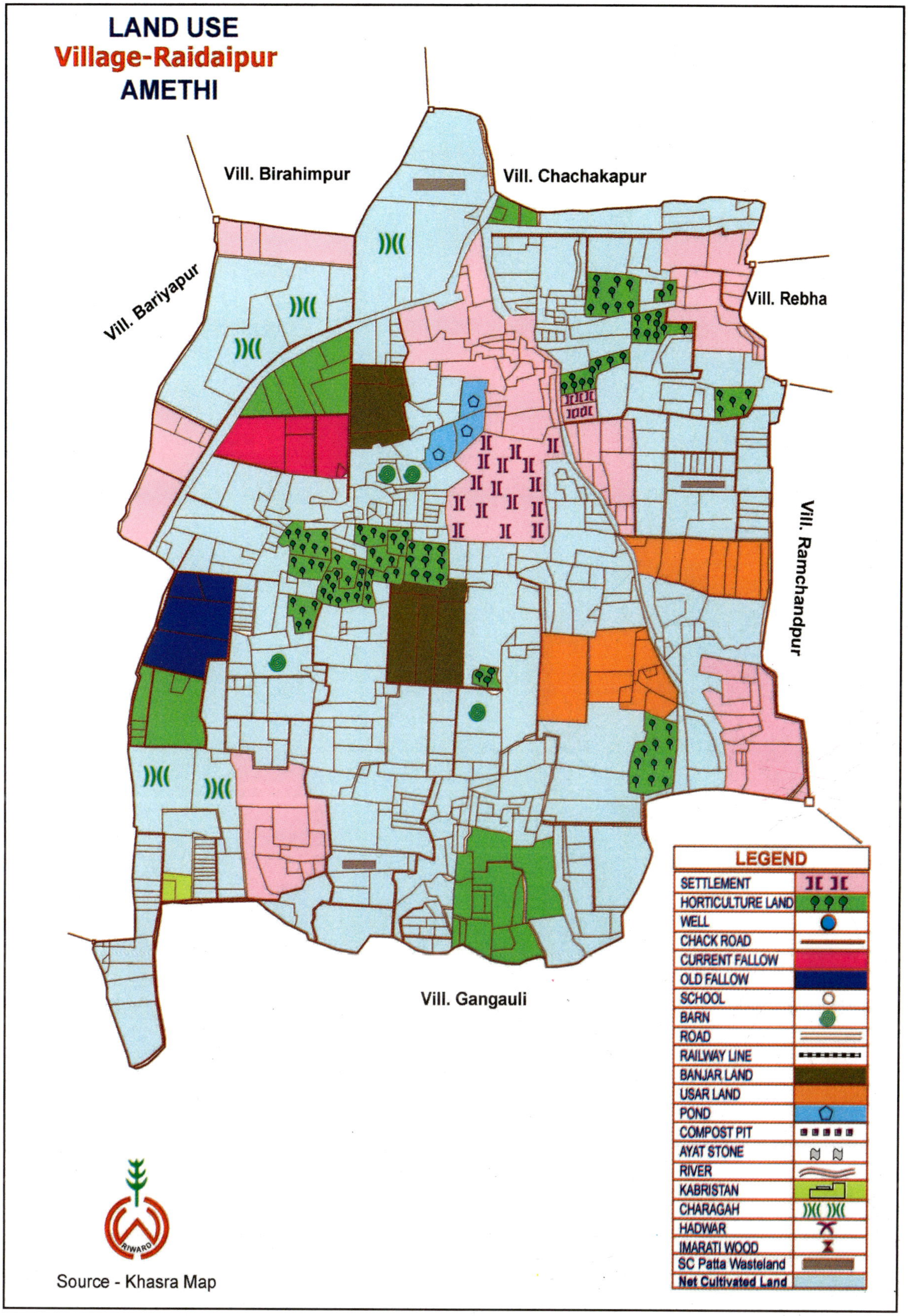

Map 6.37: Land Use of Village Panchayat Raidaipur

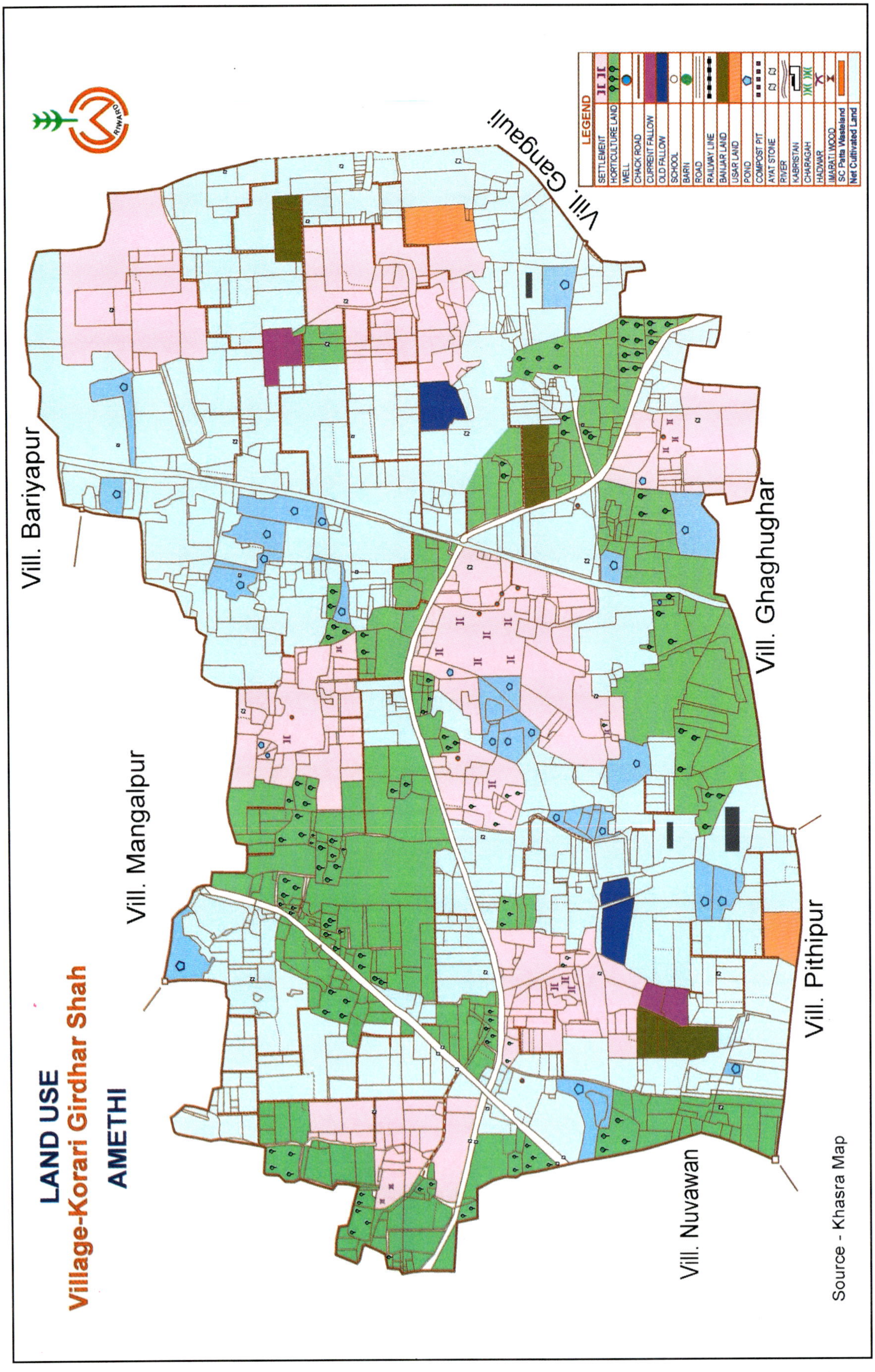

Map 6.38: Land Use of Village Panchayat Korari Girdhar Shah

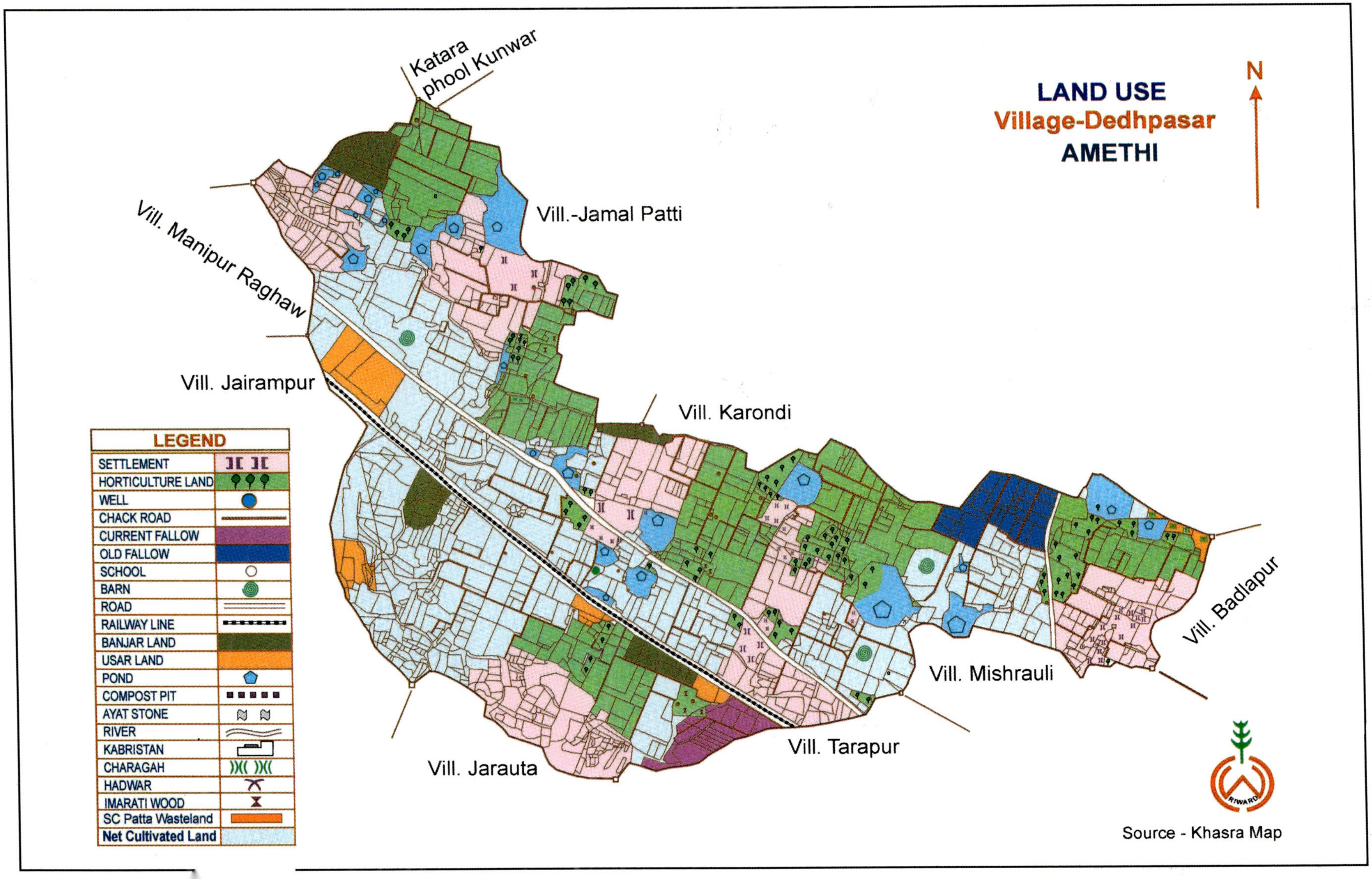

Map 6.39: Land Use of Village Panchayat Dedhpasar

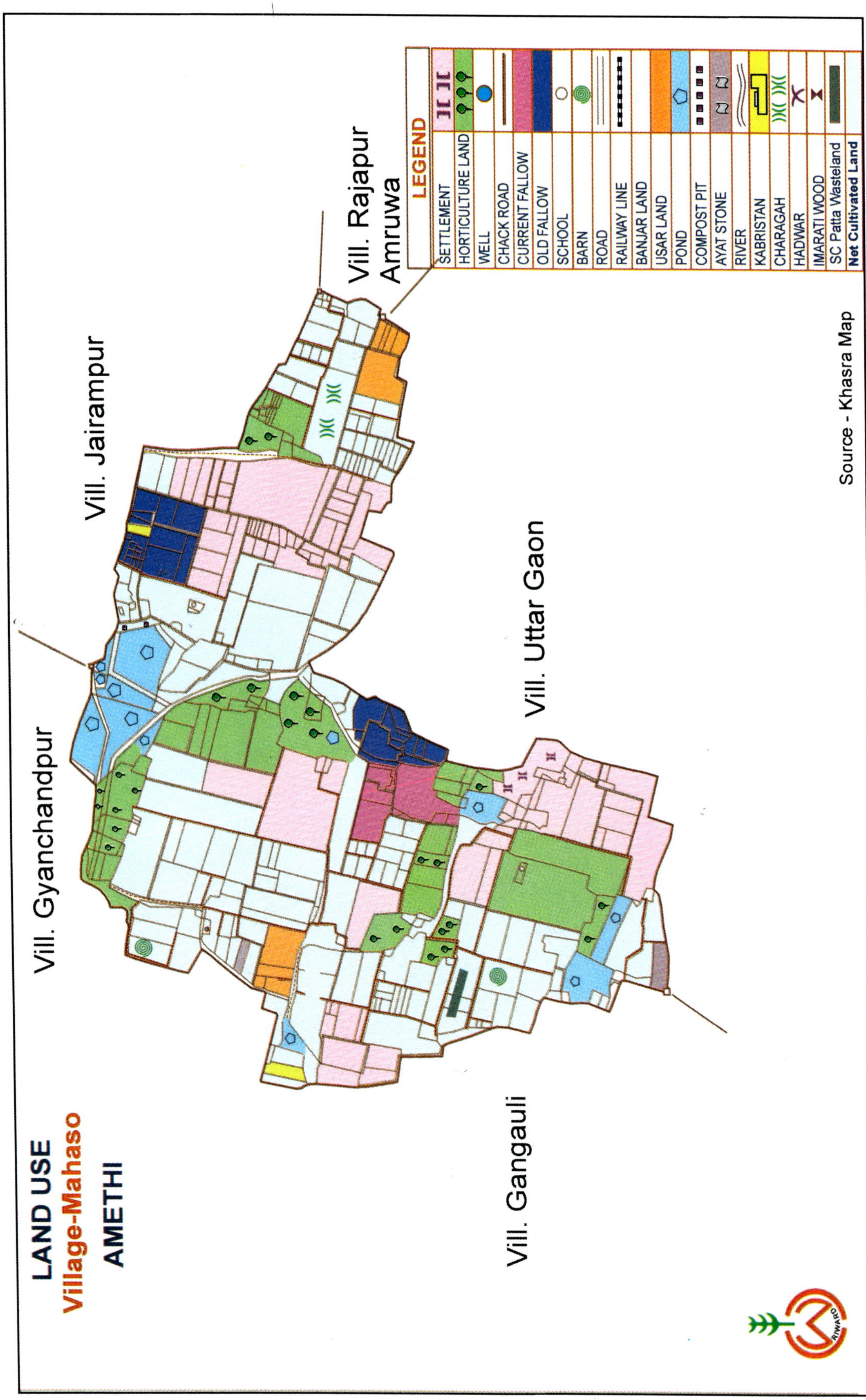

Map 6.40: Land Use of Village Panchayat Mahaso

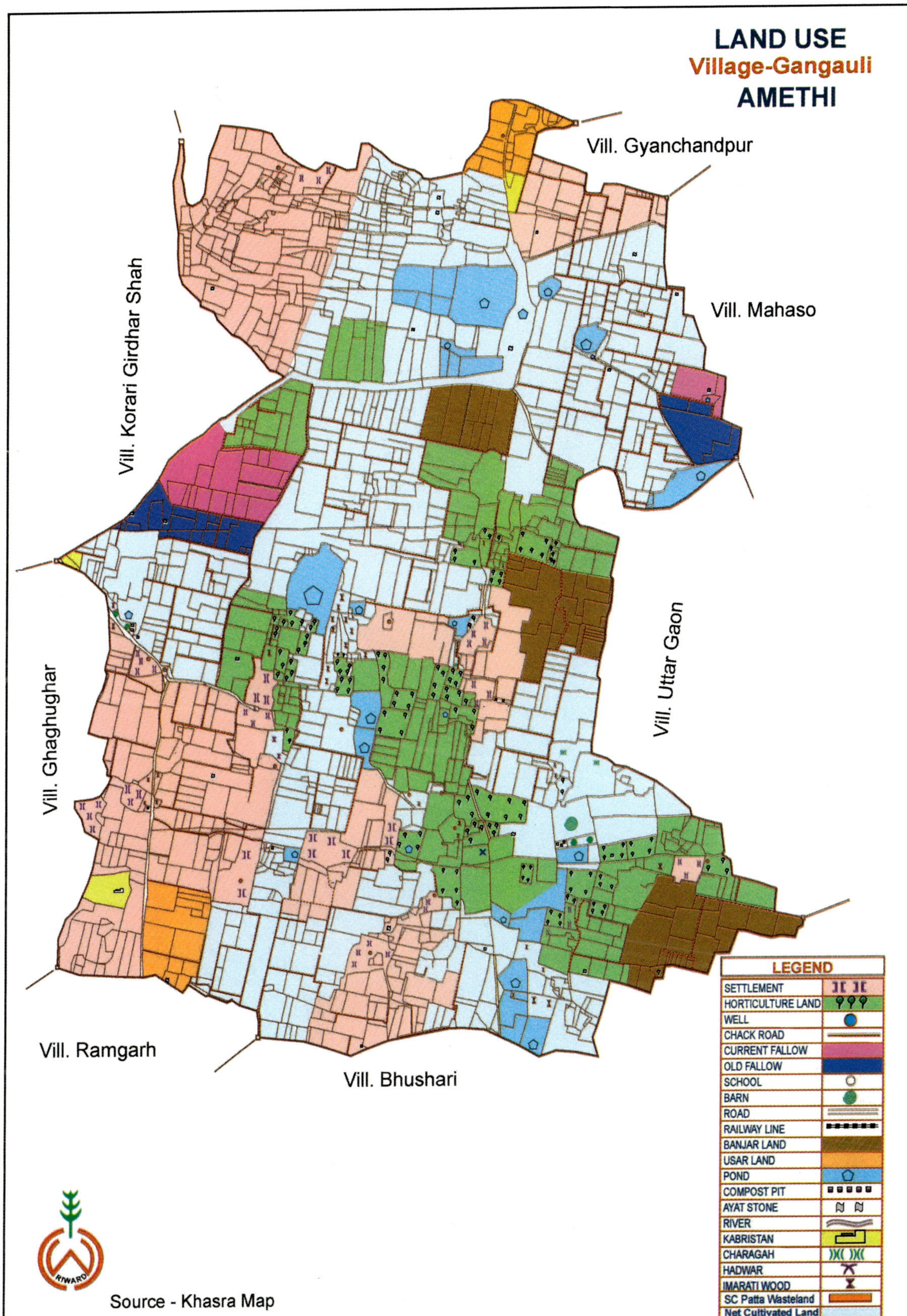

Map 6.41: Land Use of Village Panchayat Gangauli

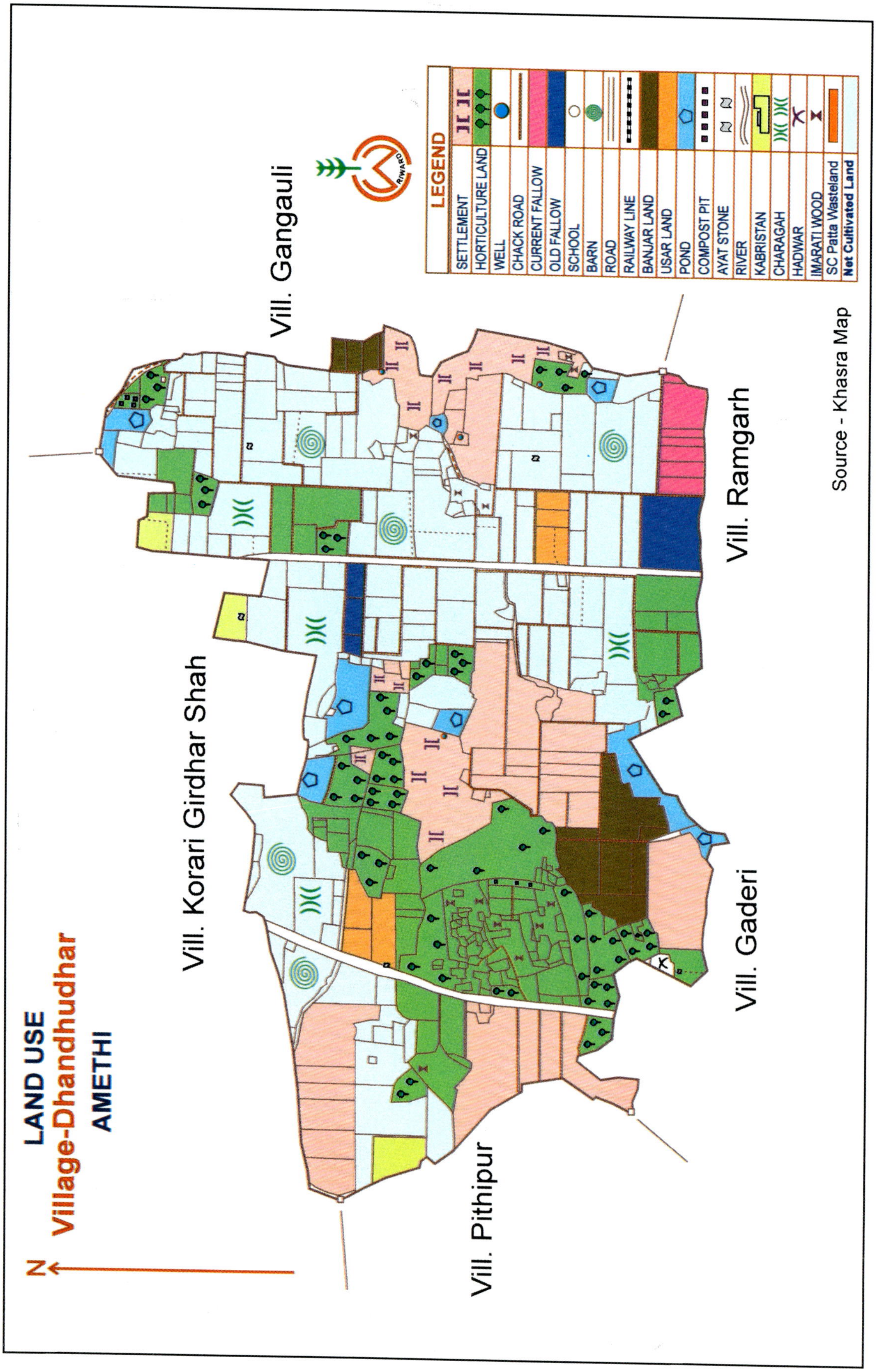

Map 6.42 : Land Use of Village Panchayat Dhandhudhar

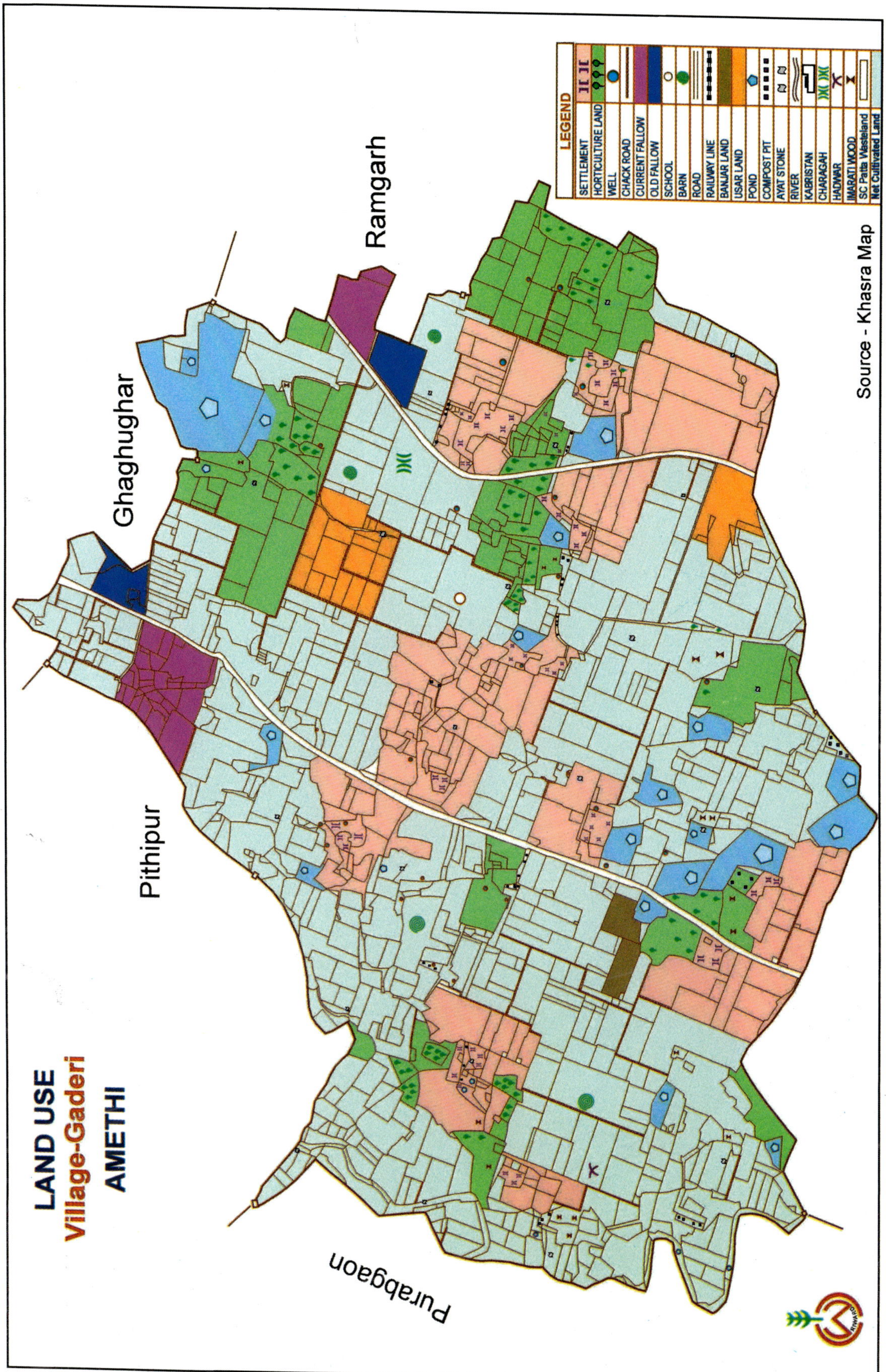

Map 6.43: Land Use of Village Panchayat Gaderi

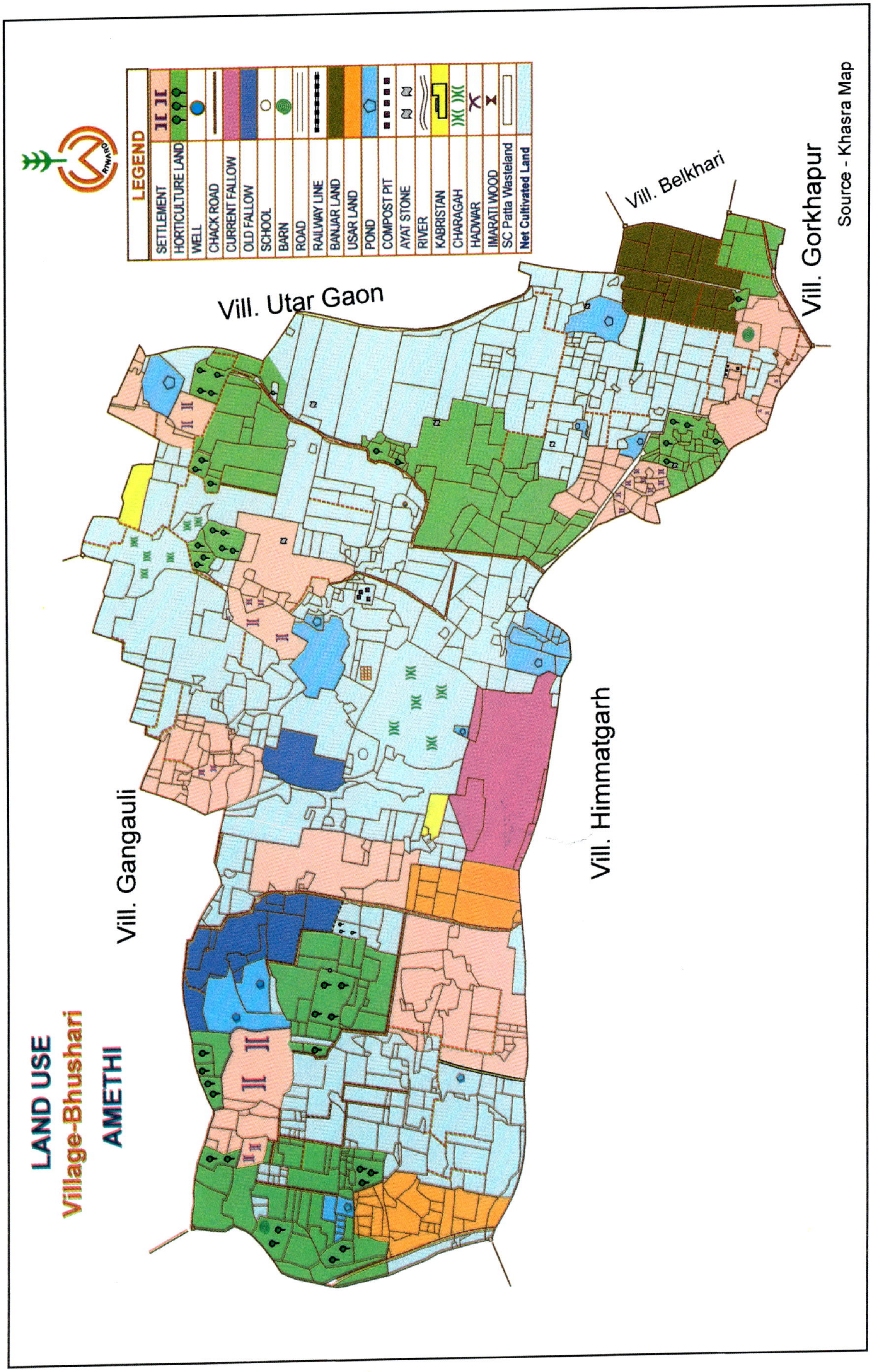

Map 6.44: Land Use of Village Panchayat Bhushari

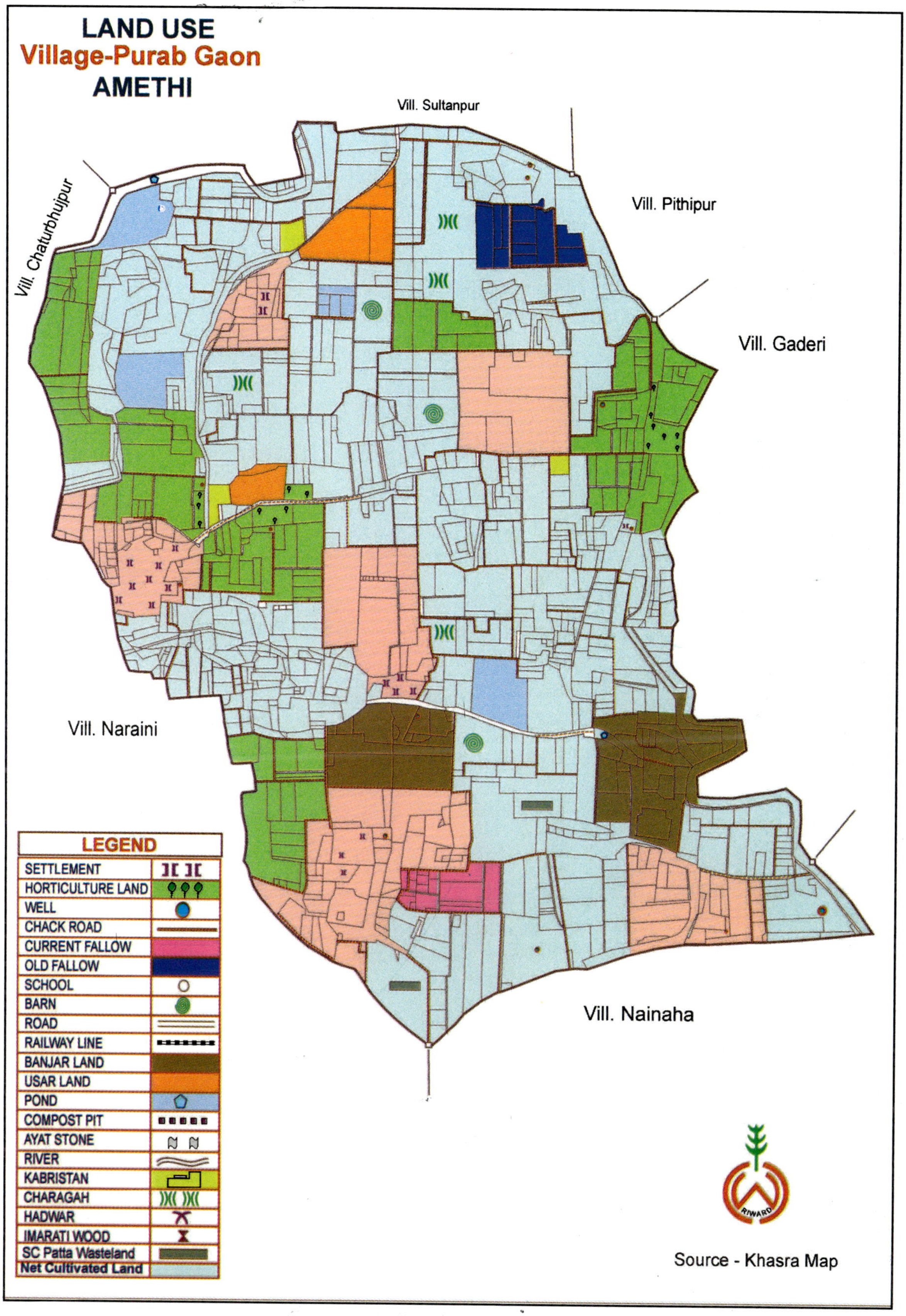

Map 6.45: Land Use of Village Panchayat Purabgaon

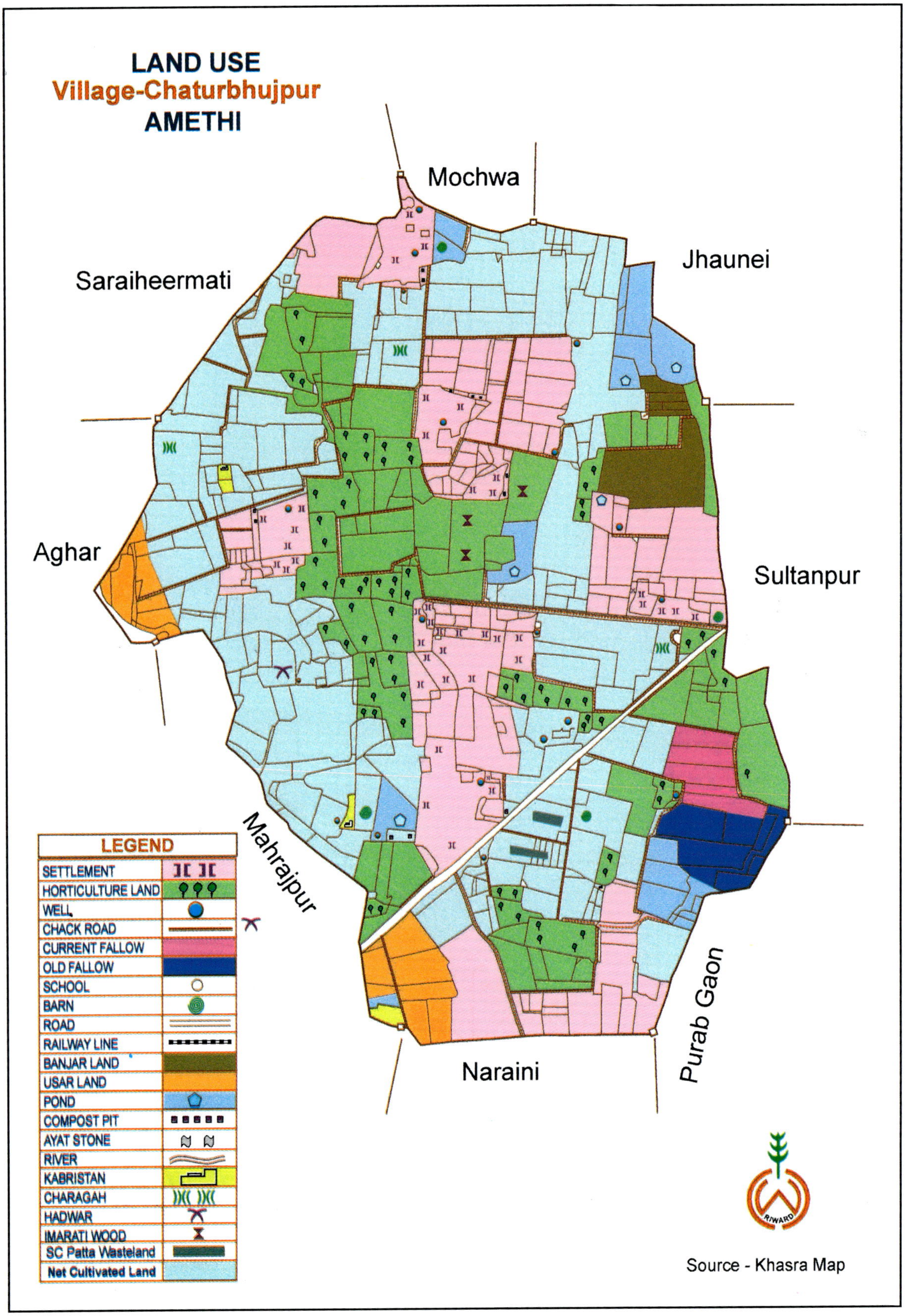

Map 6.46: Land Use of Village Panchayat Chaturbhujpur

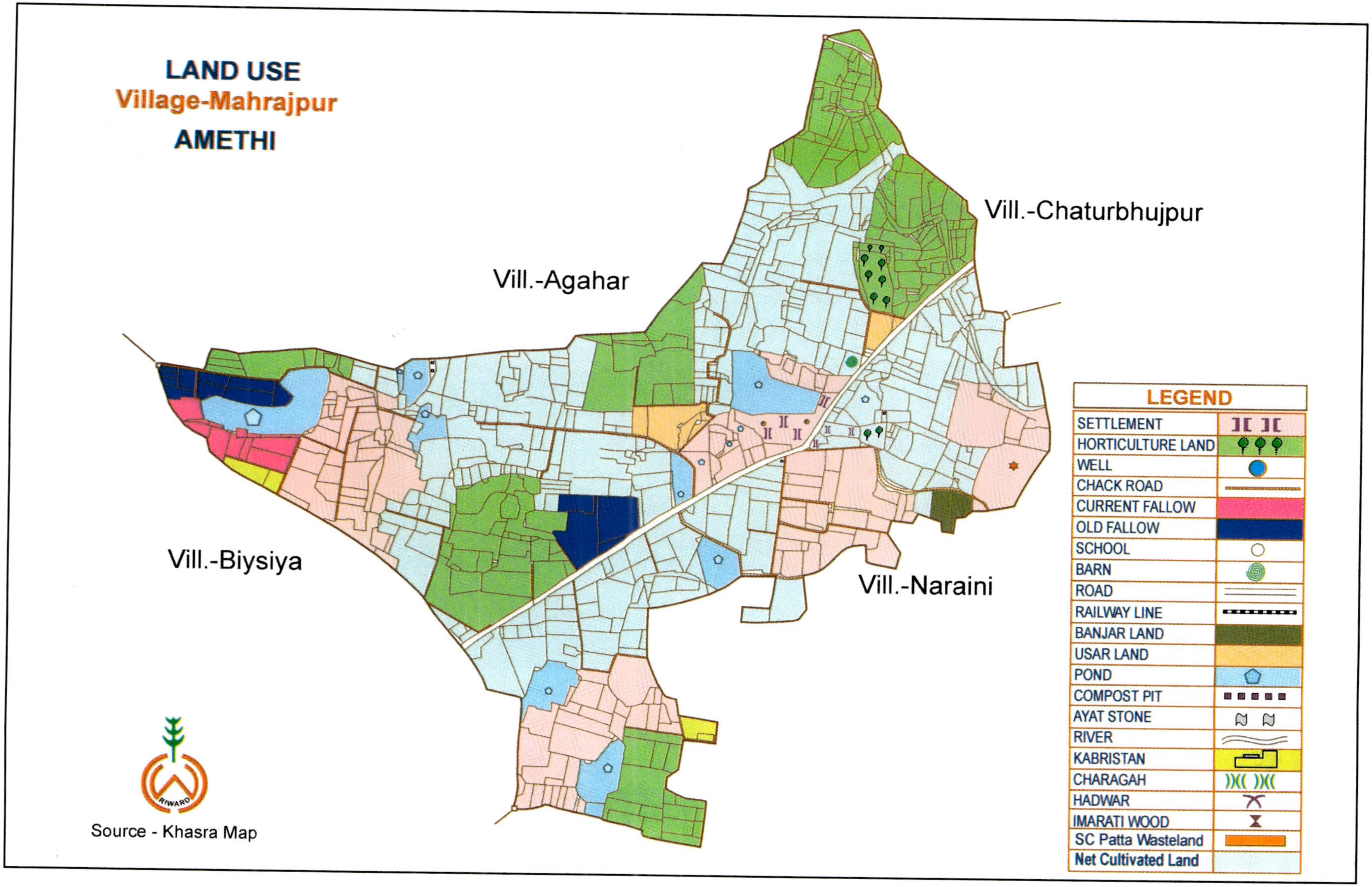

Map 6.47: Land Use of Village Panchayat Mahrajpur

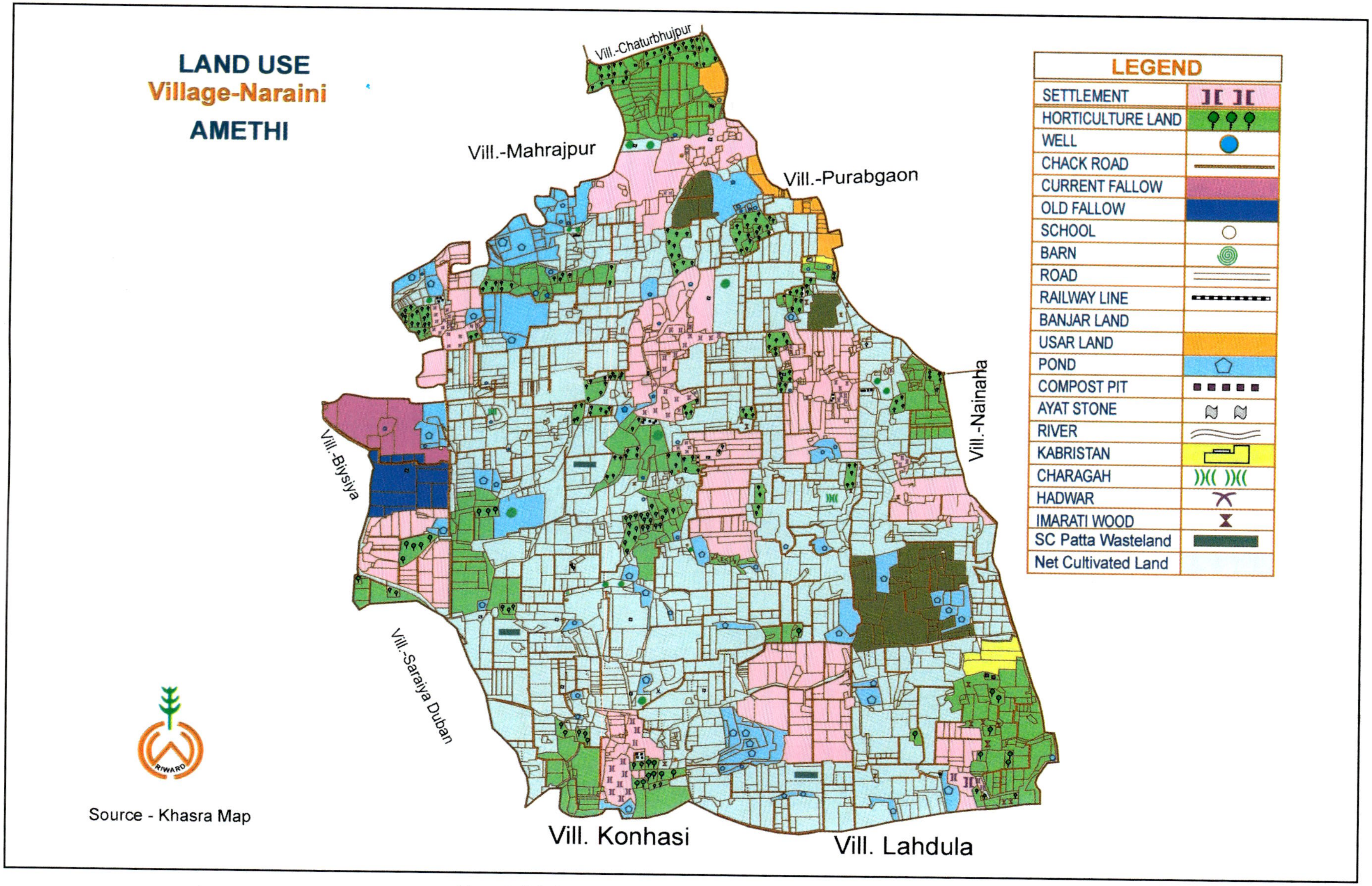

Map 6.54: Land Use of Village Panchayat Naraini

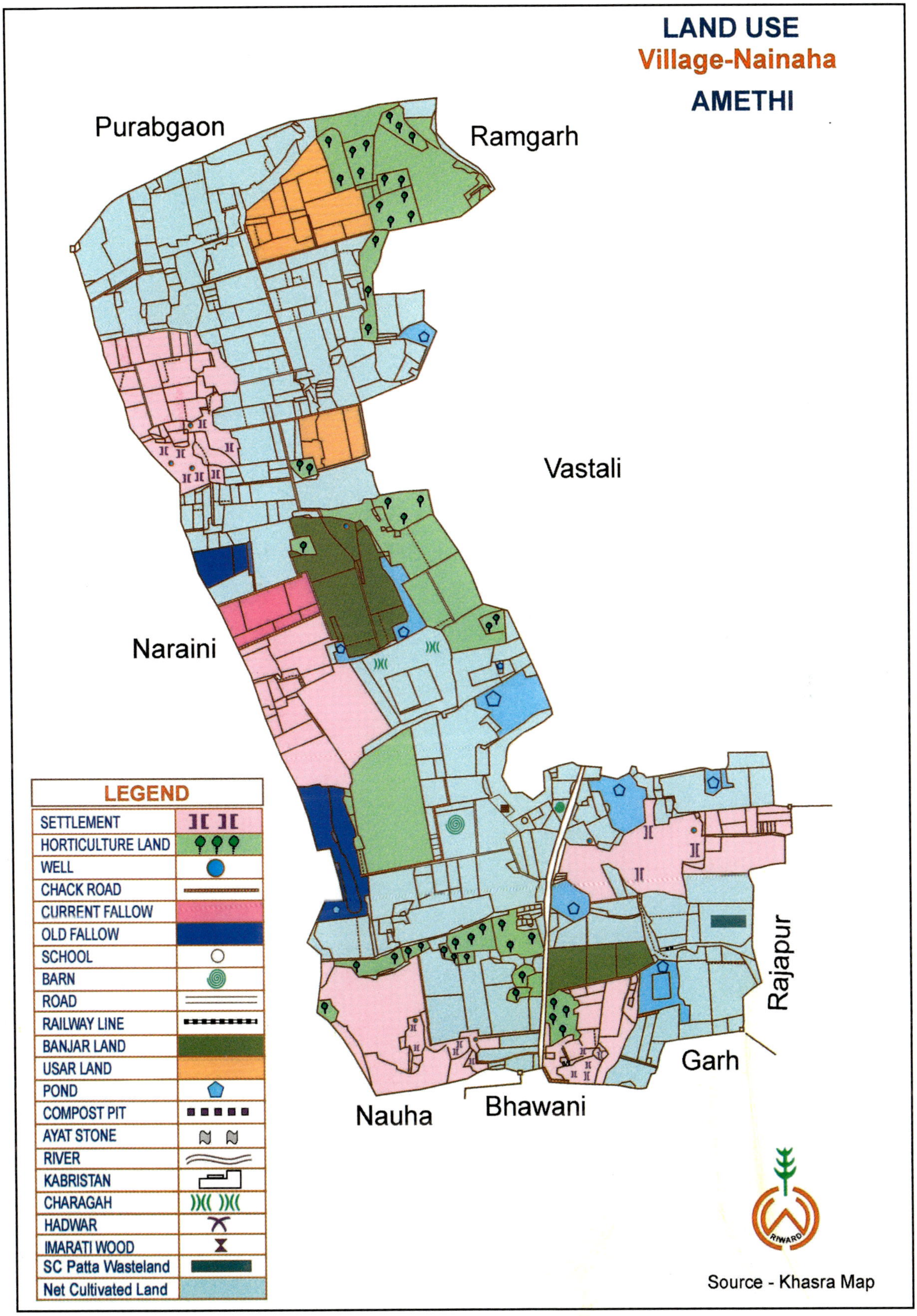

Map 6.55: Land Use of Village Panchayat Nainaha

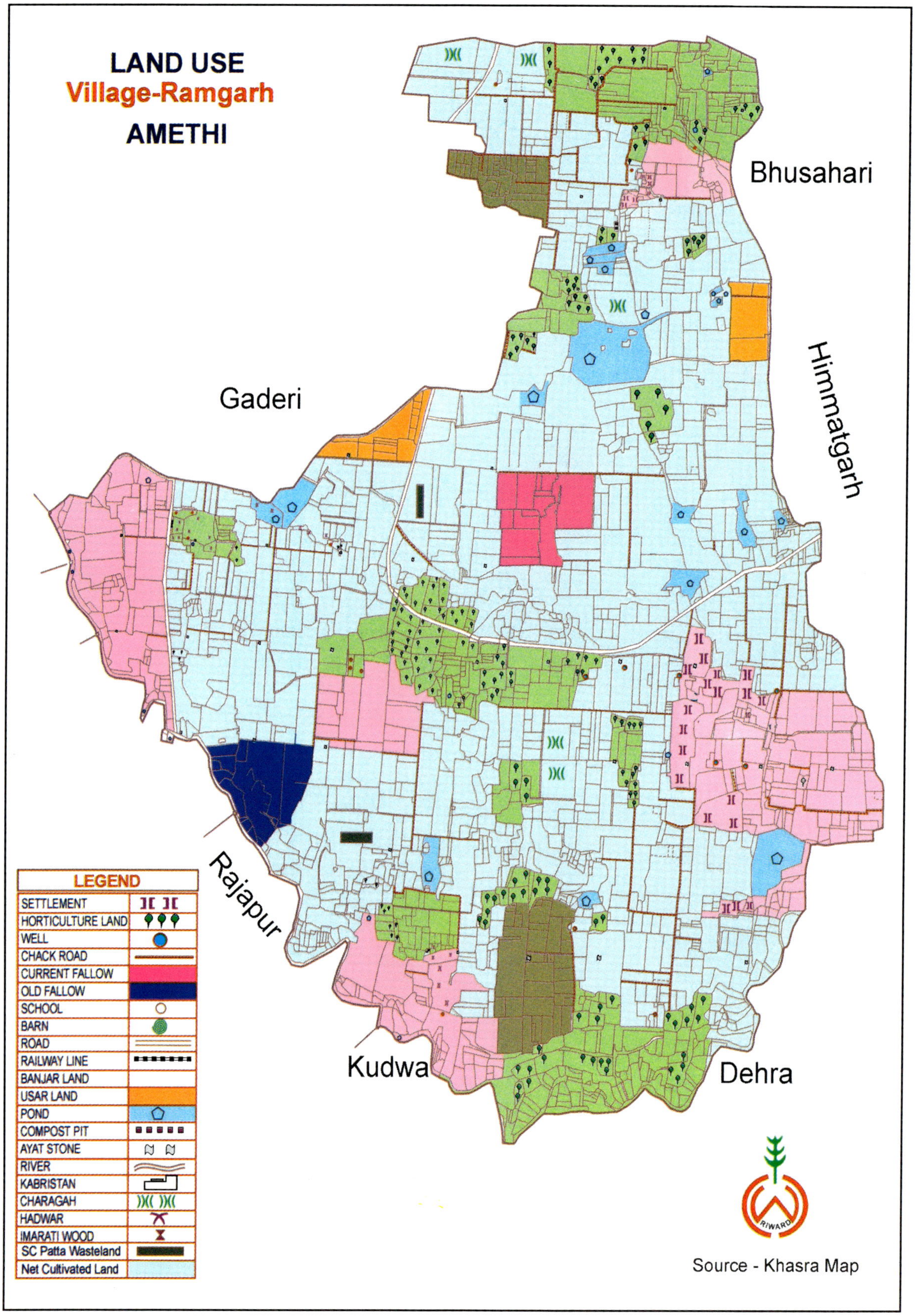

Map 6.56: Land Use of Village Panchayat Ramgarh

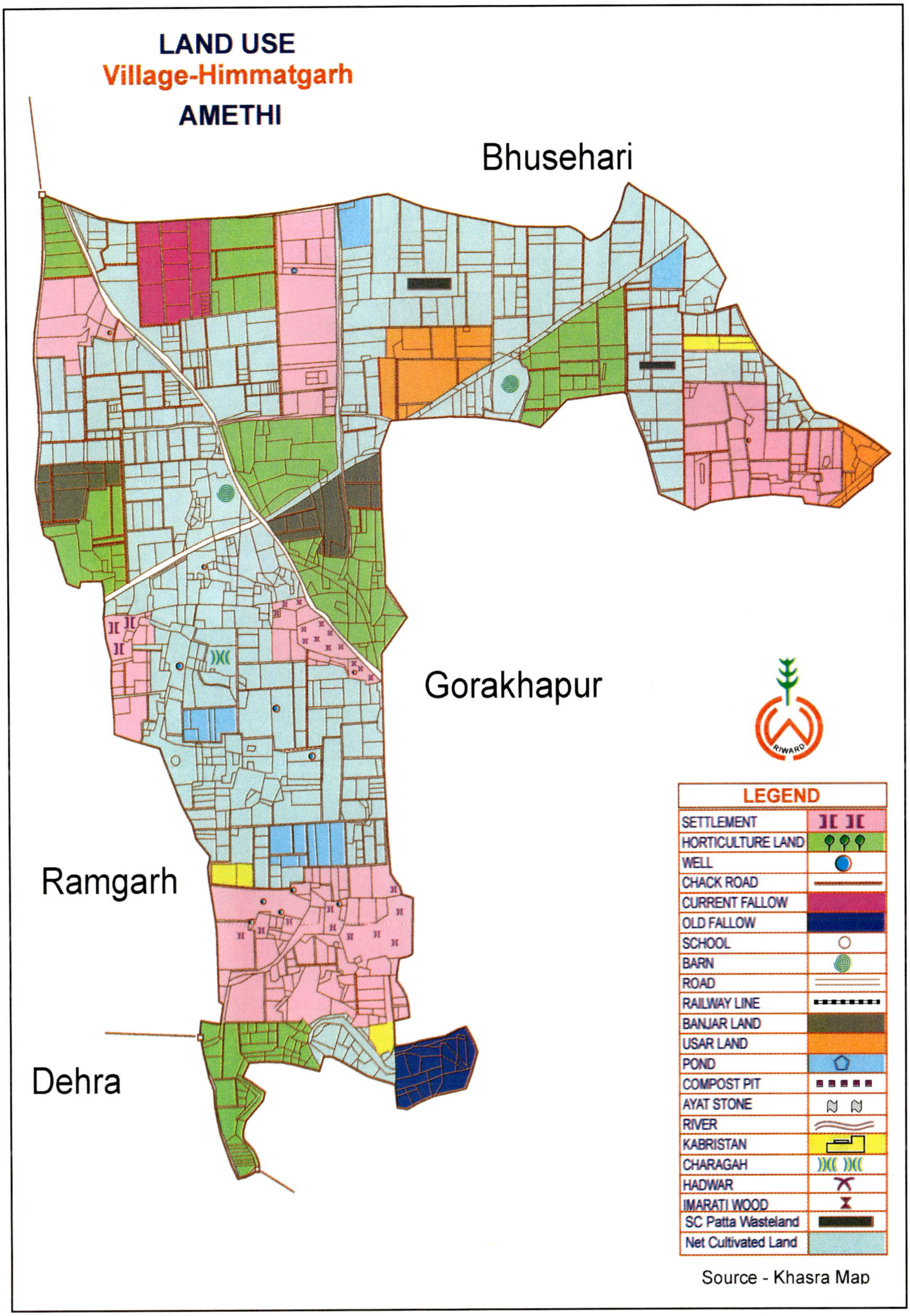

Map 6.57: Land Use of Village Panchayat Himmatgarh

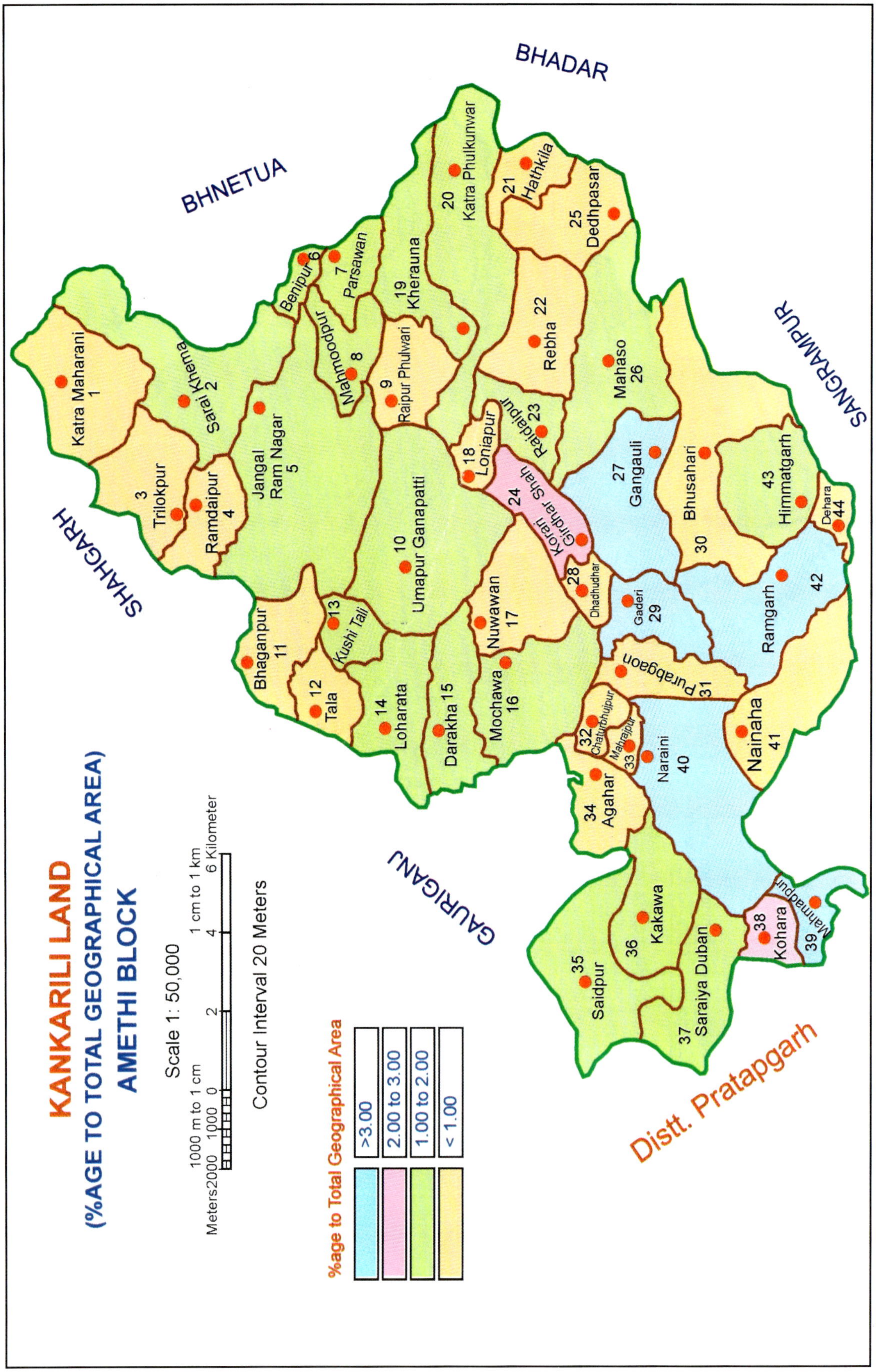

Map 7.60: Kankarili Land (%age to total Geographical Area)

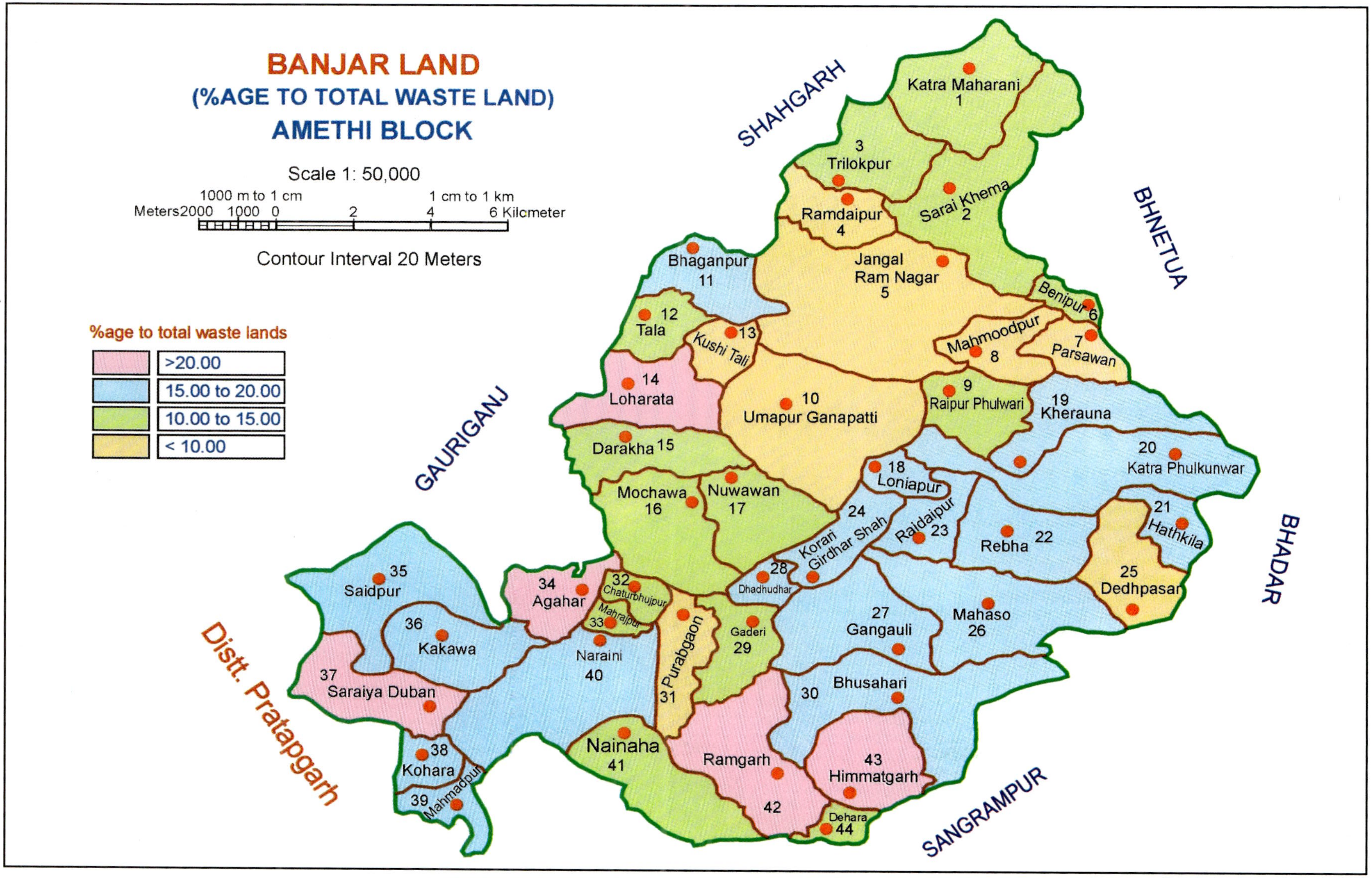

Map 7.61: Banjar Land (%age to total Waste Lands)

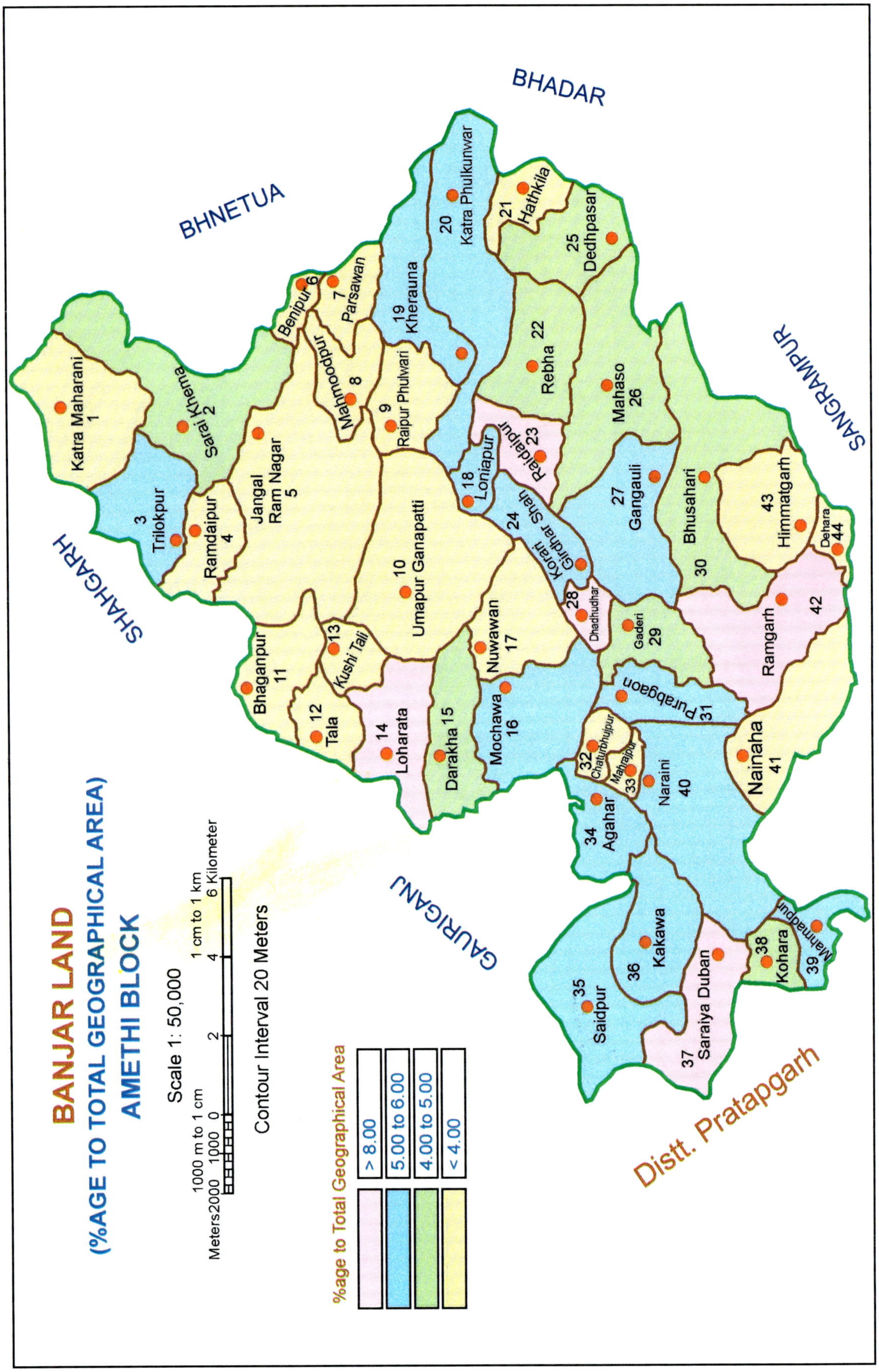

Map 7.62: Banjar Land (%age to total Geographical Area)

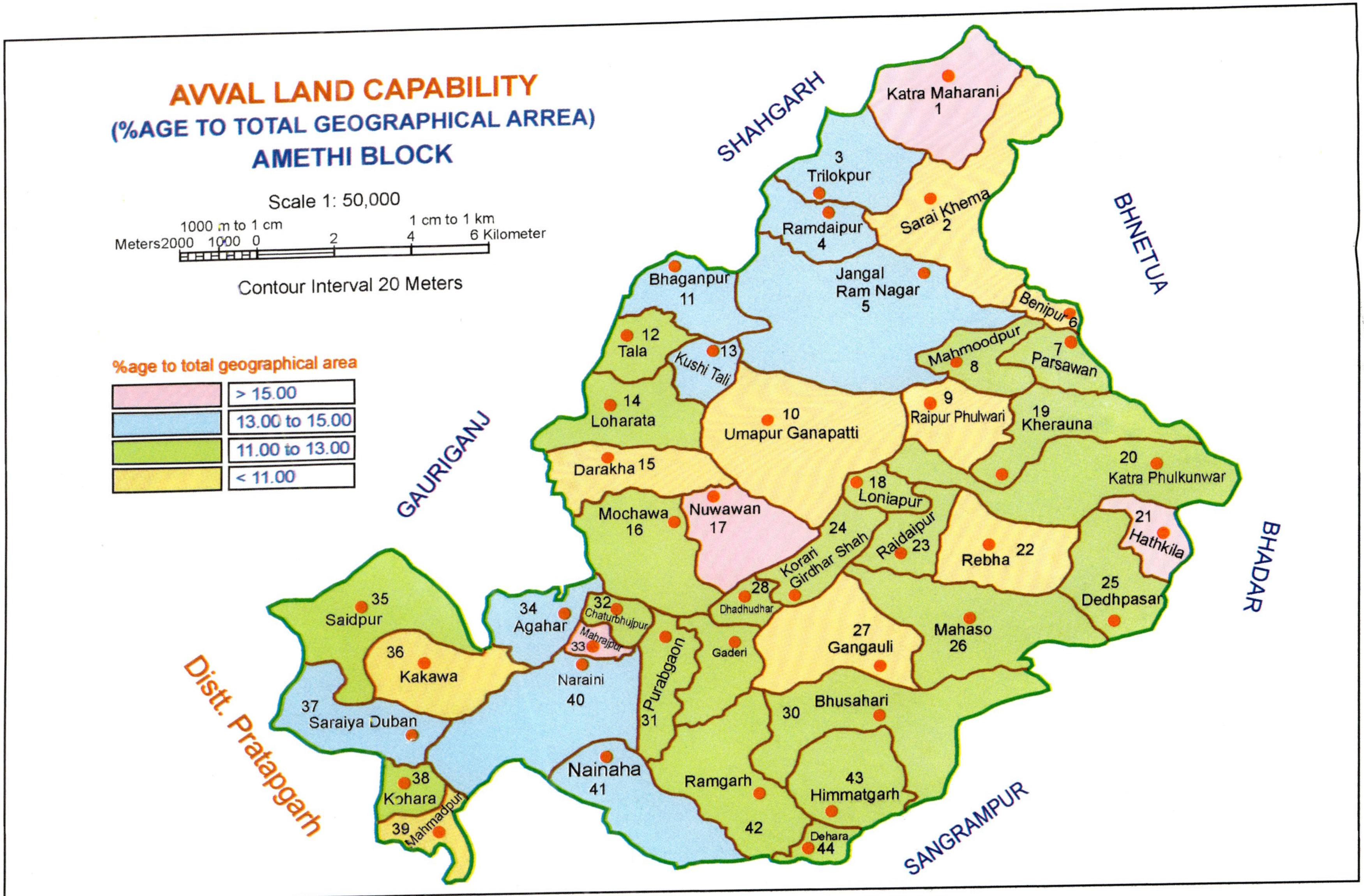

Map 8.75: Avval Land Capability (%age to Total Geographical Area)

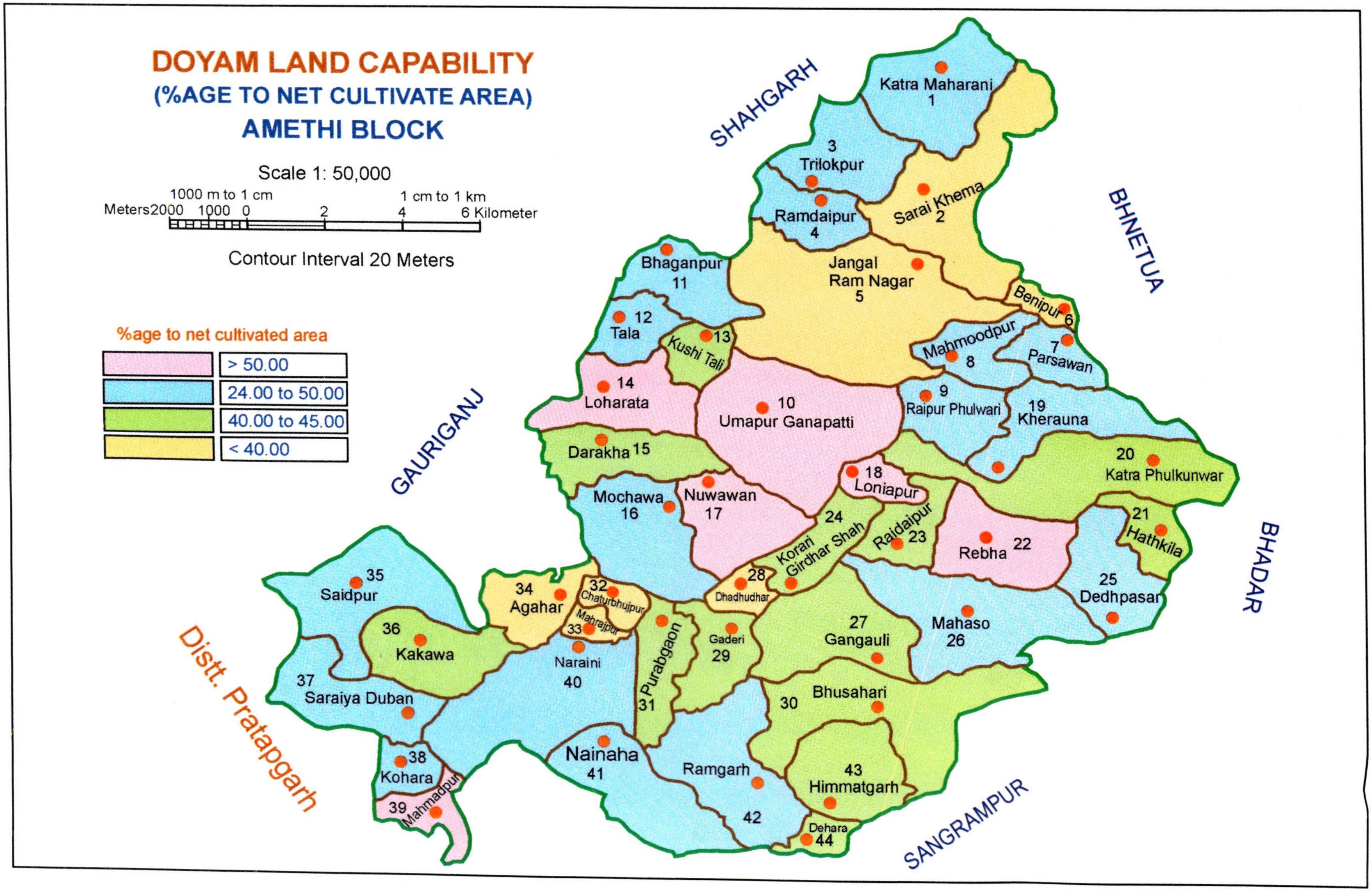

Map 8.76: Doyam Land Capability (%age to Net Cultivated Area)

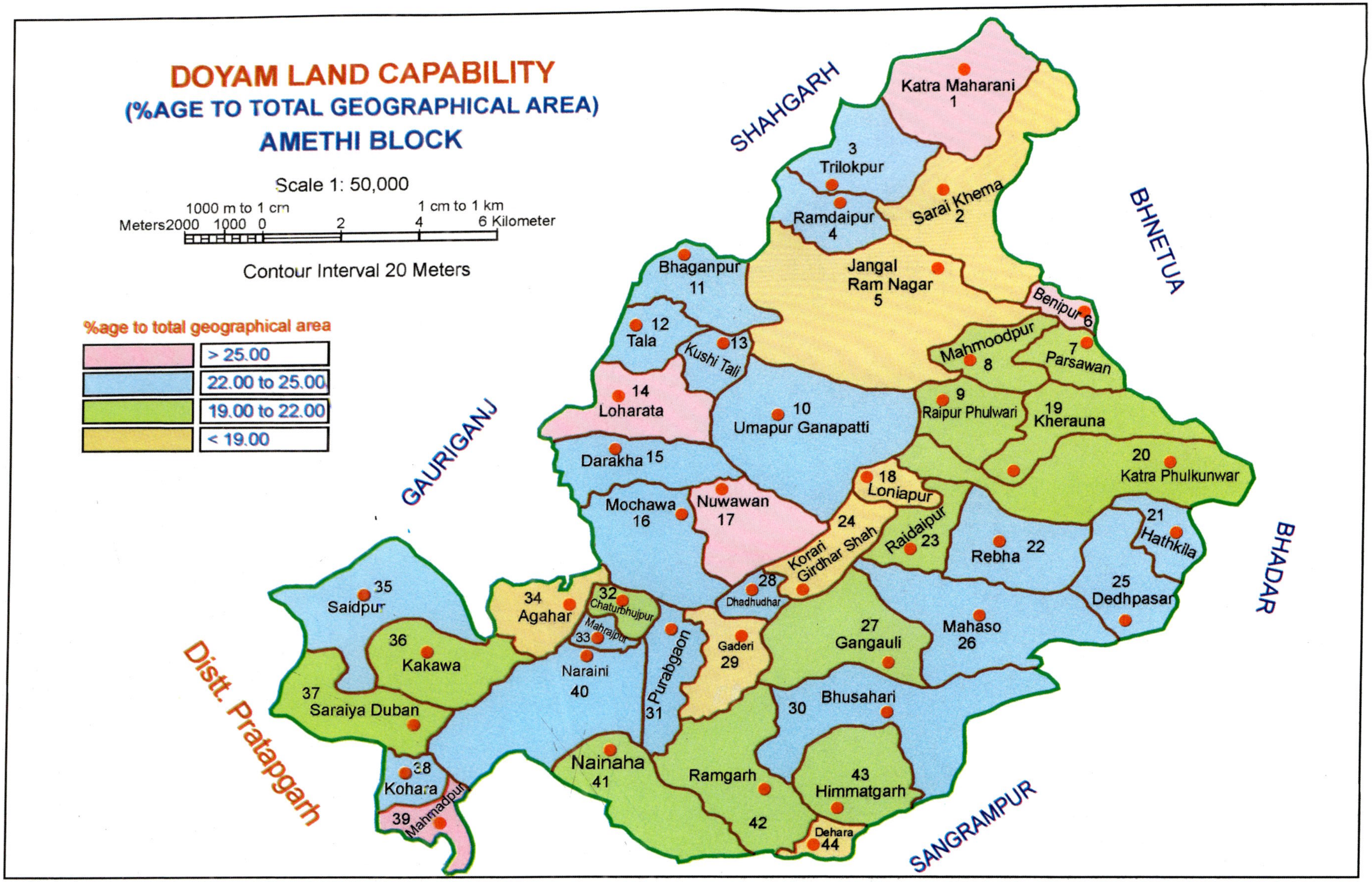

Map 8.77: Doyam Land Capability (%age to Total Geographical Area)

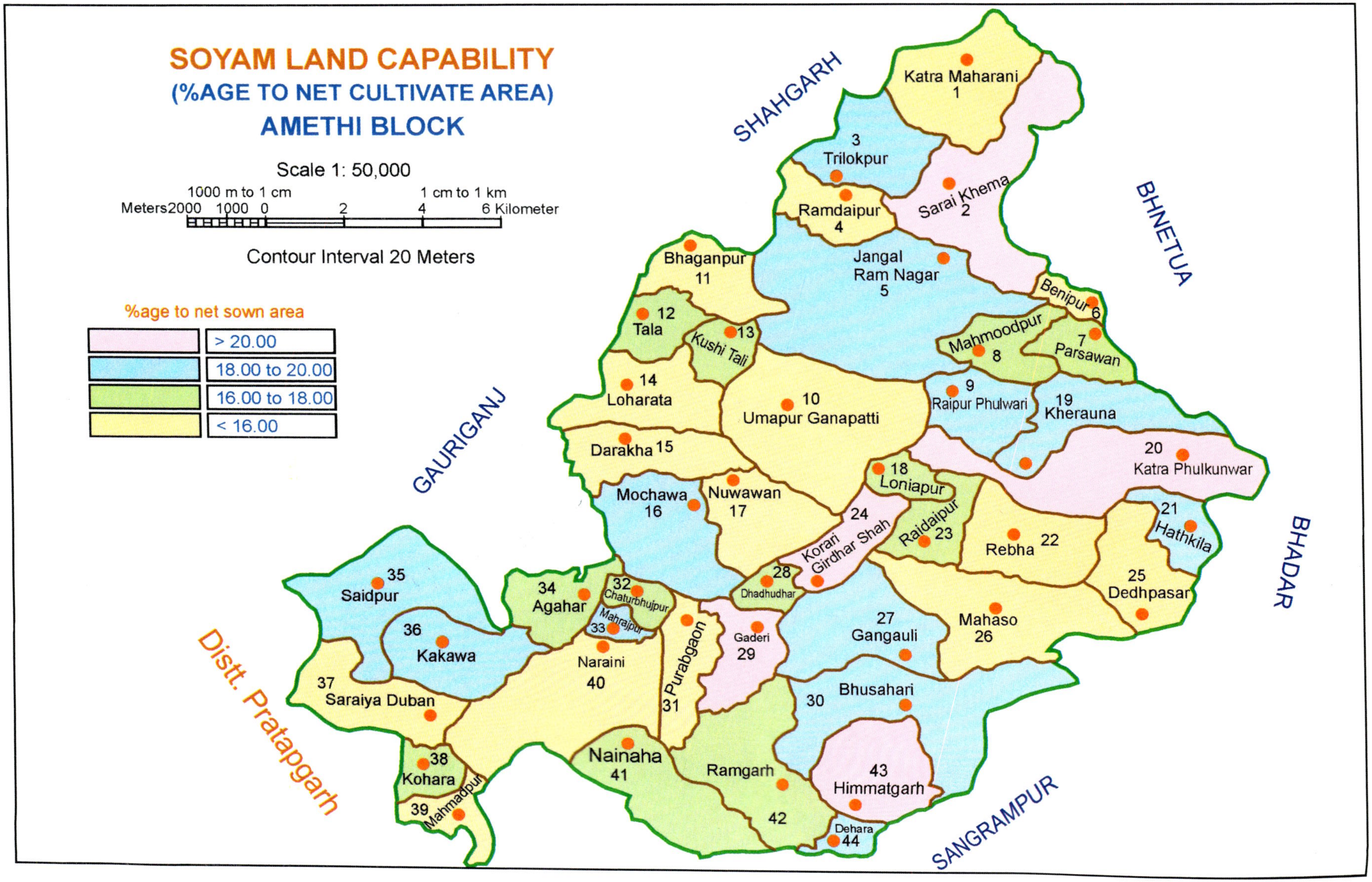

Map 8.78: Soyam Land Capability (%age to Net Cultivated Area)

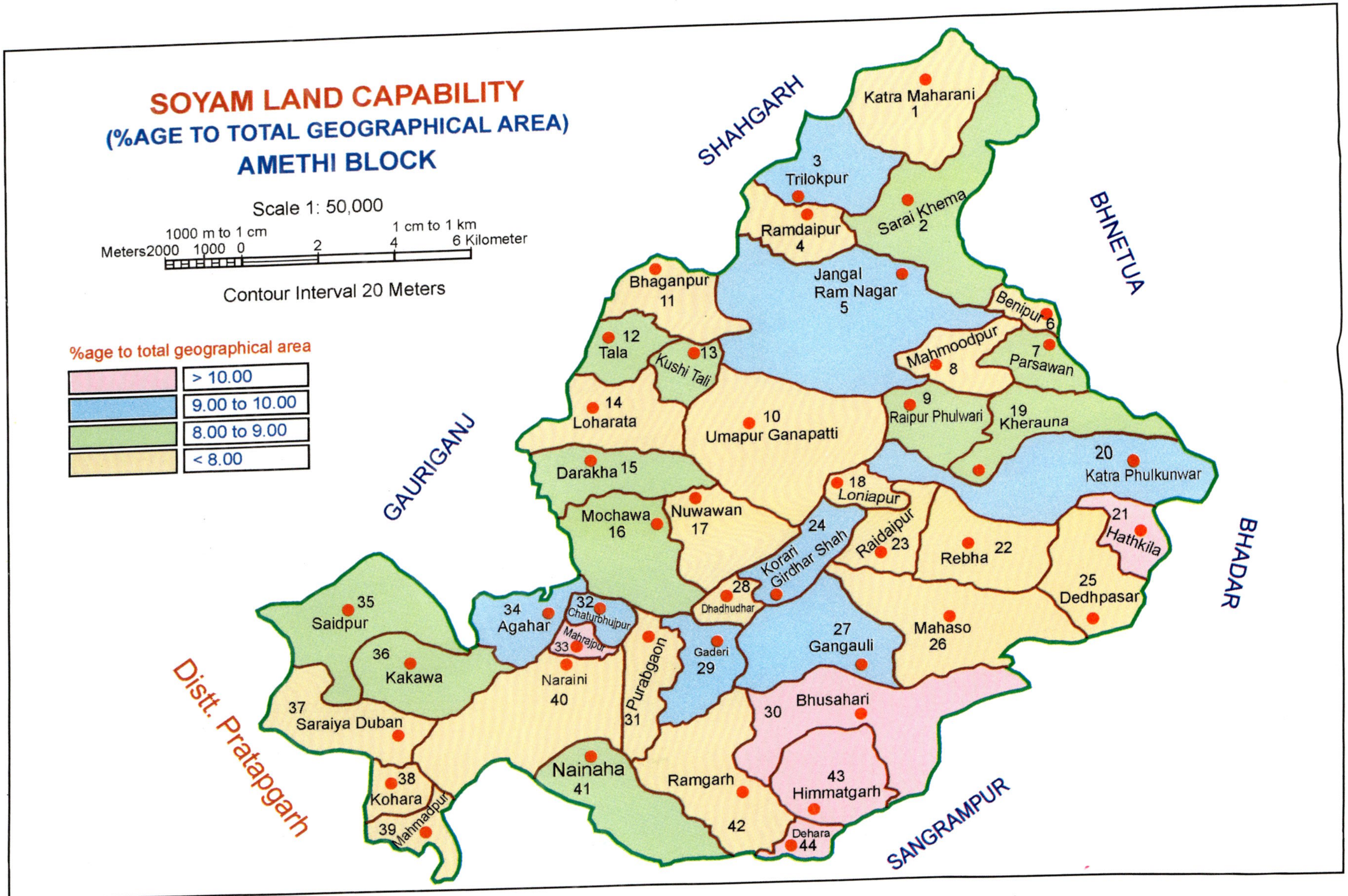

Map 8.79: Soyam Land Capability (%age to Total Geographical Area)

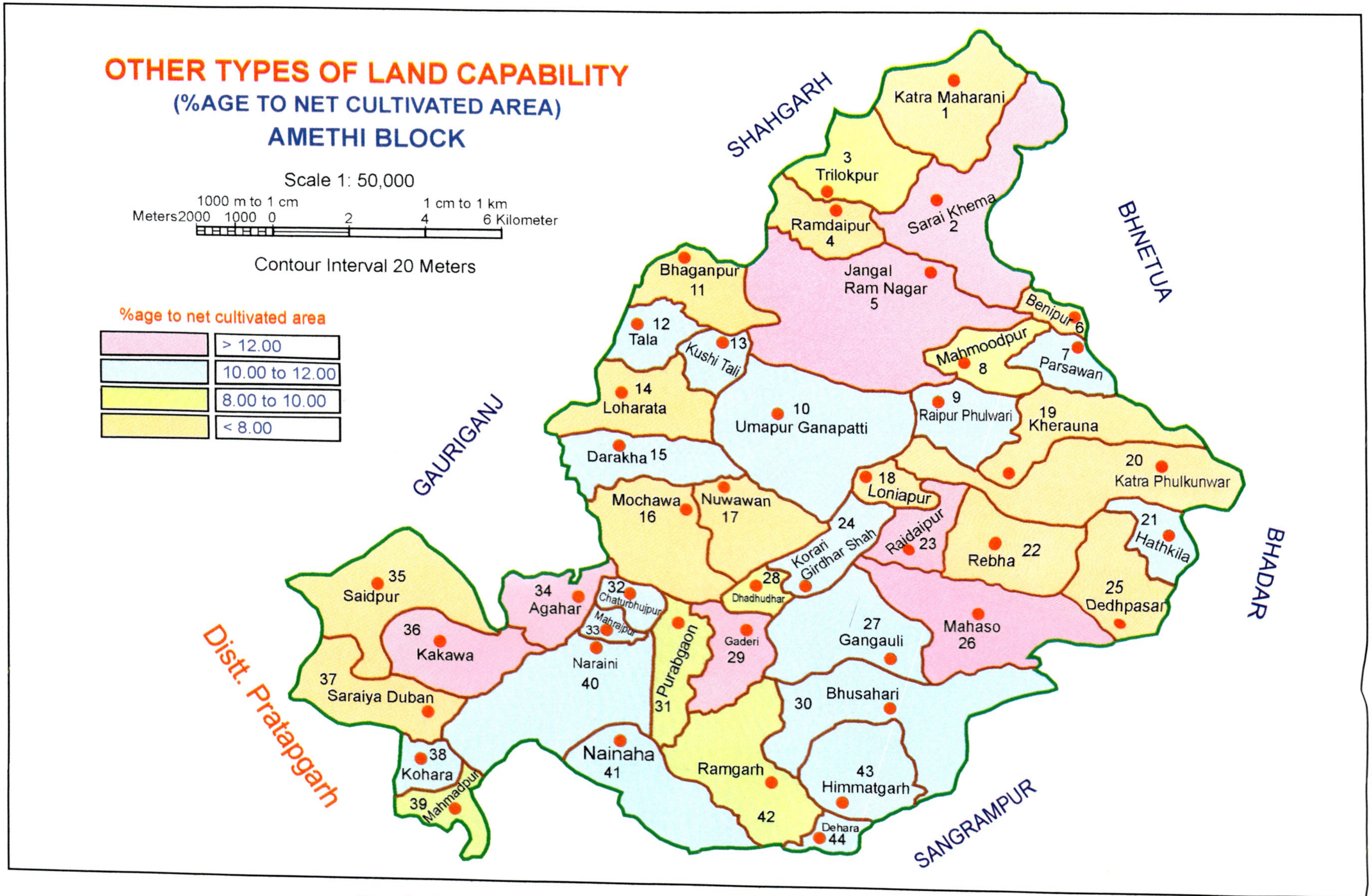

Map 8.80: Other Types of Land Capability (%age to Net Cultivated Area)

Index